Harmonics Evolution

(1897)

The Philosophy of Individual Life, Based
Upon Natural Science, as Taught by
Modern Masters of the Law

Florence Huntley

ISBN 0-7661-0522-9

Request our FREE CATALOG of over 1,000
Rare Esoteric Books
Unavailable Elsewhere

Freemasonry * Akashic * Alchemy * Alternative Health * Ancient Civilizations * Anthroposophy * Astral * Astrology * Astronomy * Aura * Bacon, Francis * Bible Study * Blavatsky * Boehme * Cabalah * Cartomancy * Chakras * Clairvoyance * Comparative Religions * Divination * Druids * Eastern Thought * Egyptology * Esoterism * Essenes * Etheric * Extrasensory Perception * Gnosis * Gnosticism * Golden Dawn * Great White Brotherhood * Hermetics * Kabalah * Karma * Knights Templar * Kundalini * Magic * Meditation * Mediumship * Mesmerism * Metaphysics * Mithraism * Mystery Schools * Mysticism * Mythology * Numerology * Occultism * Palmistry * Pantheism * Paracelsus * Parapsychology * Philosophy * Plotinus * Prosperity & Success * Psychokinesis * Psychology * Pyramids * Qabalah * Reincarnation * Rosicrucian * Sacred Geometry * Secret Rituals * Secret Societies * Spiritism * Symbolism * Tarot * Telepathy * Theosophy * Transcendentalism * Upanishads * Vedanta * Wisdom * Yoga * *Plus Much More!*

KESSINGER PUBLISHING, LLC
http://www.kessingerpub.com
email: books@kessingerpub.com

TO STUDENTS, FRIENDS AND INSTRUCTORS

The page numbers as arranged in the new edition of "The Great Work", published by R. F. Fenno and Company, are not identical with those of previous editions. It is therefore suggested that you renumber the pages of the new edition to accord with the following schedule of previous editions, should you desire to use references indicated in literature submitted by Friends of the Work, or those found in the "Key to Natural Science". Leave the present page numbers as they are and put the old numbers above them.

CHAPTER		PAGES
1.	Evolution in Operation	9
2.	Classification of Data	11 to 25
3.	Truth and Light	27 to 36
4.	The Lineal Key	37 to 71
5.	The Conflict of "Authorities"	73 to 93
6.	What Constitutes "Scientific Demonstration"	95 to 109
7.	Nature's Constructive Principle	111 to 124
8.	"Spirituality", Constructive and Destructive	125 to 138
9.	The Basis of Constructive Spirituality	139 to 165
10.	What is Morality?	167 to 174
11.	A Standard of Morals	175 to 186
12.	The "Ethical Section"	187 to 209
13.	Consciousness	211 to 232
14.	Will	233 to 236
15.	Desire and Choice	237 to 244
16.	The Law of Compensation	245 to 269
17.	The First Great Mile-Post	271 to 301
18.	The Spirit of Work	303 to 320
19.	Vanity of Vanities	321 to 336
20.	Psychological Phthisis	337 to 357
21.	Lions on the Way	359 to 362
22.	The Second Great Mile-Post	363 to 388
23.	The "Technical Work"	389 to 412
24.	Meat and Morals	413 to 423
25.	The Mark of the Master	425 to 433
26.	The Passing of the Master	435 to 456

Paste this sheet in book for reference

HARMONICS OF EVOLUTION

The Philosophy of Individual Life, Based Upon
Natural Science, as Taught by Modern
Masters of the Law

By
FLORENCE HUNTLEY
Author of
"The Dream Child"
Editor of
"The Great Psychological Crime"

Harmonic Series
Vol. I.

R. F. FENNO & COMPANY
16 East 17 Street
New York

Entered according to Act of Congress by
FLORENCE HUNTLEY,
In the office of the Librarian of Congress, at Washington, D. C.
1897.

Addressed
to
The Progressive Intelligence of the Age

HARMONICS OF EVOLUTION

HARMONICS OF EVOLUTION

Addressed to the Progressive Intelligence of the Age

TABLE OF CONTENTS

CHAPTER		PAGE
I.	Preliminary Statement	9
II.	There Is No Death	15
III.	Life After Physical Death is a Fact Scientifically Demonstrable	32
IV.	Life Here and Hereafter Has a Common Development and a Common Purpose	59
V.	The Scope, Method and Purpose	76
VI.	The Genesis of Physical Life	86
VII.	The Spiritual Basis of Evolution	115
VIII.	The Law of "Natural Selection"	146
IX.	The Natural Law of Selection	160
X.	A Question in Science	179
XI.	The Completion of an Individual—The Question Answered	143
XII.	Masculine Will and Feminine Desire	220
XIII.	The Struggle for Happiness	230
XIV.	Masculine Reason and Feminine Intuition	253
XV.	The Spiritual Basis of Love	279
XVI.	Physical Science Corroborates Natural Science	299
XVII.	Natural Marriage	321
XVIII.	Legal Marriage	337
XIX.	Divorce	353
XX.	True Marriage—The Mathematics—The Harmonics—The Ethics	367
XXI.	The Individual Solution—The True Altruist—To Know, To Dare, and To Do	432

HARMONICS OF EVOLUTION

"Fools deride. Philosophers investigate."

CHAPTER I.

THERE IS NO DEATH.
LIFE AFTER PHYSICAL DEATH IS A FACT SCIENTIFICALLY DEMONSTRABLE.
LIFE HERE AND HEREAFTER HAS A COMMON DEVELOPMENT AND A COMMON PURPOSE.

These propositions are laid down with due appreciation of their importance. They are presented as facts of Nature clearly demonstrable by scientific methods.

The writer is aware that these statements directly challenge both dogmatic theology and scientific skepticism. It is possible that they may provoke the hostility of the one and the ridicule of the other. If, however, such hostility and ridicule finally pave the way to honest investigation, the object of the writer will have been accomplished.

These positive initial declarations, be it understood foreshadow the positive character of the philosophy to be presented. The authority to state these truths in unconditional terms is derived from a school of science which transcends in scope and investigation that school commonly known as the school of modern physical science. This authority is reinforced by direct teaching and a personal experience covering a period of more than ten years.

Having made this statement, any further discussion as to the relation of the writer to her authority is obviously out of place.

Natural Science, which is the basis of this philosophy, is an exact science and not a theory. This means that its propositions are based upon the theorem of any one of our exact physical sciences, viz., the study of natural phenomena and the classification of facts in Nature, together with experiment along the lines of natural law and the demonstration of the principles involved.

The propositions of Natural Science are susceptible of demonstration with the same certainty as are those of physical science.

It will be observed that this is an unqualified statement. Positive assertion invariably accompanies either deliberate falsehood, profound ignorance, or a personal and definite knowledge as to the question involved.

In this, as in any other science, the investigator or would-be student is confronted with certain definite propositions and is given a working formula for their solution. In this, as in any other science, successful solution depends chiefly upon the individual ability, capacity and character of the student.

Over ten years of personal inquiry, instruction, experiment and experience constitute the substantial basis upon which this work rests. It is true that many of the declarations made by more advanced authorities transcend the writer's personal knowledge. Personal demonstration has, however, been carried far enough to logically bear out their broader declaration concerning the operations of natural law. It is also true that within the limitations of their opportunities, these years of personal inquiry have verified those fundamental propositions upon which this philosophy rests. These verifications constitute a reasonable basis of faith in other propositions which, as yet, transcend the writer's personal demonstrations.

In this particular instance the slow but complete transformation of an avowed skeptic was not accomplished without proof. The writer's "conversion," if such it may be called, has not been a blind process. It is not a mere matter of faith. It is not merely an intellectual opinion based upon argument or theory. The proof in question rests upon that identical character of evidence which is accepted by science as well as by the courts of law, viz.:

(1) Direct evidence, or the unqualified statements of those who claim to know, such persons being of lawful age, sound mind and irreproachable character.

(2) The direct evidences which flow from a course of personal self-development. This is a character of evidence which neither hostility, incredulity nor ridicule can affect. This is true by reason of the fact that these are evidences which flow from a personal, practical, rational and scientific inquiry and experiment. These are evidences which obtain in every department of life, physical, spiritual, intellectual and moral.

(3) The internal evidences of intelligence, consistency and truth which the philosophy itself presents to the student.

(4) The corroborative evidences conveyed to the mind of the student by exhibitions of the spiritual powers of a teacher.

(5) The corroborative evidences of the moral philosophy as illustrated in the daily life and practice of a teacher.

There are grave and serious difficulties in the path of any student who would transmit his slowly acquired knowledge to the world. Indeed, the presentation of all the proofs is not possible. The best that may be done in dealing with the facts of Natural Science by publication, is to present certain fundamental, spiritual principles in Nature and show their relation to the already proved and accepted facts of modern physical science.

The writer does not claim that her instructors are infallible authority, nor that these pages contain a faultless presentation of the teachings received. To claim for any instructor an infallible teaching of natural laws would be to push him beyond the limitations of finite intelligence and power. To claim that any student can transmit either his instruction or his knowledge without shade or shadow of inaccuracy, would be a character of presumption that no sane person entertains of his own ability.

The foremost claim made for this philosophy is its conservation of natural law. Its chief claim, therefore, upon rational intelligence, lies in the fact that it distinctly recognizes the limitations of finite intelligence and of finite endeavor.

It is claimed, however, that this entire work is the result of a definite, scientific instruction, investigation and demonstration. It is also declared that the writer will present the results of her instruction and investigations as accurately, as clearly, and as honestly as her limitations will permit.

There are other requirements which enter into a proper understanding of this philosophy than the truth of the philosophy itself or the wisdom of a teacher or the accuracy of a writer. Those other requirements rest with the reader, and comprise, in fact, a degree and quality of intelligence on the part of the reader which are commensurate with the science and philosophy presented.

It will be observed that the term "Natural Science" is used to designate the science upon which this philosophy is based. It will help the reader very materially if he understands the full significance of the term thus employed.

(1) *Physical Science* is universally understood to mean that science which has to do with physical matter only. It is, therefore, limited in its scope to the substance, functions and phenomena of physical matter.

(2) *Spiritual Science,* (as commonly understood), is that science which has to do with only things spiritual. It is, therefore, also limited in its scope to a distinct and separate field of causation.

(3) *Mental Science,* (as commonly understood), in like manner, is limited to a knowledge of the mind with its functions and phenomena.

Thus, it appears that each of these sciences is limited to knowledge of only a part of the facts of Nature. But what term shall be employed to properly designate that science which includes knowledge of the facts of physical science, spiritual science and mental science combined? What term is broad enough to cover the facts of all the departments of Nature?

Evidently no better term could be found than "Natural Science." Since this philosophy is based upon facts in all the departments of Nature, the term "Natural Science" properly designates the science upon which this work is founded. Whenever, therefore, the term "Natural Science" is used in this work it must be understood in that broad sense which includes all other sciences and departments of science.

In projecting this work another line of criticism is anticipated than that of theology or physical science. That criticism, if made, will probably be based upon the assertion that the philosophy here presented is not an exposition of Oriental philosophy. This may be anticipated from the fact that we have among us today representatives of certain sects of Oriental religions. These scholars are teaching their several doctrines which upon their face may appear at variance with the deductions laid down in this work.

In a certain sense such criticism would not be without justification.

The teachings of our eastern friends are based upon the Vedas and kindred and contemporaneous philosophies written or promulgated thousands of years ago. It is admitted that the philosophy here presented is not the philosophy of *ancient* India. That is to say, while the facts of Nature and the principles of truth are changeless, both knowledge and the comprehensive powers of man have been continually increasing.

There are facts concerning the school of Natural Science which are not taken into account by our Oriental visitors. Indeed, they are facts wholly unknown even to the great body of Oriental teachers and priests. The Vedas were written ages ago, but in the interval between that time and the present the school of Natural Science has gone steadily forward in its research, experiment and demonstration, as to Nature's finer forces. It has also advanced in its methods of teaching as well as in its methods of obtaining knowledge.

In other words, the school here represented is the progressive school which has kept abreast of the most advanced scientific thought of the age. It is that body of scholars which recognizes the fact that the scope and methods of the ancient school do not meet the demands of this age and this people.

Natural Science approaches the world at this the close of the nineteenth century, having but one object in view, viz., the transmission of some portion of its accumulated knowledge to the best intelligence of the age.

This present work is the initial volume of the first exposition of an entire system of philosophy based upon Natural Science. It is a radical but necessary step in advance of the philosophies heretofore presented. It therefore constitutes a proper and natural link in the chain of evolutionary progress. For this reason, not only this work, but those to follow, must appear as departures from previous methods of dealing with the spiritual side of Nature.

This departure represents the attempt to bridge the gulf between ancient Oriental mysticism and modern western science. It represents the effort of modern intelligence to connect the scientific knowledge of the spiritual schools with the demonstrated facts of physical science.

It is believed by representative intelligence of both systems that the time has arrived when it is possible to rationally explain the actual existence and correlation of forces in two worlds of matter, of life and intelligent activity. To this end the Modern Masters of the law have combined the knowledge of the ancient spiritual schools with the knowledge and methods of the modern physical school. By such combination they rationally establish that correlation of life and principle in two worlds which has been so bitterly denied by physical science and so long deemed a non-essential by the ancient schools.

This present writing, therefore, marks out a new path in the treatment of the so-called "occult" in Nature. It represents the attempt to explain rather than to mystify. It undertakes not merely to declare, but to elucidate and illustrate the correlation of spiritual and physical forces in Nature.

It represents, in brief, a philosophy of life deduced by modern authorities from the demonstrated facts of both the physical and spiritual worlds; which worlds constitute the natural home of man here and hereafter.

CHAPTER II.

THERE IS NO DEATH.

By death is here meant the extinction of the individual self-consciousness, personal identity and intelligent activity of a man when his physical body ceases its functions and activities.

If life after physical death were not a fact demonstrable and demonstrated there had been neither reason, motive nor excuse for this volume.

This tremendous truth, however, has for ages been in the possession of a few patient, self-denying scholars whose knowledge of the continuity of life has been a matter of daily, rational experience, just as is their knowledge of this physical existence.

Except for the zeal of these few scientists and their eagerness to present their knowledge to the world and their desire to teach others how to personally demonstrate the fact, neither this effort nor certain other historical attempts had been made to reach the public mind and heart.

"If a man die shall he live again?"

Throughout the ages man has put this question to Nature.

History, tradition and experience go to show that this question rises with the dawn of individual self-consciousness.

The world's history is made up of the issues of life and death. All of the world's activities are shaped by this expectation of death. The uncertainty as to what lies beyond the grave more or less affects every life. No individual nor community nor nation escapes the shadow. It colors individual acts. It enters into national policies. This certainty of death is the drop of gall in the cup of pleasure. It is love's terror. Childhood fears it. Old age dreads it. Even disease, poverty and crime shrink from release by death.

The certainty of physical death conditions all life to restlessness. It shadows all human endeavor with a sense of impermanency. It deflects the soul from purposeful living by bringing into life the continual prospect of reaching the end. Anticipation of death increases the apparent value of time. It creates haste. It engenders a feverish hurry and struggle for immediate satisfaction and happiness.

The controlling passion of man's nature, the passion for conquest, is strengthened by the uncertainty of life. The earth has been and is a slaughter-house. Ambition, lust, greed, and vanity have set the mark of Cain upon the human race. With the hope of a present material gain and an immediate personal satisfaction, but in ignorance of the penalties involved, men have ruthlessly inflicted death upon each other. They kill each other in open battle, in secret encounter and by a barbarous "Sanction of the Law."

Nothing but an actual knowledge of future penalties and fruitions can properly check the suffering and injustice which this passion for conquest entails.

Love of life inspires every living thing. It is, however, man alone who hopes for immortality.

It is safe to say that all men desire to live after physical death. Most of them hope for such a life. Many have faith. There are, however, more whose hope and whose faith alternate with misgiving and doubt. For hope is not faith, nor is faith knowledge, yet both are inspirations to life. Hope is but a fleeting intuition, while faith is the steady expectation of the soul.

Hope for and expectation of life beyond physical death appear to be almost inseparable from human intelligence. In this desire and expectation the savage, the seer, and the child find a common ground.

Except for this natural hope and expectation of a life to come man could not properly work out his destiny upon this physical plane. Faith is a perpetual inspiration, while skepticism clouds the best efforts. A creed of annihilation saps the springs of human energy. It thwarts the finest possibilities.

Do not these facts testify to the importance of the subject?

Do they not justify a vigorous search through Nature for actual knowledge upon this question of life and death? Who can doubt that such knowledge would fix and ennoble life's purposes as no fitful hope nor wavering faith can do?

Literature that has longest survived is that which has been based upon a desire for, or upon an expectation or actual knowledge of a life to come. The sacred writings of the older nations antedate secular history. The greatest of profane writers have speculated upon the immortality of the soul. The works of Plato represent a great intelligence inspired by a hope of immortality, while the Psalms of David represent that same hope strengthened by faith. The Sermon on the Mount, however, represents not merely hope and faith, but instead, the doctrine of the Nazarene testifies to a personal and exact knowledge of the spiritual side of life.

It is as natural to desire life after physical death, to hope for it, to seek knowledge of it, as it is to desire food, light and air. It is an unfortunate man who does not hope for life to come. It is a diseased or abnormal one who does not desire it. A man without hope or desire merely exists. He can scarcely be said to live.

He who gives heed to his own spiritual intuitions is never without hope. He who has hope may acquire faith. He who has both hope and faith may acquire actual knowledge, provided he have the INTELLIGENCE, the COURAGE and the PERSEVERANCE to prove the law.

The expectation of life after physical death comes first as an intuition. That purely spiritual intuition is as strong in the savage as it is in the civilized.

Physical science when called to note this fact dismisses it as "superstition."

This universal expectation of life arises out of conditions that are distinctly not physical. If the human mind had depended upon only physical facts and rational processes of the brain for the development of faith in a life to come it had never developed. No man who had looked upon a dead body could have conceived the idea of a future life.

In spite of the fact that physical life is a veritable house of decay and death, the expectation of, and faith in, a life to come have increased with the higher evolution of man.

It is therefore, evident, that this faith and expectation are based in the spiritual intuitions of all men. It is also evident that such faith and such expectation are not the mere superstition of savages since they increase with the higher stages of intelligence and moral life of man.

It was this spiritual intuition of the primitive man that laid the foundation of Natural Science.

Man is a rational as well as an intuitional being. Man alone is capable of reasoning upon his own intuitions. Man alone has the intelligence to seek a rational explanation of those intuitions. Man alone demands that Nature shall yield the secrets of those mysterious hopes, fears and expectations which alternately inspire or terrify the soul. The spiritual intuition of the savage establishes an expectation of life after physical death. Later on the higher grade man attempts to verify his own intuitions by rational means.

How well he has succeeded the progress of religion and philosophy shows. It was this natural desire for exact knowledge that inspired human intelligence ages ago to begin an investigation of natural laws.

It was man's determination to solve the problem of life and death that evolved the school of the more liberal science.

That earliest effort and first victory are prehistoric events. That is to say, there are no records of the earliest schools accessible to the general public. Those records, however, have not been destroyed, though oceans may now cover the continent upon which that history was made.

Men, nations and continents pass away, but human effort is never wasted. Knowledge is never entirely lost.

Centuries have rolled into cycles, but that knowledge has been preserved, transmitted and enlarged. That early school of science has never disbanded. For ages, however, its wisdom has passed only "from mouth to ear." And Natural Science today is broadened and ennobled by all that past effort, gain and achievement.

From the beginning human intelligence has occupied itself with speculations, hopes and fears as to a life beyond. The finest intelligences among men have employed their powers to merely elucidate a reasonable *theory* looking to life after physical death.

This desire for life evolved and still maintains the pursuit of Natural Science. The original secrecy of that school arose merely from popular ignorance and prejudice against learning. As a result the facts concerning spiritual life were demonstrated only by the few, and this knowledge was preserved and guarded from the world. In truth, it has been only the few who have possessed the requisite capabilities for demonstrating the fact of life after physical death. For ages these few sought only to gain and preserve this knowledge. They sought only an individual experience and an individual power and control over Nature's finer forces. It was a later consideration and a broader experience that induced them to become teachers among men.

Now, however, they have come to know that their first duty and highest privilege are to impart their science and philosophy to the world as fast as the world will receive them.

These are statements which, in all probability, will be vigorously denied by both theology and modern physical science.

These critics, however, will oppose the position of this work from different points of view. While both will declare that exact knowledge of a life to come is impossible, yet each will give a different reason for this alleged impossibility. The one regards such an assumption as sacrilege. The other scouts the idea as superstitious folly.

Between the bigotry of faith without science and the bigotry of science without faith truth runs a terrible gauntlet in this world.

The great body of human intelligence proceeds along two lines of investigation. The one system is speculative and spiritualistic. The other is scientific and physical. The one represents intuition unsupported by reason. The other represents reason unaided by intuition. The one stands for only a spiritual perception of ethical principles. The other represents only rational conceptions of physical facts.

Mankind, as a whole, in its expectation of a future life, is sustained by faith and not by any actual scientific knowledge of the life to come.

The creeds of Christendom begin "I believe." Not one begins "I know," when referring to spiritual things and a life to come. By the adoption of such a creed theology takes rank as speculative philosophy. This is true of any religion that does not offer a rational means for demonstrating its dogmas.

Physical science, on the contrary, is based upon facts of physical Nature rationally demonstrated. As a science, therefore, it rejects that which is not susceptible to demonstration. It would not be science unless it adopted exactly this course. Physical science is right when it refuses to accept as science that which is undemonstrable. It errs, however, when it undertakes to dogmatize as to what is and what is not demonstrable to man.

Thus, theology, rests upon spiritual intuitions and faith in those intuitions, while physical science is built solely upon physical Nature and the demonstration of physical facts.

The weakness of theology is its ignorance of physical facts. The weakness of scientific skepticism is its contempt for spiritual religion. Both systems, being human, are narrow. Each, however, is honest, and therefore susceptible to evolutionary processes.

The educated clergy no longer deny the facts of physical science, while the foremost specialists of physical science are studying and utilizing the spiritual and psychic forces in Nature. Theology is hampered by the fear that science will unearth theological errors. Science, on the other hand, is oversensitive as to any criticism of its deductions and theories.

Intuitions of a spiritual life are not proofs even to the rational mind of any individual. They are, however, truths to his soul and are sources of consolation, of hope and of inspiration.

Intuition is not knowledge. It is, instead, a suggestion of knowledge that may be acquired. Every man and woman knows the potency and inspiration of those spiritual perceptions which are not explainable in reason. Intuition, though not knowledge, is a higher guide to human life than is cold reason when it entirely ignores those convictions of the soul.

German metaphysics demonstrate the development of reason and the suppression of intuition. Here we see to what heights and depths of absurdity human reason may travel. Through and by such "reasonable" processes man is reduced to "an erroneous proposition."

It is little wonder that the theologian rejects those self-evident absurdities of metaphysical abstraction. On the other hand, the reasonable metaphysicians and scientists cannot accept the philosophy of a supernatural reincarnation and resurrection.

For nearly six hundred years controversy has raged as between speculative theology which teaches life after physical death, and speculative philosophy which teaches that death ends all. With a Luther, Calvin, Knox and Wesley on one side and a Voltaire, Schopenhauer, Kant and Renan on the other, it is little wonder that lesser minds become confused as to this question of life after death.

Physical science augments this confusion brought about by speculative theology and speculative philosophy. It enters the debate with demonstrated facts which dismay theology and disconcert metaphysics. Part of these discovered facts contradict certain theological dogmas. Physical science, therefore, declares that theology has no basis in fact. By this it assumes that there can be no undiscovered facts which might demonstrate a spiritual side to Nature. Instead, physical science assumes that there are in Nature only physical facts and physical forces, and that these facts and forces are demonstrable by methods of physical science alone.

If the Christian world really *knew* what it now only professes to believe, how quickly the whole existing order of theological discourse would change. If physical science were only able to conceive that there *might* be facts of Nature beyond the scope and methods of its own schools, how soon would our general scientific study and experiment include the psychic phenomena of life. As it is, however, both theology and physical science agree in declaring that human intelligence cannot penetrate further into the secrets of Nature.

Who can estimate the benefits that would flow from the exchange of mere faith in a life to come for actual knowledge that such is the fact?

The church, as a whole, professes this faith. The daily life and practice of the individual professor, however, suggest more of doubt than of a fixed faith. If the individual could really know that life after death is a fact, our whole dismal paraphernalia of death would disappear. Indeed, if men entertained even an unwavering faith, their lament for the dead would be modified. The truth is that the professing Christian mourner exhibits but little greater fortitude and faith when death claims a friend than does the average unbeliever. Our Christian brothers mourn their dead with an abandon that demonstrates the instability of their faith.

If one really believes in a spiritual life to come there is neither reason nor excuse for this intemperate grief. If, however, a man could know what he but mournfully hopes rather than believes, the house of the dead would never be a house of despair. Instead, it would be a house of unselfish rejoicing whenever death released the spirit from old age, disease or sorrow. When a man knows what physical death is he will never retard the passing soul with selfish grief.

Did women possess the faith they claim, they would not swathe themselves in unsanitary crape nor visit cemeteries to commune with the dead who are not there.

To the man who *knows,* the dead body is but the discarded mantle of his friend, one that had served the uses of the soul for the time. As such, the body is entitled to due reverence and is consigned to the earth or the fire without exaggerated grief.

If theology could but rationally demonstrate a basis for its faith, life would be transformed with new and higher impulses and aspirations. If physical science, on the contrary, could but prove its own major premise, viz.: "All is physical matter and mechanical energy," the church would disintegrate in a year.

Though science drives theology from one false position to another as to evolutionary history, it does not in the least affect the basis of theology, which is faith in accepted teachers of the law and the individual spiritual intuitions of man.

Theology, however, has never made a rational effort to verify its faith in a life to come.

Physical science, on the contrary, has conducted a vigorous, determined campaign against what it is pleased to term the "superstitions" of mankind.

Physical science has made an honest effort to substantiate its declarations upon this question. The special point of attack has been Modern Spiritualism. A long and critical investigation has been made of those peculiar phenomena which attach to "spiritual mediumship." The purpose has been to show that such phenomena are the result of trickery or brought about by certain peculiar physical and mental powers resident in and natural to the alleged spiritual medium.

As the scientific and reading world knows, that effort has been a lamentable failure. That failure is demonstrated in several ways, viz.:

1. By the rapid increase of psychic phenomena.

2. By the rapid increase in the number of the believers in such phenomena.

3. By the frank confession of failure made by certain eminent representatives of physical science who set out to "expose" Spiritualism.

The insufficiency of either the charges made or the methods employed by physical science is clearly demonstrated.

Modern Spiritualism, embracing millions of believers, has risen during the past fifty years in the very face of scientific skepticism.

Swedenborg's following increases rather than decreases. Mesmerism, once derided, is now introduced as "hypnotism" and practiced by the "regular" schools of medicine. The almost simultaneous birth, rise and development of Theosophy, Christian Science and Mental Healing among intelligent people are phenomena which physical science has not explained.

The newspapers, reflecting the average mind and practice,

publish daily accounts of faith healers, of mind reading tests, of telepathic incidents and of the varied phenomena attaching to clairvoyant, clairaudiant and impressional mediumship. Persons possessing psychometric powers have been called upon to aid in the detection of crime, while the defense of the criminals on the ground of "hypnotic control" is becoming very frequent in legal procedure. The courts have been compelled to take judicial notice of the existence of hypnotism as a fact.

These facts constitute a very substantial answer to that fundamental dogma of physical science which declares that "All is physical matter and physical force."

The failure of physical science to prove this proposition is confessed by every specialist who has made a personal investigation of Modern Spiritualism. Notably among such are Professor William Crookes, F. R. S., and Alfred Russel Wallace. Both of these scientists set out upon their investigations in a skeptical attitude of mind. Every experiment and dogma of physical science known to them was opposed to the idea of super-physical forces and of physically disembodied intelligences.

Both of these gentlemen entertained the belief that the complete exposure and utter rout of Spiritualism simply depended upon a few scientific tests directed by the rational mind of a physical scientist. The results are too well known to discuss. Both of these prejudiced representatives of physical materialism confess failure. Both have publicly admitted that the phenomena of Spiritualism are facts in Nature which defy the analysis or explanation of physical science. More than this, both agree that such phenomena are due to super-physical forces and to physically disembodied intelligences.

Having employed every test known to physical science and reason, these master minds of modern science have satisfied themselves that death does *not* end all.

And how has the body of science met that patient inquiry and cheering message?

It has received that message exactly as the learned of his time received the message of Copernicus, Galileo, Harvey, Franklin, Gray and Arago. Professor Crookes and Alfred Russel Wallace have been openly ridiculed and declared to be insane by those professed scientists who have not taken the pains to investigate the facts. This is the reward of two intelligent men for reporting honestly upon their investigations.

At this point of controversy and recrimination as between theology, metaphysics and physical science, is brought forward the testimony of another investigator of natural laws.

At this point a broader science comes forward with its cheering "I know," and rationally explains the ground of its affirmation. It does more than this. It directly challenges representatives of theology, metaphysics and modern physical science to offer themselves as students and demonstrators of the fact of life after physical death.

Just here Christian theology may find a reinforcement in the higher philosophy, and modern physical science will find its borders enlarged by contact with the more liberal school of science. Natural Science has at command the facts of Nature which support the real doctrines of Christ, which dispel the vagaries of metaphysics, and will aid in the extension of scientific knowledge and research.

Its pursuit is something more than a study of ancient creeds and Oriental Philosophy. It is the study and demonstration of those natural laws which govern the body, spirit and soul of the individual man. Natural Science has an object to gain among the western nations. That object, however, is not to arrest the aggressive material and intellectual development of this people. It seeks merely to throw additional light upon our own Christian philosophy and to stimulate western science to an investigation of the spiritual facts in Nature.

THERE IS NO DEATH, declares Natural Science.

In laying down this proposition as a scientific fact it is, of course, understood that the reader is probably not in a position to either immediately accept or reject this declaration. Before attempting to do either, the inquirer is asked, first, to consider in brief those lines of evidence which go to support the fact of life after physical death.

The most significant historic evidence as to the fact of another life is the personal testimony of the few master minds which have dominated human intelligence for ages. The millions come and gone, together with the living hosts of mankind, have acknowledged, and continue to acknowledge, these few great teachers as messengers of Truth and exemplars of natural law.

This fact alone testifies either to the wisdom and truth of those master minds or to the imbecility of human intelligence.

The great world religions testify to the universal faith that has been placed upon the individual honesty and intelligence of such as Gautama, Prince of India, and our Christian teacher, the Nazarene.

What does this faith of the world indicate? Does it point to trickery or insanity of those teachers? Does it indicate that the world has accepted knaves or lunatics as its moral guides?

Does it not rather indicate that the great consensus of human intelligence selects its best instead of its worst representatives as its standards of wisdom and virtue?

So far as history, sacred or profane, informs us, the individual lives of those accepted teachers were examples of honesty and wisdom. So far as the history of their personal lives and their accredited doctrines can show, they are the worthy guides of humanity. Their lives and their doctrines are wholly irreconcilable with any theory of either deliberate fraud or emotional insanity.

These accepted teachers of the world gained ascendency over mankind by force of two conditions:

1. They had something to teach.

2. Humanity responded to that teaching by reason of the fact that it appealed to the spiritual nature of man. Each of these "Masters" claimed to have proved the fact of life after death. Without this basic knowledge both their lives and their doctrines become meaningless.

Is it not easier to conceive that Mr. Darwin and others of our own modern scientists may not have been in position to demonstrate *all* the facts of Nature, than it is to doubt *all* of the spiritual philosophy in the world? Is it not as logical to concede that physical science may draw erroneous conclusions as to insist that these great teachers of spiritual law were either charlatans or lunatics?

If there is no life beyond the grave these teachers lied to the world and the world is without guidance.

That is, in fact, the position of scientific skepticism.

It is, however, only fair to our modern school of science to say that this radical position is due to theological dogma rather than to the doctrines of Christ. Theology unconsciously discredits Christian doctrine in the critical judgment of modern science.

The orthodox Christian church declares Christ to have been a *super*-natural being.

This so-called "super-natural" is the fatal error of theology. It is this that offends reason and creates heretics. It is this that so directly antagonizes physical science which very properly declares "there is no supernatural." It is the use of this word "supernatural," and the unthinkable propositions involved, which drive the purely rational mind to the other extreme. It is this that goads the scientific skeptic into characterizing the loftiest perceptions of the soul as mere superstitions. The Agnostic is the product of church dogma.

Between the "supernatural" of theology and the "superstitions" of physical science the broader Science has sought vainly for a judicial hearing. Theology offends against reason and is offended by reason. Physical scientists offend the soul and are offended by the intuitions of the soul.

Natural Science can find votaries only where the system or school or individual is far enough advanced to know that both reason and intuition are differing phenomena representing separate functions of intelligence. The more comprehensive Science demonstrates that there is no supernatural. On the other hand, it demonstrates that there is a spiritual side to Nature and that man lives on self-consciously after physical death.

Natural Science, and the philosophy of life founded on that science, accept Christ as an exemplar of truth. It regards Him as one among the greatest of masters who have proved the fact of life after physical death by scientific and natural means.

Every student under a Master of the Law is taught to thus prove the fact of continued life. By such personal demonstration he corroborates and sustains the doctrine of the Nazarene upon that question. That is to say, he proves that it is possible for any son of man possessing the proper qualifications to prove the spiritual side of Nature and the fact of a life to come. Thus, the specialist of the larger Science is prepared to give theology a natural basis for its faith. He is prepared to lead physical science along those rational lines of investigation which will enable him to demonstrate the errors of skepticism.

The claims made by Natural Science are supported by two lines of evidence, the one direct, the other indirect.

The first class of evidence includes the direct personal testimony of such as have proved the law, or of such as have come in contact by any means with the spiritual plane of life. The second class includes the testimony of such as have witnessed spiritual phenomena under test conditions. Under the first-named class we have:

1. The direct testimony of the world's chosen teachers, among whom are Moses, Buddha, and Jesus Christ.

2. The direct testimony of the personally instructed pupils of a Master. Among such were the Disciples of Christ.

3. The direct testimony of so-called seers and prophets and the oracles of all ages, races and religions. Among such are Abraham, Daniel, Isaiah, and Jeremiah.

4. The direct testimony of vast numbers of "psychics" covering the history of all ages, such persons being the hypnotized subjects of stronger disembodied intelligences.

5. The recorded testimony of the members of all the ancient schools of spiritual knowledge.

6. The direct personal testimony of their living members.

7. The direct personal testimony of countless living psychics, or spiritual mediums, many of whom hold daily communication with physically disembodied persons.

8. Many of the priesthood of the Catholic Church could testify to the conscious and intelligent communication between

men in the body and men out of the body. This, in fact, is the secret of the power of the Catholic Church.

9. Countless monks and nuns of the same church could testify to the same facts. The seclusion and austerities of the monastery and the convent originally had this object in view. The Roman Catholic Church preserves its direct touch with the spiritual side of life through its truly celibate priests, monks and nuns.

10. To this could be added the direct testimony of countless living Oriental yogis, fakirs and dervishes who acquire spiritual insight and spiritual powers through barbarous physical self-torture and physical self-suppression.

11. The direct testimony of countless Oriental priests and Oriental philosophers and students. Such testimony is found especially in India, where climatic conditions, and native temperament and dietary habits foster the spiritual development of man.

12. Lastly, the personal testimony of the writer who is neither a master, a seer, nor a prophet, nor yet a spiritual medium, nor a devotee of any church. The knowledge thus far gained of a spiritual plane having resulted from a certain degree of experiment and demonstration under the formula for self-development.

This brings us to the second class, or to indirect evidence of life after physical death. Under this head we have:

1. The great world religions which are the responsive bodies of faith supporting the declarations of a few great Masters of the Law. Almost the whole of mankind subscribes to one or the other of these religions.

2. The testimony of the individual intuitions. Every normay individual, risen to the plane of intelligent life, enjoys intuitions and experiences which, at times, suggest another plane of life. Such intuitions and experiences cannot be converted into rational proof. The individual concerned, however, knows them to be facts. He recognizes a silent, subtle, interior guide and critic that he names conscience. He finds that this conscience speaks through the intuitions alone. It is a thing independent of reason. In fact, it too often defies reason when reason longs

for approval. It is "conscience," we say, which warns, chides and desires. Conscience is that which tells us we love or hate, or are happy or miserable, irrespective of either reason, convention or legal codes.

Conscience, in fact, is the voice of the intelligent soul.

When the intelligent soul of man exercises itself upon the physical plane, man enjoys a rational conception of physical things. When it exercises itself upon the spiritual plane, the physically embodied man enjoys spiritual intuitions of spiritual things. How often that inward monitor declares "there is no death." How often the rational mind denies this intuition of the soul, one who has been a skeptic can testify.

3. The spiritual intuitions of man declare "there is no death." The soul of man rejects the idea of annihilation. Instead, it universally entertains faith or a hope in a life to come. This expectation of continued life characterizes the savage, the seer and the savant. Because these intuitions of man are so universal they are therefore natural. Natural impulses imply a natural law of fulfillment. A universal desire implies a natural means of accomplishment. Universal tendencies are always based upon adequate principles.

4. The indirect evidence presented by thousands who have witnessed the phenomena of the seance room or have come in contact with psychics under other circumstances. It is unreasonable to hold that these large and intelligent classes of persons are either entirely deceived or are deliberately deceiving others as to what they have witnessed.

5. Perhaps the strongest indirect evidence in this line is that which has been furnished by certain eminent representatives of physical science. Especially does the recently published work of Mr. Alfred Russel Wallace deserve attention and demand respect. This is not a hastily written opinion. It is not merely the result of a few desultory observations in the seance room. It is a carefully prepared summary of scientific tests and investigations of Modern Spiritualism covering a period of fifty-three years.

After more than half a century spent in the critical study of

Spiritualism, one of the greatest intelligences of the modern school of physical science declares such phenomena to be the direct result of super-physical (spiritual) laws and of physically disembodied intelligences.

Thus it is that six centuries of the methods of physical science have not equipped it to disprove the claims of Modern Spiritualism alone. It leaves one of its foremost representatives to declare for the truth of those claims.

In short, it leaves the collaborator of Charles Darwin to reaffirm what was declared ages ago by Buddha and later by our own acknowledged Master, Christ, viz.: THERE IS NO DEATH.

CHAPTER III.

LIFE AFTER DEATH SCIENTIFICALLY DEMONSTRABLE

The evidences of life after physical death, as obtained by men of science, are evidences which flow from a personal and purely rational course of development.

The formula for this course is based upon exact knowledge of certain fundamental elements and principles in Nature. It is a formula which has been successfully demonstrated by the special students of all ages. It is a far more exacting course of discipline than those prescribed by our great universities.

Modern systems conserve almost entirely a mental and muscular development. The broader system, on the other hand, includes development in the physical, spiritual, mental and moral departments of life. This formula for self-development has been known only to the few. Hints of this definite, tried, tested and fully demonstrated method may be found in a few existing publications. The authors, however, writing for the general public, have so veiled their true meanings in allegory, poetry and mystical symbolism as to conceal rather than reveal the correct principles and the true method of development.

A notable instance is the Bulwer literature. His "Zanoni" and "Strange Story" have been at once the puzzle and delight of the past half century.

This scientific formula remains a secret for the best of reasons:

1. By reason of both public ignorance and indifference to the fact that there is need of such knowledge.

2. The impossibility of teaching it except under the proper conditions.

3. The danger of imparting it to the ambitious and the unscrupulous.

This scientific formula, however, like almost all special and personal knowledge of spiritual things, remains undiscovered to the world at large, more because of public indifference than because of its guardianship. Whatever man has achieved man may achieve independently of any school or sect or order. Other men may work out any scientific facts that have been demonstrated in the past if they but possess the INTELLIGENCE, the COURAGE and the PERSEVERANCE required.

Bulwer truly says, "Truth can no more be seen by the mind unprepared than the sun can rise in the midst of night."

Those who possess this knowledge cannot impart it to unprepared minds. It were a crime to openly proclaim it to the vicious. It were folly to thrust it upon the foolish, the apathetic and the ignorant.

The reader, however, will recall the fact that some form of asceticism distinguishes the records of all teachers, seers, prophets and holy men of all ages, peoples and religions.

Asceticism, to some extent, is an important factor in the development of spiritual insight and of spiritual powers.

The fasting, solitude and silence set forth, with more or less prominence, in all sacred writings have a rational and scientific explanation in natural law. The long fast maintained by Christ was not a theatric episode. It was a matter of pure science. The Apostles were taught the value of asceticism. John the Baptist attained spiritual vision partly through such means. The same is true of St. Paul.

No eater of meat and drinker of wine, nor one given to undue physical indulgence, has thus far in the history of spiritual development demonstrated the fact of life after death, *scientifically*.

There is but one purely rational and scientific course of spiritual self-development. There are, however, many unnatural, revolting and dangerous practices which enable man to reach spiritual vision. These unnatural and forcing processes are as far from true development as are the forcing processes of our public schools from true education.

The horrible Yogi practices antedate Buddhism. Buddha rejected and condemned that revolting formula. To us, however, Buddha's asceticism appears needlessly harsh and barren of the natural joys of life. Christ's life, so far as we know, is in this respect a far more pleasing example. His one long fast is about all we hear of austerity. He appears to have had in view for the masses not so much the scientific development of spiritual insight, as the right performance of earthly duties. He taught temperance in all things. His purpose was to furnish the principles of a perfect earthly life. It was not to publicly impart scientific facts.

Natural Science today declares that spiritual knowledge and spiritual powers gained at the expense of a natural, purposeful, human life are useless knowledge and power to both the individual and the world.

Merciless self-torture is strongly condemned. The modern man who aspires to such knowledge may obtain it without torture to the physical body, without the suppression of every human desire and ambition and without crucifixion of the natural affections.

All those practices belong to the past. Those past rigors have, however, an explanation in natural law. That explanation (as will be seen more plainly later on) lies in the fact of evolution itself. The physical organism of man six thousand years ago was vastly coarser in texture than that of a nineteenth century man. His passions were stronger. He was more animal in his desires and appetites. He was scientifically farther from the spiritual plane. Hence the necessity for the more forceful methods of those long past ages.

If these rigors have been necessary in the past, they are no longer so to the same extent. The natural processes of evolution have refined the physical body of man and keyed it to higher vibrations in Nature. In the past, men thought it necessary to use severe means to eliminate their appetites and passions. That severity was probably necessary at that time. Even in Christ's day rigid asceticism was regarded as a necessary part of holiness. He was accused of gluttony in that he often ate with the common people and observed their common customs.

Illustrations of unnatural asceticism are found among the Yogi, the Indian Fakirs, howling Dervishes and Catholic monks and nuns. All these differing classes, however, are inspired by different motives.

The Yogi seek continual absorption in spiritual things. The Fakir trades upon his acquired powers. The Dervishes, monks and nuns are influenced mainly by the belief that they are purchasing a heavenly favor by a life of self-torture.

In one respect our western would-be teachers of "occultism" are making grave mistakes. These zealous but uninstructed instructors are suggesting and advocating certain austerities as to diet, exercise and meditation that are simply Yogi practice. They, of course, do not know that their instructions may induce mediumship, nervous prostration or insanity.

The exact and scientific formula for self-development rests upon that fundamental principle in Nature which is commonly termed the law of Motion and Number.

The law of Motion and Number is the law with which physical science is now dealing. The modern school, however, knows it popularly as the law of Vibration.

The law of Motion and Number (so termed by the ancients), or the law of Vibration (so termed by the moderns) is, in fact, the spiritual principle of polarity. The law of polarity has to do with the positive and receptive vital energies in Nature acting upon matter.

This law of vibration or polarity has to do (primarily) with the refinement of matter and its rate of vibratory action. Science has made certain important discoveries, viz.:

1. All matter is divisible.
2. Every particle of matter is in a state of motion or vibration.
3. Coarse particles vibrate slowly.
4. Fine particles vibrate rapidly.

All matter is, therefore, in a certain stage of refinement and is vibrating at a corresponding rate or degree, as it is called.

To illustrate this principle the writer quotes from an article written by her instructor and published in November, 1894, in a leading magazine:

"Not only is the physical universe a universe of matter, but "the same is equally true of the world of spirit. Both are material "in the most exact and literal meaning of that word. The spirit "of an individual is as truly a material organism as is the physical "body which envelops it. Both are matter, the one physical and "the other spiritual. 'Physical material' and 'spiritual material' "are, in truth, the identical terms employed by the masters to "distinguish between the two worlds of matter.

"But if it be true that both are, in fact, material worlds, the "question may properly be asked: Wherein exists the difference, "and what is the necessity for any such distinction? The one be-"longs to the world of purely physical things, and is, therefore, "designated by the very appropriate term, 'physical matter.' The "other belongs to the world of purely spiritual things, and is "therefore designated by the equally fitting term, 'spiritual mat-"ter.' For a similar reason we designate that which belongs to "the mineral kingdom as 'mineral,' and that which belongs to the "vegetable kingdom as 'vegetable.'

"In this case, however, both belong to the world of physical "material, and are but subdivisions of it.

* * * * * *

"There are certain distinguishable differences existing in phys-"ical and spiritual organisms which enable the spiritual scientist "—or the master—to determine with equal accuracy to which "world of matter any given organism or body belongs. What "are some of these distinguishable differences?

"1. One which may be mentioned is, the degree of fineness, "or the relative size of the individual particles of which a body is "composed.

"Let me see if I can make this clear. Suppose, for illustra-"tion, you take an ordinary gallon measure and fill it to the brim "with marbles of the ordinary size used by children at play. "Now, it is not difficult for you to understand that, although it "will hold no more marbles, the measure is not, in fact, full.

"There are many vacant spaces between these marbles, which "may be filled in without running the measure over, provided "we select a substance, the particles of which are fine enough "to sift into these vacant spaces. Now, suppose you try number "six shot. You will find that you can put into the measure sev-"eral handfuls of shot without running it over. Why is this? "Because the shot are smaller than the vacant spaces between "the marbles. You have now poured in all the shot the measure "will hold, but you can readily understand that the measure is "not yet full. There yet remain smaller spaces between the shot "which are still vacant. Now put in ordinary white, dry sand, "and you will find that the measure, though full as it will hold "of marbles and shot, will still receive several handfuls of the "sand. Why? Because the vacant spaces between the shot are "larger than the grains of sand. But you have now put in all the "sand the measure will hold. Is it full? No. You may now pour "in over a pint of water. Why? Because the particles of which "water is composed are much finer than the vacant spaces be-"tween the particles of sand, and the water has only run into "these vacant spaces.

"It now begins to look as if the measure were, in reality, full; "but not so. Now take a very high grade of finely distilled "alcohol and you will be able to drop slowly in three or four "spoonfuls of the alcohol without overrunning the measure. "Why is this? Because there are still vacant spaces, even be-"tween the particles of water, large enough to receive the finer "particles of which alcohol is composed. But how now? Have "we reached the limit? No. There is yet another fluid compound "known to chemists whose particles are so much finer than "those of alcohol that a teaspoonful or two of this may be added "without seeming to increase the aggregate contents of the meas-"ure, thus proving that even between the particles of alcohol "there are spaces unfilled. But what shall we say now? Is the "measure full? No; not yet. We will now turn into the vessel "a current of electricity, and we find that we still have room for "an amount sufficient to charge the entire contents of the measure. "But what is electricity? The finest and most subtle element "known to the physical universe. We are now just upon the "borderland of the spiritual universe. We have approached it "along the line of 'the degree of fineness, or the relative size of "the individual particles of which a body is composed.' The "next step takes us across the border line of purely physical ma- "terial into the land of spiritual matter.

"2. Another distinguishable difference between physical ma- "terial and spiritual material is found in the rate of vibratory "motion of the atom in the compound.

"Let me see if I can make myself understood on this point. "Take a piece of granite, set it before you, look at it carefully "and see if you can discover any vibratory movement among the "individual crystals of which it is composed. No; you are ready "to declare that so far as you can discover they are absolutely "at rest; but not so. Science has discovered that the individual "particles of which a stone is composed are in a constant state "of vibratory motion, one upon the other. But this vibratory "motion of the atom in the compound is, in the case of stone, "at such a low rate that it is not perceptible to the physical sense "of sight, and as a result the piece of granite appears to be a solid, "immovable, impenetrable mass of dead matter.

"Now take a piece of growing wood. Examine it as carefully "as possible with the naked eye. You are still unable to observe "any movement among the particles of which it is composed; "but if you place it under a powerful magnifying glass you will "be able to distinguish a very slight vibratory movement among "the individual cells of which it is composed. But notwithstand- "ing that the rate of this vibratory motion is much greater than "that in the case of the stone, it is still not great enough to dis- "turb the physical sense of vision. The result is that wood, like "stone, appears to the naked eye, a solid, dead substance.

"To save both time and space, we will now pass over several "intermediate substances, such as animal flesh, gelatine, etc., and "examine a drop of water. Here we find that the vibratory "motion of the atom in the compound is at a rate many times "greater than that in either stone or wood. The particles of "which water is composed move with such facility and rapidity,

"one upon the other, that to a certain extent they elude the phys-
"ical sense of sight, and the result is that water is transparent to
"the naked eye.

"Let us take another step forward and we come to the gases.
"Here we find that the vibratory motion of the atom in the com-
"pound is at a rate so much higher than in the water that the
"physical sense of vision is entirely eluded. In other words, a
"gas is invisible only because the atoms of which it is composed
"vibrate so rapidly that the physical sense of sight is unable to
"follow them. To make this fact so clear that none may mis-
"understand it nor fail to grasp it as a fundamental principle in
"science, why is it that when a gun is discharged we are unable
"to see the bullet speeding on its way? It is only because its rate
"of movement is so rapid that the physical sense of sight cannot
"follow it. It has simply eluded the eye. Again: Look at an
"ordinary carriage wheel when it is at rest and you can see every
"spoke with perfect distinctness; but place it on a spindle and
"set it revolving at a high rate, and the higher the rate the less
"distinctly you will be able to see the spokes, until they finally
"disappear.

"We now come to the last and highest grade of physical mat-
"ter, viz., electricity. The vibratory motion of the atoms in this
"compound is at a rate higher than that in any other physical
"substance.

"And here we stand again at the border line which bounds
"the physical universe of matter and separates it from the world
"of things spiritual. The only difference is that in this case
"we have approached from an entirely different direction, viz.,
"along the line of vibratory motion. The next step takes us
"beyond the physical into the world of spiritual matter. There
"are other distinguishable characteristics of physical material and
"spiritual material which enable the advanced scientist immedi-
"ately to classify and locate in its proper world, any given ma-
"terial organism, with as much certainty and precision as the
"physical scientist or physicist of the great colleges of the world
"is enabled to classify and locate the purely physical substances
"with which his science has to do; but it is unnecessary to mul-

"tiply these illustrations. What I desire to make clear is the "fact that the physical scientist, or physicist, using only physical "means, is limited in his investigation and demonstration to the "world of physical matter. He stops at the border line between "the two worlds of matter and is forced to say: 'I can go no "further; the instruments at my command are not fine enough, "nor sufficiently subtle, to test the properties and qualities of "that which lies out beyond. It eludes the methods of physical "science and all the means at my command.'

"At this point, however, the spiritual scientist—the master— "takes up the thread of science and carries it forward past the "border line of physics into the land of Psychics. In his ability "thus to view the subject from both worlds, his great advantage "is inconceivable to one whose sense of vision is limited to the "world of purely physical things. At this line running between "the two worlds of matter, he sees every law of physical matter "joined to its correlative law of spiritual matter. The chain of "law is thus unbroken. It runs from one universe of matter "directly across into the other without interruption; and in this "splendid continuity he recognizes the majesty, the power and "The glory in this, the universality of law."

The student of Natural Science undertakes his own development in conformity with the law of vibration. He undertakes it having in mind the triune nature of man. That is, he accepts as a working corollary a certain statement laid down by an instructor.

He accepts the proposition that man is composed of body, spirit and soul. This means that man has a physical body and a spiritual body which are controlled and operated by the highest entity, the intelligent ego, the soul. Accepting this proposition as a working principle, he proceeds to demonstrate for himself by purely rational and scientific means and methods.

The law he evokes to carry out that demonstration is the law of vibration, that law which refines matter and increases its vibratory action. The physical body is composed of physical matter. The particles which are coarse in texture move at a correspondingly low rate of vibratory action. The physical body is provided with physical sensory organs. Nature conditions these organs to receive and register the vibrations of physical matter only. These vibrations are registered upon the physical brain, through which instrument they become cognizant to the intelligent soul.

The physical sensory organs are not adapted to all of the vibration of even physical matter. Their combined powers only embrace a limited range of vibration. This range includes only the vibrations of physical matter which lie upon the same plane of refinement and vibratory action as the physical body itself. By aid of these organs the intelligent ego or soul becomes cognizant of different external physical objects, elements and conditions. The recognition by the ego of these external physical objects, elements and conditions constitutes what we term physical sensation.

Each one of the physical organs of sensation receives and registers a different range of vibration. The whole surface of the physical body itself is so supplied with sensory nerves as to become a medium of vibrations. The general sense of touch is experienced when any portion of the physical body comes in contact with physical matter that is made up of the coarser particles moving at the lower rates of vibration.

The physical eye, on the contrary, is the most highly specialized organ of physical sensation. When the physical eye is brought in contact with rays of physical light (which are, in fact, only fine particles of physical matter moving at a high rate of vibration), the individual experiences the sensation of sight. As with the eye, so with the ear, another specialized organ. When the physical organs of hearing are touched by physical atmosphere moving at certain rates of vibration, the individual hears physical sounds. As with the ear, so with the other special physical organs of smell and taste. Both of these organs represent still other and differing rates of vibration which obtain in external physical matter.

Thus, in fact, every physical organ of sensation is an organ of touch. The general sense of touch that obtains upon the entire surface of the physical body differs from physical sight only in degree. The one registers the slow vibrations of coarse material particles, the other registers the rapid vibrations of finer material particles.

Through the operation of these several physical organs, each registering a different range of vibration, the intelligent ego is brought into conscious relations with a very wide range of vibratory activity of physical matter.

It must now be clear that the physical sensory organs and the physical brain are adapted to receive and register only the vibrations of physical matter. It must also be clear that they are adapted to receive only a limited range of physical vibrations.

This fact is proved by physical science.

For example, the animalculæ that swarm upon the earth, in air, in water, and in all living bodies, are intangible and invisible to the physical sensory organs. They affect neither the sense of touch nor of hearing, taste, smell nor sight. Except for the microscope, physical science would declare a belief in animalculæ were a superstition. The recently discovered "X-ray" makes no impression upon the highest physical organ of sensation. None the less, the "X-ray" is a vibration of physical matter.

The photographer's art best illustrates the feeble powers of our highest physical organ of sensation, the eye. The sensitized plate reveals the existence of stars which the human eye, aided by the finest telescope, fails to discern. These vibrations from remote stars are, however, plainly registered upon the chemicalized plate. Physical science accepts these photographs as pictures of realities. It thus accepts as a fact that which man can neither feel, hear, taste, smell nor see.

When one realizes the limitations of the physical senses in a physical world, it is easy to understand why these organs fail to report the vibrations upon a higher plane of matter. It is now comparatively easy to understand how the vibrations of an infinitely finer grade of matter must entirely elude these physical organs which are intended for physical vibrations only.

For example, a physically embodied man cannot (by the aid of his physical sensory organs alone) in any manner "sense" the presence of a disembodied spiritual man. The spiritual man is neither physically tangible nor visible.

These facts of Nature, demonstrable to Natural Science, are already foreshadowed by modern physical science. The claims of the older school are substantiated by a current series of lessons in science written by Elisha Gray. Speaking of the physical sensory organs of sound and of sight, this recognized scientific authority says:

"While vibration ceases to affect our senses at 40,000 per second, as sound, we find ourselves conscious again of periodic motion when it reaches 398 trillion times per second; then we hear with our eyes or see with our ears, whichever you choose. The sensation is in all cases the effect of motion."

Again he says:

"There is much food for speculation in the thought that there exist sound waves that no ear can hear, and color waves of light that no eye can see. The (to us) long, dark, soundless space between 40,000 and 398 trillions and the infinity of range beyond 764 trillions, where light ceases, in the universe of motion, makes it possible to indulge in the speculation that there may be beings who live in different planes from ourselves and who are endowed with sense-organs like our own, only they are tuned to hear and see in a different sphere of motion."

Once more the learned professor rises to the role of prophet without retreat from his scientific basis when he says:

"The eye is more perfectly developed, but it is capable of but comparatively crude photography. The red ray comes to the eye with the lowest number of vibrations—398 trillions—four thousand billion. The eye cannot record anything with a less number of vibrations a second. The highest color visible is violet, with 764 trillion vibrations, beyond which the eye cannot vibrate in sympathy with color. But there are colors we cannot see. The universe is filled with things which to the human eye are invisible. In the same way there are things which we cannot feel, and odors we cannot smell, and flavors we cannot taste. For all that we know, this world may be the home of another race of beings who pursue their course unknown to us and perhaps we to them. It is the aim of science to come to the aid of man's imperfect organism and to lift a little the veil of mysteries."

The spiritual body of a man is composed of "spiritual material." That is, of matter much finer than the finest physical matter, and moving at a higher rate of vibration than the finest particles of physical matter moving at their highest possible rate. The spiritual body permeates the physical and constitutes the model upon which physical material integrates. The spiritual body, like the physical, is provided with five sensory organs. They are adapted to receive and register vibrations of spiritual material only, that is, of matter lying upon the same plane of vibratory action as the spiritual body itself. By the aid of these organs the intelligent ego becomes cognizant of different external spiritual objects, elements and conditions. The recognition by the ego of these objects, elements and conditions constitutes what we term spiritual sensation.

Each one of the spiritual sensory organs receives and registers a different range of vibration. The whole surface of the spiritual body is itself so constructed as to become a medium of spiritual vibration. The general sense of spiritual touch is experienced when any portion of the spiritual body comes in contact with spiritual material of the coarser texture moving at lower rates of vibratory action. On the other hand, the spiritual eye is the most highly specialized organ of spiritual sensation. When the spiritual eye is brought in contact with rays of spiritual light (which are in fact only particles of spiritual material moving at a higher rate of vibration), the individual experiences the sensation of spiritual sight.

As with the spiritual eye, so with the spiritual ear, another specialized organ of sensation. When the spiritual organs of hearing are touched with spiritual atmosphere moving at certain rates of vibration, the individual hears spiritual sounds. As with the spiritual ear, so with the other specialized organs of smell and taste.

Thus every spiritual organ of sensation is an organ of touch. By use of these special organs, each registering a different range of vibratory activity upon the spiritual plane, the intelligence or soul is brought into conscious relation with a very wide range of the vibrations of spiritual material. These spiritual organs register the vibrations of spiritual material only. They are also limited in their capacity upon the spiritual plane in a manner analogous to the limitations of the physical senses.

However, the capacities and the limitations of the spiritual senses are not a matter of present consideration.

That which is important in this connection is the fact that the spiritual sensory organs do not register the vibrations of physical matter. They are too fine, too highly attuned. The slow vibrations of physical matter fail to register through the spiritual organs of sense.

For example, a spiritual man (that is, a physically disembodied one) cannot sense the physical side of man by aid of his spiritual senses. The physical side of an earthly man is neither tangible nor visible (under ordinary conditions) upon the spiritual plane of vibration.

The seance room often illustrates these differing laws of sensation on the two planes of matter. For illustration, the physically embodied visitors often experience peculiar thrills, pleasant or unpleasant, while they neither see, hear, taste nor smell anything unusual. This "thrill" (which is unlike electricity in that it conveys an impression of intelligence) is, in fact, the general spiritual sense of touch. It is touch with the lower vibrations of the spiritual plane. It is the touch of the spiritual organism and not the physical.

On the other hand, the invisible "spiritual control" will explain that he "senses" his earthly visitors spiritually and not physically. That is, as he explains it, he sees them only in their spiritual forms.

Thus, in fact, the two worlds of matter are separated only by difference in refinement of matter and in rates of vibratory action. When the physical sensory organs act independently upon the physical plane they sense nothing of the spiritual. When the spiritual sensory organs act independently upon the spiritual plane they sense nothing that is physical. They receive and register the vibrations of spiritual material only. Under these conditions nothing lying upon the physical plane is tangible or visible to a liberated spiritual man. A liberated spirit on the other side of life is as completely out of touch with physical matter, or the earthly plane, as the earth man is out of touch with the spiritual plane.

Except under extraordinary conditions a "spirit" does not see the physical side of matter even when he moves immediately upon the earth plane. What the spiritual man sees, ordinarily, is the spiritual side of physical matter. He sees only the spiritual models upon which physical matter integrates. He sees only that part of Nature which has vibratory correspondence with the spiritual organs of sensation.

This accounts for the fact that in mediumship the physically disembodied intelligences are usually as eager to come in touch with earthly friends as are the friends who seek the services of a spiritual medium to come in contact with the spirits. In the present state of development, mediumship offers the easiest line of communication between the two worlds. It is also the most hazardous method, involving many dangers to the physical, mental and moral nature of the medium.

However, with the law of vibration clearly in mind some of the mysteries of life are cleared up. It now becomes possible to conceive of two planes of matter, life and intelligence, correlated yet separated by apparently impassable barriers. This knowledge explains how matter upon one plane is invisible and intangible to intelligence upon another. It explains how sound on one plane is silence on the other, how the light of one plane is darkness upon the other.

"The light shineth in the darkness and the darkness perceiveth it not" is a scientific fact upon which an ethical truth has been formulated.

The earth man is, therefore, the inhabitant and operator of two distinct instruments for the uses of his intelligence. One instrument is of course physical material whose individual particles move upon one another very slowly. The other instrument is of fine spiritual material whose individual particles move rapidly upon one another. Though interdependent the two bodies are not identical. Each perform functions peculiar to its own plane of matter. Neither body is more than a mere vehicle for the uses of the operating ego. Both are important. Both are indispensable to the soul which is seeking knowledge of itself and of its environment.

These are some of the facts of Nature that are taught the would-be student of Natural Science at the outset of his undertaking. With certain definite instructions concerning this law of vibration he is now prepared to retreat or to intelligently enter the path of a personal self-development.

The preparatory work of every student's life may be said to be chemical. The chemical refinement of the physical body is the foundation upon which he builds. This refinement is brought about solely through scientific knowledge of the vibratory principles. Given a healthy body, a vigorous brain, a determined will and the proper instruction and environment, and the course of this self-development will increase rather than impair the physical strength. Unnatural and unscientific methods only do injury to either the body or the mind.

Ignorant experiment is always fraught with danger.

The chemical refinement of the physical body is brought about to a certain degree by a system of diet. Fine foods in limited quantities are substituted for coarse food in unlimited quantity. By fine food is not meant rich food, but fine natural food, as grain, fruit and nuts. Supplementing this dietary course is a systematic course of exercises which may be termed breathing exercises. This is a course of training analogous to our athletic exercises. It is, in fact, a purely physical training, having in view primarily an increased regular and rapid oxygenation of the body. Supplementing this chemical and physical course of development is a purely intellectual system of training. This is a course of instruction by which the mental powers and the will are trained to the knowledge and employment of Nature's finer forces.

This scientific course of self-development involves, as will be seen:

1. A physically refined body and brain.
2. The acquirement of scientific knowledge.
3. The exercise of power.
4. The right application of knowledge and the right use of power.

The chemical refinement of the body has a specific object. That object is to so refine the physical body as to permit the ego to exercise the spiritual organs and the spiritual powers independently. This result is worked out through knowledge of and conformity to the law of vibration. It must be explained that the spiritual body during physical life is conditioned by the vibratory action of the physical body. That is, physical matter is the dominant material upon the physical plane. The coarser the physical substance, the more it hampers and obscures the spiritual powers and the spiritual senses. The coarser a human body is in particle, the slower its vibratory action, and the coarser the physical organism and the slower its vibratory action, the less independent is the spiritual organism. The vibratory action of the spiritual organism is ordinarily lowered or accelerated in just the degree that the physical body is coarsened or refined.

The independent use of the spiritual organs of sensation is measured (during earth life) by the degree of approach to spiritual material that can be obtained in the physical body. Fine food produces two mathematical results:

1. It reduces the size of the individual particles of physical flesh.
2. It increases the vibratory action of the atom in the compound.

This, as has already been shown, causes the particles to move upon one another more rapidly. This rapid vibratory action of the atoms in a compound when it has reached a certain rate produces transparency. There are also important secondary effects following this physical refinement.

1. The vibratory action of the spiritual organism is correspondingly increased.
2. The ego is enabled to exercise the spiritual organs of sensation independently.

Ordinarily, the physical body in its coarser state is opaque to its own embodied spiritual sensory organs. It has the effect of darkness to spiritual vision. When the physical body is refined to a certain stage under scientific direction, a remarkable thing occurs to the student. While still exercising his own will and rational powers he finds himself as it were, in a house of glass. He finds that he is able to exercise independently first one, then another, and finally all of the spiritual organs of sensation. He does this independently of the material composing his physical body.

He finds himself now just as consciously and as rationally in touch with the spiritual plane as he is ordinarily with the physical. He now feels the touch of spiritual objects. He hears spiritual sounds. He smells spiritual odors. He also sees objects which are reflected by the rays of spiritual light.

The spiritual plane is just as tangible and visible to a spiritual man as our physical plane is tangible and visible to the physically embodied man. The hand-clasp of two spiritual beings is just as real as, and far more magnetic than, that of two physically embodied individuals.

This free outlook upon the spiritual plane is the first great victory of the student.

Thus, by personal experiment under an exact scientific formula a man in the physical body proves the existence of a spiritual world inhabited by ex-human beings. This experiment involves the demonstration of the fact that there is no death.

This is the most important single discovery ever made by finite science. To prove that death does not end all has been the most valuable single achievement of man in the physical body.

While this is the most important single fact, it is not the greatest victory from the student's point of view. The triumph of the student's life is the hour when he clearly proves that he is a spirit inhabiting a physical body. His proof of this fact is the ability to entirely withdraw the spiritual body from the physical, to stand apart from it, to travel at will, to collect rational knowledge, and to return again to the physical body. This he does as consciously, as intelligently, as independently as he departed.

This complete temporary release of the "spirit" from the body is the most difficult experiment known to a student of this physical plane. It is a power attained only by the few. In fact, it should not be attempted except by such as expect to devote their lives to a practical teaching and demonstration of both the science and the philosophy of life. This experiment is not essential to an intelligent communication with the spiritual plane, nor is it necessary for the purposes of a practical human life.

It is, indeed, a dangerous experiment except under the personal supervision of a master of the law.

The first mentioned experiment, however, is one involving no risk. It is indisputable proof of the spiritual side of Nature. It is indisputable proof of the fact of life after physical death. It is to this point of demonstration that the school of Natural Science is prepared to carry any man who possesses the necessary qualifications of body, mind, will and motive.

The complete mastery of Natural Science necessitates:

1. A physical organism that will sustain the proper refinement.

2. An intelligence which comprehends the philosophy as a whole.

3. The will to maintain self-control over every department of individual nature.

4. The moral courage to rightly apply and practice the knowledge and powers gained.

It is true that the student encounters difficulties. They are considerable in both character and number. They are not insurmountable, however, or this science and philosophy had not been evolved by man.

It is true that there are certain other subtle and difficult experiments and processes known only to the few. The system which those few have followed, however, is opened to any determined intelligence just as our universities are open to all, viz., to such as have the necessary qualifications and to such as can make the right conditions for the prescribed course of study.

The results which flow from a course of self-development under the prescribed scientific formula are as certain as those which flow from certain combinations of chemical substances. This is speaking as to general results. There are, of course, effects which are personal to the student. Those effects vary in character as to the individual bodies, minds and souls of the several students.

The results upon a group of students show all such variations as occur when different proportions of the same chemical substances are subjected to the same conditions. The different results and effects upon individual students express the difference in their physical endurance, mental acumen, powers of will and purity of motive.

The law governing the relation of chemical particles is absolute. Each experiment, however, records the minutest variations in the quantity, quality and proportion of the substances involved. So the principle governing self-development is universal. The results, however, in each case are a record of individual INTELLIGENCE, COURAGE and PRESEVERANCE. For example, one student may incline to the life and the work of a cold scientist. The study and knowledge of material laws and the material forces of the universe may be his chief ambition. Such a man would give himself up to the acquirement of knowledge and the exercise of power for the ethical pleasure which such knowledge and such power afford the individual.

Another student may incline to a deeper knowledge of the purely ethical significance of natural law, in which case the philosophic study of mankind becomes his greatest pleasure. Still another may be impelled to utilize his science and his philosophy for the practical benefit of humanity less developed than himself. He who seeks and obtains harmonious development on all planes of life is the most successful of students.

It is such as these who are rightly termed Masters of the Law.

Under proper guidance the chemical refinement of the physical body is the natural foundation of increased mental and moral capacities. This diet of fine foods referred to rigidly excludes, for a time, flesh, stimulating drinks, narcotics and all highly seasoned foods. What is meant by fine foods are those which are fine in particle and high in vibratory action. Gradually the physical body is refined by this fine nutriment. It loses vibratory correspondence with coarse foods. It assimilates (and rises to vibratory correspondence with) the fine physical foods.

As a result the gross passions and appetites decline. Gross physical conditions are abnormal conditions. They are the result of man's ability to experiment in nutrition and with the laws governing his physical body. Animals that depend upon intuition instead of independent judgment select proper foods.

This scientific dietary system is, therefore, a simple process of correction. It is a restorative of normal human conditions. With the return to a normal demand for nutrition, gluttony and intemperance die a natural death.

The return to normality is also the end of lust. It increases the powers to love just in proportion as abnormal and diseased conditions are eliminated. Gluttony, intemperance and lust, the lowest and coarsest of vices, are essentially human errors. They are the results of over-feeding, over-stimulating and over-indulging the natural physical functions.

The chemical refinement of the body, therefore, substitutes temperance for intemperance and thus restores normal physical conditions and removes all abnormal physical appetites and desires. Gradually those vices so largely dependent upon food and drink are eliminated. Temperance becomes natural, both as to digestion and to physical passion.

This course of development should not be termed the suppression of the animal nature. It is simply the restoration of the human nature to natural conditions.

Naturally the intellectual powers are benefited by this experiment. The physical brain is quick to respond to finer and more normal conditions. Relieved from the congestion of over-stimulation it is free to exercise normally. The physical brain-stuff is refined in particle. Its vibratory action is increased. It becomes a finer instrument for the uses of the intelligent ego, or soul. As a result the intelligent soul of man is inspired to greater activities, and the acquirement and application of knowledge become a delight instead of a task.

Thus the stores of knowledge are increased and the powers of the soul quickened after the student has been physically prepared for the highest exercise of his intelligence.

The moral effects of this course of self-development are equally great. When the physical body ceases its abnormal demands the natural tendency is temperance and chastity. When the mind becomes enriched with knowledge and the soul becomes conscious of the advantages of normal living, the desire is to so live as to enjoy Nature's beneficences. A calm, clean and philosophic life is the natural result of a scientific course of self-development.

It is, however, in his final use and application of all this knowledge and power that the student proves himself a worthy or an unworthy representative of this philosophy. In this he will demonstrate either the selfish inclinations of the school of Black Magic, or the altruistic principles which distinguish this school of White Magic.

"White Magic" and "Black Magic" have no meaning other than to signify the right and the wrong uses of spiritual knowledge and power.

Christ's life and philosophy typify the principles of the school of White Magic. Mohammed's life and acts, on the contrary, proclaim the method of Black Magic. Christ applied both knowledge and spiritual powers to save men from error. Mohammed used his knowledge and spiritual powers to control men in their temporal affairs. The one taught regeneration and development through love. The other taught conquest and domination by the sword.

A man trained under Natural Science, who is governed by the principles of fraternity, equality and love, becomes a Messiah (or messenger of truth) to the world. When, on the contrary, he is impelled by personal ambition and thirst for personal power he becomes a selfish priest and a scourge to mankind. The practical results of Christian and Mohammedan principles may be read in the governmental character of the United States as compared with Turkey.

A man very highly developed under spiritual law is freed from the coarse temptations of the flesh. Such a man, however, is not above temptation. In fact, to such as these come the most subtle and powerful temptations that can assail the ambitious soul of man. To one equipped with great knowledge of Nature's finer forces and the power to operate and control them comes the strongest temptation known to man. It is no longer the clamor of the gross appetites and passions. It is an imperious demand for power, a demand that rises in the soul itself.

The desire for a personal, temporal supremacy and aggrandizement is the most powerful temptation that can move an intelligent being who is equipped with unusual powers. The desire to exercise sovereignty over the temporal affairs as well as over the minds of men has been the stumbling-block of most great men. Thirst for power causes greater suffering in the world than do the mere fleshly appetites and passions of men.

This is the temptation over which Christ triumphed. It is that to which Mohammed yielded. It is this which has been, and is, the fundamental error of Catholicism.

Bulwer was in a position to know when he said: "Rightly is the fundamental principle of our Order that we must impart our secrets only to the pure. The most terrible part of our ordeal is the temptation our power affords."

Given a healthy body, a normal brain and a determined will, and proper instruction will develop the use of the spiritual faculties and establish control over spiritual forces. Except, however, such a man be deeply imbued with a sense of personal, moral responsibility for his acts he becomes a dangerous power in the world. Such a person actuated by selfishness becomes a menace instead of a blessing to mankind. For what incalculable evil is Mohammed responsible? His terrible doctrine has almost entirely wiped out a great Christian nation.

The misuse of spiritual power in another direction is leaving its impress upon our own civilization. It would require a volume to explain the dangers which lie in the modern practices of spiritual mediumship and of hypnotism. A large per cent of our insane are merely weak-minded and weak-willed individuals who have yielded to the control of vicious intelligences on the other side of life. Already a limited knowledge of hypnotic control has become a dangerous medical, commercial and social practice. These experimenters with unknown forces, whether they be physicians, showmen or entertainers, are exponents of Black Magic.

It is to be regretted that men destroy the body with narcotics. How much more to be deplored is paralysis of the rational will power such as occurs in hypnotism. No greater evil menaces modern society than this ignorant experimenting with hypnotism. When once that danger is understood, the public will not condemn the few who would guard the secret of such powers from the ambitious experimenters, the avaricious, the thoughtless and the ignorant.

True spiritual development involves correct knowledge and correct use of power. True spiritual development is, after all, merely an accelerated development obtained through a rapidly extended knowledge of certain principles and elements in Nature. In the course of time, by the very gradual steps of evolution, the whole mass of humanity will attain spiritual development now enjoyed by scholars only.

Even to-day the physical refinement of the western nations has brought upon us the dangers of mediumship and hypnotism. Nothing but a rational knowledge of the principles involved in those practices will avert serious evils to large numbers of innocent people.

The past half-century has been a turning point in the history of the highest races. Certain groups of the Anglo-Saxon race have reached that stage of physical refinement which makes access to the spiritual plane comparatively easy. To such as these the ancient rigors of Yogi are unnecessary. Nature has already done the greater part of the work.

There are, however, dangers to be encountered by reason of this near approach of the two worlds. The daily press is filled with accounts of hysterical mediums and hypnotic crimes. Everybody is investigating. Most of these investigators are walking blindly into danger.

The chief difficulties to be overcome by the student of Natural Science are not dietary. He meets something more exacting than the mere demands of the physical appetites and passions. The intellectual and moral demands of this philosophy constitute the severest test to which he will be subjected.

Neither torture of the physical body, the renunciation of material comfort, nor the suppression of the affections is required of the modern student. Holiness in the modern sense does not mean a life of isolation, introspection and subjective ecstasy. Instead, it means a practical life in the midst of men. It means a natural, wholesome, human life lived out in conformity to the spiritual principles in Nature and the requirements of an intelligent soul.

It means a practical share in the world's activities, benefits and accomplishments. It means an exemplification of natural, physical law upon the physical plane of life, as well as natural spiritual laws through physical conditions.

It will be observed that no attempt is made here to give the scientific formula for self-development. The underlying principle alone is given. There are good and sufficient reasons for this seeming oversight, viz.:

1. The writer is a student and not a Master of the Law.
2. The scope and purpose of this work exclude such discussion.
3. The proper scientific course must be laid down and personally directed by a Master.
4. The entire formula could not be put in books. General rules may be published, but individual cases demand individual treatment.
5. The instruction requires a personal and continued relation between Master and student.
6. Were it now possible to publish this formula it would be a crime to do so. Formulas for extracting poisons and compounding explosives are not published in the daily press. Drugs and explosives have their uses only in proper hands. They become deadly weapons in the hands of the ignorant and the vicious. No man, except he prove his moral, as well as his physical and intellectual soundness, can become an accepted

pupil in that school which primarily conserves the welfare of humanity.

To such a man as this, however, Natural Science opens the way to this formula far enough to demonstrate that this philosophy has a scientific basis. The individual may do this without detriment to health, to business, or to any earthly relation or ambition. More than this, he is assured that he would be able to fulfill the last requirement of that formula without violating "his duty to his God, his religion, his country, his neighbor, his family or himself."

To such a man, having passed all tests, is finally awarded that degree which distinguishes him as a Master of the Law.

In summing up these propositions which relate to the demonstrability of life after death, the writer turns to modern physical science for corroboration. Natural Science declares this demonstration to be a scientific possibility. It declares that the principle involved is that which is known as the "Law of Motion and Number," and also as the "Law of Vibration."

This declaration is supported by the authorities of the modern school of physical science. The principle involved in the scientific demonstration of life after physical death is the same principle to which Mr. Edison refers when he says: "As a matter of fact, all matter is in a state of vibration, and any force containing the same vibration affects all classes of matter tuned to that particular key."

Thus, modern physical science recognizes and accepts that fundamental principle in Nature which guides the student in all his experiments and forms the basis of this philosophy. Upon this agreement of both the ancient and modern schools Natural Science declares:

1. All matter is in a state of motion or vibration.

2. All matter is composed of individual particles of larger or smaller size.

3. Large particles vibrate slowly, small particles vibrate **rapidly**.

4. By a purely scientific process a man may obtain knowl-

edge of higher vibrations than those received and registered by the physical sensory organs.

5. This scientific formula simply accelerates the natural processes of evolution and is, therefore, a natural process.

6. The practical results of this formula are:

(a) To discover to man a higher plane of matter.

(b) To show that this higher plane is inhabited by ex-human beings.

THIS PERSONAL EXPERIMENT, GOVERNED BY EXACT RULES AND IN CLOSE CONFORMITY TO NATURAL LAW, CONSTITUTES THE SCIENTIFIC DEMONSTRATION OF THE FACT OF LIFE AFTER PHYSICAL DEATH.

CHAPTER IV.

LIFE HERE AND HEREAFTER HAS A COMMON DEVELOPMENT AND A COMMON PURPOSE

It is with an increasing interest that the writer turns to the third and final proposition upon which this philosophy rests.

It is with this final proposition that this present work is mainly concerned. However interesting may be the mere enumeration of the cold facts of Nature, the deeper interest always attaches to man's relation to those facts and to the ethical principles involved.

Reference to the scientific course of self-development is introduced in this work for one purpose only. The ethics of life are built upon this scientific foundation.

That brief outline contains but a necessary hint as to the scientific method of personally proving the spiritual facts of Nature and personally demonstrating the ethical principles underlying all activity.

This work is not intended as a purely scientific treatise. It does not include either the chemical formula or the ethical code for self-development. It does not embrace definite instruction as to the control of spiritual elements nor give direction as to the development of the mental forces.

Instead, the reader at this time is asked to consider mainly the ethical effects which flow from the operation of these principles, elements and forces which the student of the law demonstrates for himself. He is asked to consider the ethical effects rather than the scientific processes.

When the student of natural law demonstrates the fact of life after physical death, his education is only begun. He is now confronted by other problems. He has pushed himself into a new and unfamiliar world of matter, life and intelligence. He is now in the midst of those conditions of which he had formerly only speculated. He is now called to study and analyze this new world under the laws which directly govern spiritual phenomena. He finds analogies to physical life everywhere. However, he finds no conditions that are identical with earthly life.

He finds instead a world of superior material refinement, whose vibratory activities respond to the spiritual senses only. He is oblivious to the physical side of matter. This new world appears to be wholly independent of the old. There is nothing in it that would impress the physical senses. The suddenly liberated student is very like the man born blind and deaf suddenly restored to the normal use of those senses. He sees that which he has no language to describe. He finds it difficult to comprehend those new conditions, and he has neither words nor ideas by which he can report the effects even upon himself. He is simply dazzled, delighted, bewildered. He confesses that all efforts to translate that experience to man in the body are necessarily inadequate. He finds, as literal truth, that statement which declares "It hath not entered into the heart of man to conceive" what this higher world means, either as to its appearance, or its intelligence, or its activities.

The first great fact that forces itself upon the intelligence is the universality of matter. In this one fact he corrects the common error of regarding the spiritual world as an immaterial world.

This dogma of an "immaterial world" is one of those absolutely unthinkable propositions which discredits theology. It is also one that naturally offends modern science.

The spiritual world is as truly material as our own. It is simply a world of matter finer in particle and more rapid in vibratory action than our physical world. Because of this fact Natural Science uses those distinguishing terms "physical material" and "spiritual material." The spiritual world is just as real and tangible and visible to a spiritual man as is the physical world to the physical man.

Finite intelligence has never yet succeeded in passing the

limitations of matter. As far as science has penetrated (thus far) it finds both matter and intelligence in their natural relations. That is, it finds intelligence everywhere manifesting itself as the positive force in Nature acting upon matter. It finds everywhere matter as the negative property of Nature being acted upon by intelligence.

When the student is able to consciously and intelligently release his own spiritual body from the physical he proves another fact. He proves that statement of St. Paul which has been the subject of controversy for nearly nineteen hundred years. He proves that there is a "natural" or physical body and that there is also a "spiritual body." He finds that this spiritual body is a material body which in form and expression is but a finer representation of the physical organism he has temporarily quitted.

Whether the ego, or intelligent soul, ever discards this spiritual body is a question not germane to this present writing. It is a question, however, that is much debated in spiritual life.

Another important fact reveals itself to the investigator. This spiritual world has locality. It encircles this planet like a vast girdle. How far outward and upward it extends is a question not involved in this discussion. In appearance that world is analogous to this. That is to say, it has a similar distribution of land and water. There are oceans and continents. There are mountains, valleys and plains. There are forests, lakes and rivers. The same activity in the material world exists there as here. There is movement of all waters. There are magnetic changes of matter. There is growth in vegetation.

Whether there are other spiritual worlds than the one correlated to our planet is another question beyond the limitations of this work.

The world is inhabited just as this world is, by intelligent beings capable of moral improvement. They are real people; in fact, the same people who have previously lived here. They are simply spiritually embodied intelligences instead of physically embodied individuals. They preserve their identity as certain individuals from this plane. They continue to follow in the same general lines of intellectual and moral activity which engaged them in this world.

The student who learns these facts for himself demonstrates the correlation of two worlds of matter and the continuity of life, intelligence and activity.

The spiritual plane is divided into many planes or spheres of life and action. The divisions are local as well as intellectual, social and moral. Scientifically speaking, the spiritual material of this higher plane is subject to the law of polarity or vibration. By reason of this law its several zones or spheres are regulated by what is known as the spiritual law of gravitation. The coarser material moving at lower rates of vibration very naturally constitutes the immediate stratum encircling the coarse physical earth. By the same law of spiritual gravitation the remote regions of the spiritual world are those whose material substance is finest and whose vibratory action is highest.

The physically released inhabitants of the spiritual plane occupy the several zones or belts composing the spiritual world. They find their spiritual homes in one or another of those strata representing lower or higher states of refinement and vibratory action. They select as their homes that particular sphere or locality to which their own vibratory condition impels them. This means that spiritually embodied men differ in degree of material refinement and vibratory action just as do men in the physical body. It means that spiritual people, like earthly people, seek those localities and that social environment which correspond to their own stages of development.

The spiritual law of polarity or vibration is again illustrated by the separation of spiritual beings into many social castes.

For example, the spiritual plane immediately surrounding this earth consists of the least refined spiritual material moving at the lowest rates of vibration. Hence that zone is the negative state of spiritual material. It therefore does not reflect light. It appears (to spiritual vision) darker than those localities above and beyond.

Another important fact confronts the traveler. He finds that mere liberation from the physical body does not in the least change man in his essential nature. He finds that it does not mean an instantaneous absorption of universal knowledge. It does not effect a sudden revolution in the moral nature. In fact, the change called death leaves a man very little wiser and morally no better than he was immediately before. He finds himself somewhat in the same position of an ignorant American who has been suddenly transported to the center of some European civilization. Certain additional facts of Nature are literally thrust upon him. He cannot, however, obtain any definite information as to the nature of those facts without study and investigation.

His own limitations, spiritual, mental and moral, curtail his comprehension and enjoyment of that which confronts him, just as ignorance curtails a man's enjoyment in this world. He finds in the spiritual world, as in the physical, that a man enlarges his store of knowledge and capacity for enjoyment through the slow processes of education and self-development. If ignorant and vicious when released by physical death, such a man will follow the same impulses and passions which governed him in earthly life. He neither appreciates nor seeks that which is refined, intelligent and noble. Instead, he seeks that which is in natural or vibratory sympathy with himself.

The physically disembodied man discovers that it is his own acts, thoughts and motives, which have conditioned his spiritual body to one or another of the spiritual zones or localities. If his earth life has been intelligent, chaste and purposeful he finds himself attuned to the higher planes and the higher circles of spiritual life. Under such conditions he passes outward from the earth plane by the law of spiritual gravity and dwells in that sphere and among such people as are harmonious to himself. If, on the contrary, his life has been vicious, ignorant, criminal and impure he finds that the "spirit" has been coarsened by that previous life in the body. He finds, therefore, that he is in touch with only the lower stratum of spiritual material and spiritual society.

Under these conditions the spiritual body cannot rise. It remains in the negative regions of spiritual existence. It is therefore an integral part and natural representative of negation or darkness. It cannot rise to the light. It appears, therefore, as darkness to itself and to others.

By reason of this natural law and this actual condition in the spiritual world have arisen those well-known but mysterious allusions to "earth-bound spirits," to "angels of darkness," to "regions of darkness," to the "outer darkness" and to the "darkness of ignorance."

In this physical world darkness and evil and ignorance are linked rather in a figurative sense. In the spiritual world they are linked in a literal as well as in a figurative and ethical sense.

What we term the law of spiritual gravity is, in fact, that universal spiritual principle of polarity which governs evolution upon both planes of matter. This is the principle which underlies all of the propositions of Natural Science. It is the principle of spiritual affinity which constitutes the text of this present writing.

Reducing this principle of polarity to a general proposition it would read as follows:

There is in Nature a universal principle which impels every entity to seek vibratory correspondence with other entities of its kind.

Under this universal principle the spiritual world divides itself into many material and social regions. By the same principle spiritual beings seek that zone or locality whose material refinement and social development correspond to themselves. It follows, therefore, that the wise and the ignorant, the good and the evil, the active and the idle, group themselves in the order of their affinities.

When the student sees these conditions as facts in Nature he proves for himself that Nature embraces a principle which continually tends to bring together all things of the same degree of refinement and vibratory activity.

This same principle of polarity operates in human society upon the physical plane. While caste in this world appears to depend almost wholly upon external advantages and physical conditions, this is not the fact. In reality all human organizations and social integrations are dominated by the spiritual sympathies rather than the physical conditions, professions or advantages. A common expression in our own society illustrates this principle. We hear it said that such or such person is "out of his sphere," meaning that the individual is not mentally or morally equipped for fellowship in a certain circle.

There is one radical difference between the physical and spiritual worlds in this respect. On the physical plane we only know by intuition when a person is spiritually out of touch with his social environment. In the spiritual world, however, the law of vibration sets its ineffaceable sign upon every man. That is to say, one who remains in the lower stratum of the spiritual world is himself in that negative state which causes him to appear dark. On the contrary, he who rises by reason of his own refinement vibrates at a higher tension. He is in a positive state of activity which may appear first as color and next as light.

"Hell" and "darkness," therefore, are not mere figures of speech. The word "heaven" has a literal significance. Darkness is both locality and condition, as well as appearance. Light is both locality and condition, as well as appearance.

Every individual, whether physically or spiritually embodied, throws off spiritual magnetism. During earthly life those magnetic waves are invisible to the physical eye.

In spiritual life, however, they are distinctly visible to the spiritual eye. The spiritual man appears to his fellow men as veiled in darkness, or he gives off magnetic waves so rapid in vibratory action as to produce the effect of either color or light to the spiritual eye. Thus a spiritual man appears in an aura of darkness, or of color, or of light, according to his degree of development, viz., according to the degree of his material refinement, vibratory action and magnetic power.

Persons familiar with the seance room will recall how frequently a clairvoyant medium will refer to the "aura" of certain visitors. At one time the medium may allude to a physically embodied visitor, at another time to a spiritual visitor. In both cases the medium sees spiritually those waves of spiritual magnetism thrown off by the person in the physical body or out of it, as the case may be.

In this we find explanation of the "halo" or the shining cloud surrounding "angels" that have for all ages been written about or reported by mediums or pictured by artists. The finer the spiritual condition the more brilliant and magnetic becomes this aura.

Saul of Tarsus was prostrated by what he could only describe as a great light. There was a very excellent reason why Moses' spiritual visitors were concealed by a cloud, from the uninstructed Israelites. The effect of sudden contact between a highly developed spiritual being and a physically embodied man wholly unprepared for such contact is a dangerous encounter for the earthly man. It would be as fatal as the live electric wire.

Thus the law of vibration governs the material conditions of the spiritual world. It is a natural law that assigns man to darkness or envelops him in light in accordance with his own essential nature.

It must be understood, however, that it is the intelligent soul itself which controls the vibratory action of both bodies, physical and spiritual. The vibratory action of the spiritual body is, in fact, but the reflex action of the soul itself. It is, therefore, the soul or the ego which is coarse or fine, weak or forceful, dull or active. To be dull and heavy of soul is to be coarse in material texture, slow in vibratory action, negative in condition, dark in appearance. To be active of soul or intelligence is to be fine in material particle, rapid in vibratory action, positive in condition and luminous in appearance.

Thus, after all, it is the soul which drags the body down. It is the activity or inactivity of the intelligent soul which determines the local habitation of the spiritual body and thus its own social environment.

Caste in the spiritual world means more than it does here. People there are not so often found out of place. In a literal sense, men show their "true colors" in the spiritual world. The natural spiritual law leaves the individual little opportunity for simulation. He appears as he is, stupid or active, dull or intelligent, evil or good. He appears selfish or cowardly, noble or exalted, just as he is in fact. He is clothed in darkness or light according to his own self-made conditions. In short, he is "known" in the spiritual world.

At the hour of physical death the released ego, invested with its ethereal body, may rise rapidly from the earth or it may cling indefinitely to its former earthly haunts. It may condition itself to the coarser and darker regions close to the physical plane, or it may be able to rise rapidly to those finer, lighter and more positive regions lying far from the physical world. It is the soul of man which holds his spirit earth-bound, or impels it to higher planes when once released from the physical body.

Except a man knows this law he can form but the faintest conception of earth's immediate spiritual surroundings. It is only the student who realizes that humanity as a whole is in closest touch with the lowest stratum of spiritual life and intelligence. He preceives that mankind is assailed by evil spiritual influences more frequently than he is approached by the higher and better influences. The too often demoralizing results of the seance room are particularly due to the easy approach of vicious disembodied intelligences.

It must be understood, however, that the physically disembodied man is not permanently bound to any one locality nor to any particular social environment. Nothing but his own condition binds him either to place or to people. When that condition changes he releases himself. In that life, as in this, the individual is the arbiter of his own destiny. He rises or sinks by and through his own efforts or his own failures. In that world, as in this, love of knowledge, together with the courage to do and the strength to persevere, will gradually raise a man from a lower to a higher plane of existence.

The spiritual man overcomes unhappy spiritual conditions through educational processes. As a result his spiritual body becomes finer in particle with increased vibratory action. He is thus changed from the coarse, negative and dark state of being to the fine, positive and luminous state.

Thus there is a basis for our popular belief in halo-enveloped angels who go down into the dark places and among the fallen.

In the spiritual world the student finds an increase of purely intellectul activity as well as of philanthropic effort. Released from the exactions imposed by the physical functions of life man finds leisure for the intellectual pursuits he may have been denied here. Release from physical life means increase of time and opportunity for higher work. When the strain of this planetary life is over, the intelligent soul is free to follow its highest aspirations.

The scholar, statesman, artist, and poet, as well as the philanthropist, is now free to move forward in his chosen lines. In that world, as in this, there are facts to be learned. There are people to govern. There are beauty and love to be translated. There is music to be written and there are songs to be sung. There are romances to be lived. There is work to do.

In short, the higher life furnishes opportunity and means which most of us vainly seek here.

In another respect that life is analogous to our own. The same differences of opinion and intellectual controversy exist there as here. It is true that men no longer dispute as to the fact of life after physical death. It is true that they no longer regard the spirit world as a supernatural world. There is, however, no abatement of discussion and dispute and speculation over other facts in Nature. Men there, as here, debate the ultimate issues of existence. They speculate upon the immortality of the soul, the nature and character of God, and the probabilities of rebirth upon this planet.

In yet another respect these two worlds are analogous. The inhabitants of the spiritual world are men and women. The sex division is as distinct on that side of life as on this. More than this, upon that plane as upon this, man is the intellectual master. He is the organizer and ruler of spiritual mankind. On the other hand, woman is there, as she is here, his companion, co-worker and mate. In that world, as in this, man is the aggressive factor, while woman is the pacific factor, typifying the spirit of love. Though man continues to dominate the larger public life, the influence of woman continues to penetrate and ennoble the entire social organism.

Men and women continue to occupy the same relative position in that life as in this. In spiritual life, as in this life, man particularly represents law, order and knowledge, while woman particularly represents peace, love and all the æsthetic and ethical activities.

The student learns another still more important fact. He settles for himself a question that has been debated for ages.

He discovers that men and women in spiritual life continue the individual love relation. Between men and women on the spiritual side exists the same irresistible attraction which leads them into individual love relations.

This individual love relation must not be understood to mean merely an impersonal and altruistic friendship. It means instead, a personal and exclusive love and partnership based upon the spiritual law of affinity. This relation, therefore, is a more permanent one than the average marriage here. The union of a spiritual man and woman is unlike marriage upon the physical plane in that it lacks the physical functions, passions and sympathies growing out of the purely physical nature and physical conditions. Instead, it is a much closer bond based upon the spiritual principle of affinity. It is a union based, not upon physical passion, but wholly upon spiritual, intellectual, æsthetic and ethical sympathies.

In short, the student of Natural Science discovers that sex is an immutable spiritual principle.

He discovers that the bond between men and women outlasts all earthly and physical ties and relationships. The mutual love of man and woman transcends the physical functions and passions. It includes all that which goes to make up the higher man and woman. On earth we cherish an ideal of true marriage. That ideal contemplates not only the physical relation but a perfect sympathy in the higher intellectual and moral nature. How seldom that ideal is realized, let each one judge from observation and from his or her personal matrimonial experiences.

These facts concerning sex in spiritual life are accepted by the student as proof that it is a fundamental principle in Nature, and that the spiritual destinies of man and woman are correlated to each other. They are accepted as proof that the sex office transcends the mere physical functions of generation and reproduction. They are accepted as proof that the uses and purposes of sex are not exhausted in physical life. Instead, on the best authority of Natural Science it is declared that this question of sex is bound up in the highest development of intelligent moral beings.

It is natural, perhaps, that physical science should claim that sex represents nothing but a set of physical organs. This is to be expected of a science which studies man wholly through the organs of generation and reproduction. Upon the spiritual plane, however, science clearly demonstrates that sex represents a set of principles.

Thus, one by one, the advanced student of Nature demonstrates principles and gathers facts which directly controvert certain popular dogmas of physical science.

A scientific acquaintance with the spiritual side of life establishes the fact of two correlated worlds of matter, life, intelligence and love. Man occupies each of these worlds in turn. The investigator discovers that both are governed by one set of general principles and that causes in one world may produce effects in the other.

There is no death. Instead, a man has one life in two worlds. When he leaves the physical body he simply takes up life on the other side as would any stranger suddenly transported to some strange and unfamiliar country. He takes up life under new conditions while remaining in essence the same man he was on earth. He is released from physical exactions and physical activities, nothing more. He continues to feel the same impulses, passions, appetites and desires that he had encouraged here. He is moved by the same hopes and aspirations which governed him here. He is released from physical toil but not from activity. He does not suffer pain through physical disease. He is not, however, exempt from pain.

All of which means that an ex-human being is the identical individual who passes from this life to that, be he wise or foolish, good or evil.

Spiritual life is an inevitable sequence of physical life and development. An intelligent, purposeful and happy spiritual life depends upon the substantial basis of an intelligent, purposeful and chaste human life. Man is, therefore, the arbiter of his own destiny. Nature furnishes the time and the opportunity. Man is left to either improve or waste his time. He is left to accept or ignore the opportunities which Nature offers.

What may appear to be adverse conditions in this life may, in fact, be the very conditions which best develop the individual spiritually and morally.

With this array of important facts Natural Science looks upon the evolution of man from a very different point of view from that taken by either scientific skepticism or dogmatic theology. These conclusions represent the combined research, experiment, experience and judgment of all the great Masters of the Law, past and present. Thus it follows that the whole philosophy of life is based upon that proposition which constitutes the text of this chapter, viz.:

LIFE HERE AND HEREAFTER HAS A COMMON DEVELOPMENT AND A COMMON PURPOSE.

This is one of those propositions which, though contradicting physical science, yet explains those facts of Nature which physical science cannot explain. Scientific investigation, extending through all the ages in both worlds of matter and life, clearly demonstrates that the evolution of man is a correlated physical and spiritual process. It goes further. It shows that this correlated and ceaseless movement of life and intelligence involves a stupendous plan and purpose.

Evolution is not a matter of chance nor of mere physical adaptation to physical ends. Instead, evolution moves in given lines according to certain immutable spiritual principles. Nature's plan and purpose are discovered through and by the demonstration of one principle, fundamental in Nature, which ceaselessly operates to improve Nature's products. This plan of progress is traced through the refinement of matter, the increase of intelligence, and the development of morality. The study of this plan begins with the study of the mineral atom upon the physical plane. It embraces the plant and the animal. It includes man upon this physical plane. It reaches out to the study of man in the spiritual world.

Nature's plan involves a purpose. The universal activities look to something more than a physically improved or even a spiritually improved being. The principle which operates to improve all of Nature's products has an ultimate purpose in view. That purpose is faintly foreshadowed in the mineral and in the plant. It is revealed in the animal. It must be consummated in man. Natural Science, therefore, declares:

1. That evolution is the expression of intelligent spiritual principles.

2. That evolutionary processes conserve an intelligent purpose.

3. That Nature seeks to fulfill that purpose under one general spiritual principle which refines matter and increases its vibratory action.

This work will be devoted to an exposition of this universal plan and purpose and of that particular principle which conserves this plan and purpose.

The philosophy to be presented rests upon Natural Science. It, therefore, consists of deductions drawn from demonstrated principles and facts in Nature. The theories of physical science rest entirely upon certain proved facts of physical Nature. The philosophy here discussed rests upon certain proved facts of both physical and spiritual Nature.

Modern physical science traces the individual living entity from a single life-cell to the higher state of man, a rational being with physical functions, necessities and passions. The higher science goes further. It also deals with the individual living entity from the life-cell to the higher estate as a physical man. It does more than this. It deals with man, first, as a rational, spiritual being physically embodied, and next as a rational, spiritual being upon the spiritual side of life. Natural Science deals with the individual entity, from cell to physical man, not merely as a physical entity, but as a spiritual entity about which physical material integrates.

Natural Science goes further. It declares that man is the

highest expression of Nature's great plan. It declares that he is the nearest approach to Nature's ultimate purpose.

Man is found to be the highest product of Nature on either plane of life and activity. It is found that it is he who accelerates or impedes his own progress under Nature's plan. It is man alone who hastens or delays individual fulfillment of Nature's ultimate purpose.

Having demonstrated all this, the broader Science seeks to assist humanity in working out the plan and purpose of existence. Such assistance may be rendered only by teaching mankind the spiritual laws governing the physical world and the natural functions of the soul of man. The first lesson to be taught includes those few fundamental principles which govern the evolution of man. By such exposition the individual man is enabled to more intelligently, rationally and more directly move toward his own best good.

The student, coming in touch with both planes of life, is at first overwhelmed. He is absorbed by his own sensations growing out of visible and tangible spiritual phenomena. Later on his soul demands higher satisfaction than the æsthetic pleasures of sense. He must unravel the mysteries. He must know the elements, principles and forces which go to make up the spiritual plane. He must unfold the plan. He must seek the causes of things. He determines to know the meaning and the purpose of it all.

Gradually he formulates certain definite questions. He appeals to Nature for reply. It is true, however, that the questions which finite intelligence formulates depend for reply upon the endeavors of the finite questioner.

He is confronted by an Infinite Intelligence expressing itself through this evolutionary process in two worlds of matter and life. In his attempt at solution he nowhere finds higher authority than the individual intelligence of man himself.

The student questions Nature. Nature leaves him to analyze and demonstrate for himself.

What is the principle which governs these correlated worlds of matter? What is the exact relation of these correlated lives?

How far does physical life condition the after spiritual life? What is sex? What is the meaning of "masculine" and "feminine"? Why are man and woman immutably bound in flesh and spirit? What is love? What are the final meaning and purpose of an individual existence that ranges over two planes of matter, life, intelligence and love?

These are a few of the questions asked by the student who penetrates the spiritual plane. Ages of research, experiment and demonstration have furnished some satisfactory answers. That knowledge, however, has involved centuries of secret study. It has involved either silence and sacrifice or persecution and humiliation on the part of those who outstripped the popular intelligence. All religious and mystical so-called revelations have a basis in natural law. Our own Scriptures, together with other sacred literature, represent the honest efforts of honest men to impart to an unprepared world some portion of their own knowledge.

With this the reader is prepared to judge as between the theory of scientific skepticism and a philosophy of life deduced from exact knowledge of spiritual Nature. He is left to decide which of these schools is the more likely to be right, the one which illogically explains a small part of the physical facts of Nature, or the one that consistently explains a very wide range of physical facts and metaphysical or spiritual phenomena also. The reader must determine as between two conclusions which are diametrically opposite. He must accept or reject the fact of a life to come. He must perceive or fail to perceive an intelligent plan in Nature. He must find or fail to find life's purpose. He must recognize or fail to recognize the duty he owes to himself in working out that purpose.

This work undertakes to do no more than outline that universal principle in Nature which conserves Nature's plan and Nature's purpose.

The knowledge and the demonstration of this principle enable science and philosophy to declare that LIFE HERE AND HEREAFTER HAS A COMMON DEVELOPMENT AND A COMMON PURPOSE.

Briefly summarizing the foregoing pages, we find:

1. There are two correlated worlds of matter, life, intelligence, morality and love.

2. The difference between these two worlds, as to their material substance, is a difference mainly in the degrees of their refinement and the rate of their vibratory action.

3. Man has but one life and one course of development ranging successively over both planes of existence.

4. Nature provides but one evolutionary plan.

5. Nature develops that plan under one universal spiritual principle.

6. Nature's plan involves a definite purpose.

7. Man, as Nature's highest product, may accelerate his own development and hasten the fulfillment of Nature's purpose in the individual life.

8. He does this by and through exact knowledge of and rational conformity to Nature's plan of progress.

9. Such knowledge is obtained by and through investigations and experiments which cover spiritual as well as physical life.

10. The study and analysis of one world without reference to the other yield only partial results and a partial science.

11. The individual who best conserves his own development is he who seeks knowledge as to both states of being. He it is who will soonest fulfill and enjoy Nature's higher purpose.

CHAPTER V.

THE SCOPE, METHOD AND PURPOSE.

The following pages are intended as a partial exposition of the philosophy of individual life based upon Natural Science.

This work must be understood as departmental in character. It therefore has limitations. The first of these limitations is as to scope. The work is confined, primarily, to the consideration of man as an individual. Secondarily, it is confined to the individual man physically embodied. Lastly, it is concerned with the individual man in his individual relations upon this physical plane.

This philosophy holds that the individual man and the individual human relations have their distinct places in the higher evolution of man.

The vital issue to man in the physical body is not the nature and character of ultimate conditions upon the spiritual plane. On the contrary, it is to discover how he, as an individual, may best fulfill the duties and responsibilities of this present life. It is to discover the secret of individual attainment and happiness in this present world. Mere knowledge is valueless except it be made to serve a present and practical purpose. Knowledge that does not have a direct bearing upon the present daily life of the individual is an incumbrance and not an acquirement. It does not matter how unusual or wonderful that knowledge, in itself, may be.

Knowledge of the life to come is valuable to an earthly man, only in so far as it furnishes the motive and inspiration of a higher and better earthly life. This present work is not intended to divert the mind of the reader from earthly life or living. It is not intended to substitute a purely intellectual vision of future spiritual possibilities for a practical knowledge that will conserve present life upon this physical plane of action. This philosophy wisely holds that so long as man lives in the physical body and makes his home upon this solid earth, his best energies belong to his development in this earthly sphere of action.

The physically embodied man who becomes exclusively engrossed with the spiritual side of life makes as grave an error as the man who permits himself to become completely absorbed in the physical side.

It is important that man should know that he survives physical death. It is equally important that he should know that this earth life affords opportunities which do not obtain in spiritual life. It is important that he should realize that earthly life constitutes a series of activities which appear to have endless consequences. This philosophy continually reminds man that it is the life here which determines life there. It is the motive here which determines the habitation of the spirit there. It is the act here which evokes recompense or penalty upon the soul in that other life. In brief, it is the foundation here which supports the superstructure there.

In treating of man from the standpoint of the individual, Natural Science simply follows Nature's universal plan.

Natural Science is so named, first, in that it includes all natural phenomena, physical, spiritual and psychical; next, in that it adopts Nature's plan when it comes to the study, analysis and demonstration of those phenomena. Nature universally begins with the individual. Nature works only through the individual, whether it be an atom, a cell or a man. Rocks are but aggregations of individual atoms. Plants, animals and men (physically speaking) are but aggregations of individual life cells. Species are but aggregations of individual animals. The human race is but an aggregation of individual men and women.

The preliminary work of the higher science is to outline those spiritual and psychical principles, properties and elements which go to make up the physically embodied individual man or woman. True philosophy based upon this science, therefore, seeks first to outline the individual rights, relations, duties and responsibilities of this physically embodied individual. Physical science shows that Nature constitutes the individual as the center of all physical improvement in species. Natural Science proves that the individual is the magnetic center of all spiritual forces. Its consequent philosophy shows that the individual is the intellectual and moral center of all human relations. In brief, the individual must be accepted as the center of all integrations and organizations, whether they be physical, spiritual or psychical.

By thus accepting Nature's plan this particular work is not only confined to man in his earthly sphere of action, but to man in his personal relations during this earthly life. Therefore, the work is necessarily departmental and limited in scope.

In preparing this primary work the writer has adopted the rational method of treatment. In this respect alone the discussion may be found to differ materially from all previous attempts to explain the spiritual side of Nature and the ethical duties of man.

In this respect the discussion may be found to differ radically from the great body of spiritual philosophies and "occult" works now published and widely read.

The problem that has confronted the school of Natural Science for ages has been the proper presentation of its philosophy. Some mistakes have been made. There are records of failure where intemperate zeal on one side has been met and defeated by popular ignorance on the other. There have, however, been notable successes all along the path of effort. The great world religions testify to the truth of this statement. The correspondences of all great religions show that mankind are working out the same fundamental spiritual principles and are classifying the same facts in Nature and the same ethical truths.

The disagreements of religions indicate merely varying conditions as to time, race, nation, custom, language and climate. Those disagreements also indicate the variation in degrees of knowledge, power and purity attained by a particular teacher.

Buddha and Christ were expounders of spiritual and psychic laws under widely different conditions. They differed also as individuals. They differed in personal attainment, character and method. As a result, we have two philosophies colored by the individuality of their teachers. Though both systems are based upon the same principles, they are very unlike in outer form and creed.

Each of these teachers expressed his individual concepts of spiritual principles and of human responsibilities. Each employed the method which appeared as best adapted to the intellectual and moral development of his own time and race.

Buddha founded an order of ascetics vowed to poverty and beggary. He taught a literal renunciation of material comfort and of natural human ties as the life of highest attainment. He saw in a life of seclusion and introspection what appeared to him a higher life than one which shared the common activities and the common joys of human life. The fundamental doctrine of Buddha was the *sacredness of all life*. The fundamental characteristic of the man was *reverence for law*.

Christ, on the contrary, conducted an active ministry in fellowship with "publicans and sinners." He did not found an order nor adopt a distinctive garb. While he did not exalt himself, neither did he abase himself with shaven head and begging bowl. He preached temperance without imposing austerities. He neither condemned the natural activities of men nor disparaged natural, human relations. The church he attempted to found was one not built by hands, nor was it symbolized by priestly garb nor priestly authority. The fundamental doctrine of the Nazarene was the *universal brotherhood of man*. The fundamental characteristics of the man were *compassion, pity and love*.

The method adopted by Buddha would have found no response among the Jews and Gentiles of Judea. However, the truths taught by Buddha were the same as those laid down by the later Teacher. Who that knows the spirit of this age will venture to say that the method of teaching employed by Christ would be effective in this age of scientific skepticism?

How and by what method shall Natural Science convey its message to the twentieth century?

Nineteen hundred years have wrought great changes. The natural processes of evolution have advanced the average of humanity physically, spiritually, intellectually and morally. The average physical organism among the superior races is more refined and is keyed to higher vibrations in Nature. The average of rational intelligence is immeasurably higher.

The twentieth century is confronted with purely intellectual problems undreamed of by the children of Israel three thousand or even nineteen hundreds years ago. Neither the methods of teaching employed by the Indian sage nor those of the Hebrew Messiah would meet the requirements of this age and people.

Six hundred years of the most exact methods of physical science have evolved a higher character of civilization and a higher order of intelligence. The discoveries of physical science have enriched the popular mind with facts of Nature and equipped it with a scientific vocabulary. It has done more than this. It has inspired the popular mind with a desire to search for other facts in Nature, especially those facts which have to do with the evolution of man as an intelligent and moral being. Physical science has done still more than this. It has prepared the modern mind to search for the facts of Nature by the same rational methods as are employed by the specialist of the modern schools.

Because of these facts and conditions Natural Science to-day adopts that method which is in harmony with the present stage of intellectual development.

The rational method of physical science creates the inductive mind. It inspires intelligence to seek facts rather than to exercise itself with speculation. It replaces intuitive methods with rational methods. It creates the desire for that which appears practical, useful and germane to life as we know it.

Modern western intelligence is, in method, almost the exact reverse of ancient Oriental intelligence. It therefore demands methods of teaching in conformity to itself. A practical people must be practically taught. A rational and scientific intelligence must be met with rational and scientific explanations of the phenomena of life.

It is not enough for religion to say, at this time and to this people, "God is love." Modern intelligence demands of science that it shall elucidate those principles, elements and forces which will rationally and scientifically prove that "God is love."

The needs of to-day are not those which called out a Moses and a Christ. It is neither a governmental code nor an ethical creed that is required at this stage of development. It is no part of the present purpose of this philosophy to present an amended constitution for our own Republic. It neither intends nor desires to abrogate the Sermon on the Mount.

The Constitution of the United States is an almost infinitely higher ideal of human liberty and justice than was the Mosaic code. The Sermon on the Mount still embodies an ethical creed far in advance of the world's development.

The object of science in this connection is mainly to meet the demands of an advanced intelligence for a more definite and scientific knowledge of the higher laws of life. It is rather to show the scientific relation between natural law and ethical creeds. In working out this purpose the first problem is that which relates to the mere manner of presenting the laws in question.

How and by what method shall Natural Science seek audience of this critical twentieth century western intelligence?

Within the past twenty years a great wave of spiritual inquiry has passed over the western world. During that period our own country and Europe have been flooded with "occult" literature, part of which is translation of ancient writings and part is pure speculation and theory based upon those translations. The errors into which much of this literature and teaching have fallen were inevitable. Those errors arose from causes inseparable from all such sudden and widespread impulses. Those causes are:

1. Absence of authoritative teaching.

2. Lack of personal and definite knowledge obtained in a scientific course of self-development.

3. Confusion as to the differences and distinctions between Indian and Egyptian philosophies, their methods of teaching and principles of practice.

4. The use of poetic and figurative language where exact terms and scientific explanations are demanded.

5. The natural tendency of the human mind to speculate.
6. The absence of pure literary art.

Confusion in the mind of a writer is confusion to the mind of the reader. The intellectual grasp of a writer is conveyed to the reader just as unmistakably as his insincerity, egotism or selfishness are conveyed, no matter how lofty the theme nor how ingenuous the language.

It is just as impossible to teach the spiritual facts of Nature by intuition or by faith as it would be to teach mathematics by intuition and by faith. The practical and rational western mind has encountered many perplexities in its attempts to follow and assimilate the mystical poetry of a remote time, people, language, and development.

The attempt has resulted in a flood of mystical and poetic, but unintelligible explanation which is profusely adorned and disguised with Sanscrit and Hindoo terminology. Anything like a clear understanding of Natural Science involves the same conditions that are necessary to an understanding of physical science, viz.:

1. Authoritative teaching.
2. The presentation of that knowledge in language reduced to the demands of the rational mind.

Because of these facts and conditions this work endeavors to present natural principles governing the ethical phenomena of life, in plain Anglo-Saxon, and in conformity to the rational methods of modern science.

This work has a distinct and definite purpose.

This purpose is easily understood when the scope of the work is clearly outlined. The purpose involved is the transmission of certain scientific knowledge to the individual man and woman. This knowledge, it is believed, may serve their present efforts for an individual development and a personal happiness.

Every student of this philosophy learns in time that his own best development and happiness involve certain tasks. He finds that something more is required of him than the mere acquirement of knowledge, or the aesthetic enjoyments of the spiritual senses, or the exercise of individual powers. He realizes that the knowledge he is gaining is priceless knowledge to men who have not enjoyed his opportunities. He realizes that the pleasures and powers which he enjoys are those so deeply desired by but practically unknown to the average man. He realizes that the whole world suffers because of its ignorance of laws he is daily demonstrating.

Once realizing this, the student's duty to humanity becomes paramount. From this time forward his mind is occupied with the desire to pass on to others his already acquired gains.

Every student of this philosophy discovers that each individual owes a duty first to himself and next to the world. His first duty is to improve himself, his second to benefit humanity. The latter is his duty whether the world appears friendly or hostile to his purpose.

That which the individual owes to himself and that which Nature promises to him as an individual are very naturally the first themes selected by the student for discussion.

Science undertakes to analyze the individual. Philosophy presents that analysis and deduces principles therefrom. It becomes the privilege of the individual reader to investigate such findings and teachings. After this it becomes his duty to develop himself in accordance with Nature's plan and purpose, as they shall appear and appeal to him.

Physical science has discovered and classified a long list of physical elements and essences that are deadly poison to the physical body of man. It has also discovered and classified a still larger list that is beneficial and curative to physical man. In the same way Natural Science discovers and classifies certain spiritual principles, elements and forces which, rightly understood, lead on to development and happiness, or where ignorantly dealt with breed discord and degeneracy.

The purpose of this work is not merely to generally reaffirm those spiritual principles laid down in the past. Its particular purpose is to disclose certain other principles and relationships that have never yet been scientifically and rationally explained to the general student.

The fundamental doctrine of Buddha was the sacredness of

life, while that of Christ was the universal brotherhood of man. This philosophy recognizes and includes those fundamental truths as taught by the great Masters of the past. It declares further, however, that there is one other fundamental principle and one relationship which have never yet been properly nor publicly explained by theology, science, philosophy or law.

The principle here referred to is the universal spiritual principle of sex. The relationship meant is the individual, physical, spiritual and psychical relationship of man and woman.

This, indeed, is the paramount problem of this age. One who has kept pace with the development of the superior races will not question this statement. The sex question invades every department of modern life. The literature of the hour is almost wholly given over to theorists upon this question. The widespread "woman movement" is a part of this agitation. Neither theology, science, philosophy, law nor art has thus far intelligently or satisfactorily demonstrated the principle or explained the relationship involved. This, in truth, is a problem which only the combined effort of man and woman (aided by Naturel Science) can properly solve and explain.

True philosophy of all ages to come must include those spiritual principles and ethical truths which are laid down in Indian and Christian doctrine, for, as has been truly said by Colquhoun, "No one truth can possibly militate against another." It must be remembered, however, that schools of science as well as all things else, are subject to evolutionary laws. This being true, philosophies based upon science must extend their scope to meet later discoveries and demonstrations.

This being true, it will be found that the teaching in the twentieth century must be an advance upon that which was given out thousands or even hundreds of years ago. The philosophy of to-day imposes discipline but not austerities. It forbids greed, but it also advocates the proper accumulation, use and enjoyment of material wealth. It forbids intemperance and lust, but it sanctions a proper indulgence of the physical nature and of the human affections.

It reasserts the sacredness of life. It enjoins the universal brotherhood of man.

It does all this. It does more than this. The higher science to-day is prepared to state, to explain and to demonstrate the universal principal of sex. Philosophy based upon that science is to-day prepared to publicly and explicitly teach the value of individual life and the supreme importance of the true relation of man and woman.

CHAPTER VI.

THE GENESIS OF PHYSICAL LIFE.

It must be admitted that no satisfactory theory of the evolution of man can obtain so long as the genesis of life upon this planet is shrouded in darkness.

The real factors and causes of evolution are bound up in this question of life itself.

Once given a primordial life cell demanding nutrition and capable of reproduction, and modern physical science constructs a man. Out of this primordial cell and these physical functions and a "hostile environment" it evolves a man physically, intellectually and morally.

This, however, does not in the least explain the original appearance of the cell itself. Neither does it explain the nature nor the cause which originates it. It does not explain the original division of life into male and female. It does not explain the phenomenon of intelligence which attaches to the operations of all living things.

Indeed, physical science explains none of these things. On the other hand, it most unscientifically relegates them to the regions of the "Unknowable."

Interest in the genesis of physical life remains unabated. The inquiring mind of man is not dismayed by the "Unknowable" of other men nor schools of men.

The vital problem in science to-day is the appearance of organic life upon this globe of inorganic matter.

Physical science confesses itself baffled at every point when it would explain how life evolves from non-life, how sensation evolves from non-sensation, or why intelligence inheres in living things. It fails to explain these phenomena just as it fails to explain how intuitive intelligence rises into rational intelligence or how unmoral perceptions rise into moral conceptions.

Modern physical science has traced the man physical through the lower forms of life. It has not, however, mastered the secrets of either life, sensation, intuition, reason, morality, love or happiness. The vigilant biologist traces life to the nucleated cell. Here, however, in the department of Protozoa he becomes bewildered. He misses the connecting link. He fails to discover that subtle element which enters in and converts a simple vegetable cell into the nucleated animal life cell.

Physical science is baffled here just as it is when it seeks the connecting link between man and the ape.

The specialist of physical science finds a cell which may be either animal or vegetable. He can neither feel, see, weigh nor measure, however, that which differentiates the simplest form of the animal life cell from a vegetable cell. His science can not determine as to certain cells, whether they will germinate vegetable or animal life.

Thus, physical science fails to determine at what point a vegetable life cell is converted into an animal life cell. In the same way it fails to demonstrate where mineral activity is converted into the higher activities of vegetable life. Indeed, physical science has not as yet discovered the connecting link between any two of the great kingdoms of Nature, mineral, vegetable, animal and human.

The Darwinian theory of evolution depends upon the original hungry life cell.

This theory, it will be observed, does not account for the hunger of that cell any more than it does for life itself. There is one fact which physical science consistently ignores, namely, that the life cell seeks nutrition because of an innate and preexisting demand of some character.

Darwinism declares that everything came to be as it is because it was reinforced from without and because hunger and hostile environment forced it to do thus or so. It therefore accepts both hunger and life as ultimate mysteries. It thereafter concerns itself entirely with the physical phenomena manifested by this unexplained hungry cell after it is generated by undiscovered forces.

Thus, physical science rests upon assumptions which preclude further investigation as to the genesis of physical life. Failing to account for it by the means known to physical science, it therefore holds that it can not be accounted for in science. It contents itself with assuming that life somehow generates through mechanical and non-intelligent physical forces. It does not, however, explain either the principle or the elements underlying this mechanical and non-intelligent physical movement.

With this for a major premise physical science thereafter is bound to assume that all further evolution or organization or variation of life cells rests upon mechanical principles and non-intelligent physical forces.

Physical science is thus bound to assume that man is the direct result of a blind digestive apparatus.

The physical materialist simply assumes that the physical functions constitute the sole factors in evolution. He does not concern himself with the principle which set the hungry life cell in operation. He fails to demonstrate and explain the generation of life, the pre-existence of hunger and the original capacity for reproduction.

These assumptions, as will be seen, also ignore that principle in Nature which endows original protoplasm with the masculine and the feminine characteristics, qualities and capacities.

No school of science will satisfy human intelligence which attempts to explain evolution without reference to the principles which generate a living entity and endow it with hunger, sex and an impulse to persist as an individual.

"The present state of knowledge furnishes us no links between the living and the non-living." Thus declares the Enc. Brit. when summing up the value of all experiments thus far made by physical science as to the origin of physical life. This is true as to the research and experiment of modern physical science. It is not true, however, of the research and experiment of Natural Science. For years the world of science lived in expectancy of generating life by experiment. Two groups of scientists warmly debated the issue. One group, the experimenters, held to a theory of spontaneous generation of life from non-living substance. The other group maintained that life generates only from antecedent life.

The experiment consisted in sealing boiled water in air-tight jars.

When generation did not occur it had to be conceded that life does not generate in boiled water sealed in air-tight jars. The experimenters were right when they afterward agreed that life does not generate in a vacuum. Their opponents were also right when they declared that life generates only from antecedent life.

Upon the strength of these experiments physical science formulates what it terms the "Law of Biogenesis." While this theory fails to explain both the principle and process involved in the generation of life, yet it does undertake to explain how life can *not* generate. It claims that there can be no passage from mineral to plant life nor from plant to animal life. It declares that the doors of each kingdom are hermetically sealed upon the mineral side.

"Natural Law in The Spiritual World" is a popular work which attempts to apply this theory to the spiritual side of Nature. The argument is based upon the deductions of Huxley and Tyndall upon this subject. The author takes for granted that the opinions of these great specialists closed the case of Nature, for he says: "The organic is staked off from the inorganic by a barrier never yet crossed from the other side." Again he says: "No change of substance, no modification of environment nor any form of energy nor any evolution can endow a single atom of mineral substance with life."

If by "life" the author refers to animal life he is right. If he means vegetable life, however, he is wrong.

Physical science is not familiar with the process by which mineral substance is raised to correspondence with the vito-chemical life element.

All that physical science has really demonstrated is:

(1.) Animal life appears in a single nucleated cell.

(2.) It contains one substance not found in vegetable forms of life.

(3.) These cells are endowed with the functions of nutrition and reproduction.

(4.) These cells have an inherent tendency toward organization.

(5.) These organisms differentiate and improve in appearance, complexity and capacity.

(6.) All entities endowed with animal life display intelligence in their operations.

(7.) Animal life successively develops sensation, intuition (instinct), reason and morality.

In its last analysis physical science fails to suggest the causes which produce the original cell. It therefore relegates the origin of physical life to the region of the "Unknowable." Thus, the modern school stands convicted of insufficiency upon a question vital to scientific progress.

More than this, it is highly probable that physical science will reject any explanation which emanates from another school.

Is it not singular that familiarity with the history of science does not more rapidly promote a spirit of toleration among men of science? On the contrary, modern physical science, as a school or coterie, is almost as intolerant as were the older groups of investigators.

A certain pride of intelligence has been the stumbling-block all along the path of science. Egotism and dogmatism appear to be the weaknesses of most great specialists. It seems inevitable that when the individual intelligence finds itself completely baffled it declares to the world, "We have reached the unknowable."

For confirmation of this statement turn to the history of modern physical science alone.

When men of science will consent to substitute the word "Undiscovered" for the word "Unknowable," dogmatism and persecution will have received their death-blow. A pursuit of the facts of Nature will then replace the mere effort to sustain an individual opinion or the deductions of any certain school or coterie.

Physical science, in the person of Mr. Edison, goes far enough to declare that "All is a matter of vibration." Mr. Edison has already demonstrated that only a given range of the vibrations of physical matter can be reported and registered through the physical senses.

In the person of Professor Roentgen, physical science demonstrates that there are certain higher vibrations of matter, invisible and intangible to physical sense, which may yet be dealt with by science. The "X-ray" is neither seen, heard, touched, tasted nor smelled. It exists, however, as an acknowledged reality and demonstrated force.

Physical science is making wonderful strides in the demonstration of these higher vibrations of physical matter. It has already conclusively proved that Nature embraces both matter and forces which make no impression upon the physical senses of man. By reason of such experiments it has arrived at the point where it could naturally and readily extend its lines into the broader field of Natural Science.

It remains to be seen whether pride of opinion is still stronger than a desire for knowledge. If so, the open door to another line of research and experiment will be ignored.

Natural Science embraces a comprehensive knowledge of physical science. It goes further. It demonstrates that there is a whole world of material vibrations and forces which eludes all physical organs of sensation, all physical instruments of registration, all physical experiments, and all physical means and methods of analysis and demonstration.

Physical science agrees that the vibration of matter means the movement of matter. It agrees that this vibration or motion of matter, as a whole, is maintained by the action and reaction of individual particles moving upon each other. Physical science also agrees that by and through this ceaseless activity of matter force is generated.

Thus far physical science coincides with Natural Science. The more liberal school, however, simply goes farther along the same line of experiment.

Natural Science goes far enough to demonstrate that it is the ceaseless activity of individual particles moving upon each other which refines matter itself and increases its vibratory action.

In yet another respect it points the way for modern experiment.

Physical science is concerned at present with the vibrations of physical matter only. Natural Science, on the other hand, is concerned with the vibratory activities of spiritual material as well as physical material. It studies physical matter by physical means and spiritual matter by spiritual means.

The higher science is therefore in position to discuss the law of motion and number from two points of view instead of one. That is to say, it is prepared to estimate and compare the difference in the refinement of matter and the vibratory activities of two worlds of matter.

The present position and deductions of Natural Science represent its researches crystallized to certain definite propositions concerning this law of motion and number. As already explained, the law of motion and number refers, first, to matter in motion; next, to the degree of refinement of the atom in the compound. These deductions are presented to supplement rather than to antagonize physical science. It is hoped to establish co-operative relations with the modern school and to secure its sympathy and aid rather than to invite antagonism.

Even though physical science rejects the principles here laid down, there are still many independent thinkers and investigators open to truth from whatever source it may come. There are many difficulties in this attempt to discuss the genesis of life upon this planet. This is frankly admitted. This is true by reason of the fact that those experiments which deal with the progress of matter and the genesis of life are experiments which lie beyond the means and methods of physical science.

They do not, however, transcend the experiments of human science or they would not now be subject of record and publication.

It is, indeed, difficult to present as facts of Nature those operations of Nature which can not be demonstrated by the means known to popular science. For this reason the deductions of

Natural Science concerning the genesis of life may be mistaken as theoretical, if not purely fanciful speculation. In view of this the writer suggests to the reader as follows:

(1.) Discard as false whatever conflicts with facts of Nature already proved by physical science.

(2.) Keep clearly in mind the difference between proved physical facts of Nature and the theories advanced by physical scientists to explain those facts.

It is admitted that experiments involving proof of these principles transcend popular science. It is, however, explicitly declared that they come within the range of human science and are clearly comprehensible to finite intelligence. As already stated, Natural Science has its limitations. They are not, however, the same limitations which hamper physical science. Physical science is limited to the knowledge gained by finite intelligence operating upon the physical plane. Natural Science is limited to knowledge gained by finite intelligence operating upon both the physical and the spiritual planes.

When finite intelligence masters a universal principle in Nature and acquires definite knowledge as to the processes and effects of that principle, the limitations of finite science are apparently reached. This is the nearest approach of the finite to the Infinite Intelligence.

All that man here or hereafter (as it appears) may aspire to know of the Law-Giver is some knowledge of His laws.

Colquhoun, in his "History of Magic," has well said, "When we have once established a general law of Nature we have reached the limit assigned our faculties and must take our stand on the primitive will and fiat of the great Creator of the universe; for who would otherwise attempt to explain the cause of a general law? The true philosopher endeavors to connect the various phenomena of the universe in such a manner as to elicit one or more of these general laws; and it is in this way—and in this way alone—that we can best contribute to the completion of the sciences. To attempt to go beyond this point is an error into which no man of sound sense and philosophical tact will readily fall."

There is, however, so much obtainable knowledge of the spiritual side of Nature that Natural Science is justified in presenting certain spiritual principles as universal principles and certain deduction as fundamental in science.

The spiritual principle of polarity is the universal principle under immediate consideration. Those propositions which relate to the evolution of man rest upon this principle. Evolution involves the refinement of matter, the increase of vibratory action, the generation of life, the individualization of intelligence, the development of love and the attainment of happiness.

Those experiments which demonstrate the foregoing propositions lie beyond the present scope and methods of physical science. They are experiments possible only to such as are specially trained for such work. Further, it must be understood that these propositions are not intended to explain the principle of polarity itself. They indicate only the fact that science discovers such a principle. They are not intended to explain why this principle governs evolution. They are merely intended to state the fact that under this principle matter is refined and increased in vibratory action, that life is generated, intelligence individualized, love developed and happiness attained.

All that science can ever hope to accomplish by mere publication is to give facts, state principles and rationally elucidate them.

The principle to be stated and elucidated in this work is that principle by which matter is refined and increased in vibratory action through and by the efforts of individual entities seeking vibratory correspondences in other like individuals of opposite polarity.

Natural Science discovers something more in Nature than a universal spiritual principle of polarity or affinity. It discovers something more than physical matter in motion. It finds that a physical entity is something vastly more than mere physical matter keyed to a certain rate of vibratory action. It finds that all matter is alive, or rather that matter is animated by something which we name either magnetism, vitality or life. It finds, for instance, that a steel magnet exhibits a certain character of vitality, a tree possesses yet another, the animal still another, while it finds that man exhibits higher and more subtle energies than anything below him.

Science demonstrates that in addition to matter, Nature embraces certain subtle elements which are universal in time and space and are defined as the "Life Elements."

Finite science, be it understood, does not attempt to account for these elements in Nature any more than it does for matter and motion. It accepts them as universals and ultimates in Nature, and works along that hypothesis. Science simply classifies these Life Elements along with the other universals, namely, matter, motion, life, intelligence and love.

The universal Life Elements are four in number, defined as follows:

(1.) Electro-Magnetism.
(2.) The Vito-Chemical Life Element.
(3.) The Spiritual Life Element.
(4.) The Soul Element.

Science demonstrates that one or more of these vital elements magnetizes, vivifies, vitalizes or animates all physical matter, including the mineral atom, the plant, the animal and the man. It finds, therefore, that what we know as magnetism in metals, vitality in vegetation, and life in the animal and man are, in fact, certain temporary relations established between physical material and the finer and more subtle Life Elements.

The discovery and demonstration of the operation of these Life Elements constitute part of the occupation of Natural Science. The range of phenomena included in the operation of these elements extends downward to the unconscious mineral atom and upward to the self-conscious soul of man.

Study of the Life Elements in operation is the study of the universal principle of polarity or affinity. Through analysis of these elements is demonstrated the fact that the principle of polar attraction inheres in the vital elements themselves, and not in the solid particles of physical matter.

Knowledge of this important fact gives rise to certain other important deductions, viz.:

(1.) The union of physical matter and the Life Elements is brought about through and by the operation of that principle of polarity or affinity which inheres in the universal elements themselves.

(2.) Each Life Element displays dual and yet differing powers of positive and receptive energy.

(3.) In the union of physical matter with one or more of these Life Elements arise the phenomena of magnetism, vitality and life.

Science holds that these universal Life Elements are essentially unlike in essence and manifestation. As already stated, most of the experiments which support these statements lie beyond the present means and methods of modern physical science. It is, however, possible for the intelligent reader to gain a perfectly rational conception of the principles, properties and elements with which the higher science deals. It is also possible for the casual student to obtain an intelligent idea of the processes by which these properties and elements are converted into physical life.

The four universal Life Elements successively give rise to the four great physical kingdoms, viz., mineral, vegetable, animal and human.

Each one of the Life Elements gives rise to a distinctive form of physical activity or life. Each element governs a kingdom which is entirely distinct in physical appearance, in characteristics, capacities and activities.

Electro-Magnetism, the least potent of all the vital elements, animates mineral substance and displays merely the powers of union, cohesion and aggregation.

The Vito-Chemical Life Element, the next higher and more potent of the universal Life Elements, vivifies vegetable substance and governs a higher form of life and activity. This element displays something more than the powers of union, cohesion and aggregation. It does something more than unite already existing particles. It possesses in addition the powers of growth and organization. It generates new forms by attracting the necessary material from the universal elements.

THE GENESIS OF PHYSICAL LIFE

The Spiritual Life Element, a still higher and more potent element, generates a still higher form of life. Animal life is a distinct advance upon vegetable life. Animal organisms possess a vitality, capacity and power wholly unlike and superior to the plant. The animal cell exhibits the same unconscious union, cohesion and aggregation of the mineral atoms. They also possess the power of growth and organization as do the plants. The animal, however, as a whole, possesses capacities which far transcend those of the mineral or vegetable kingdom. These capacities are defined as an individual consciousness and the powers of sensation, intuition and volition.

The Soul Element, the highest and most subtle known to science, inspires the kingdom of man. In this fourth and highest kingdom are found every capacity, characteristic and possibility of all of the lower kingdoms. Added to these, however, is a certain character of life, of energy and of capacity never yet discovered in the lower kingdoms. In this kingdom and in this alone are found self-consciousness, a rational intelligence, morality, altruism, and a free and independent will and desire.

Thus, each one of the Life Elements dominates a particular kingdom and gives rise to a distinctive class of phenomena.

The higher science demonstrates that these elements are unlike in essence, differing in character and capacity. It demonstrates also that they are of different degrees of refinement and therefore move at different rates of vibratory action.

The two lower Life Elements, Electro-Magnetism and the Vito-Chemical, have a physical as well as a spiritual side. That is to say, these are the only two of the vital elements which move at such a low rate of vibratory action as to come within the range of physical experiment. Man, through his knowledge and control of these two particular elements, revolutionizes, from time to time, the established order of human society. Electrical, chemical and medical sciences rest upon the knowledge, control and a practical application of the Electro-Magnetic and Vito-Chemical forces in Nature.

Such men as Franklin, Edison, Morse, Roentgen, Kelwin and Gray are masters of electrical science to whom the whole world is indebted. Considering the marvelous results achieved by physical science through its knowledge of the least potent of the vital elements, it is not difficult to conceive of the still more wonderful results which would naturally flow from a knowledge and control of all the Life Elements.

A knowledge and mastery of all the Life Elements constitute the active pursuit and employment of the Natural Scientist. For illustration, it is a fact, and not a fairy tale, that through knowledge and control of the Vito-Chemical Life Element a scientific man may suddenly materialize a grain of wheat or a growing plant. It is a fact, and not a fiction, that by knowledge and control of the Spiritual Life Element a physically disembodied man may materialize a part or all of his own body to the physical sense of sight. The spiritual man does not, however, really create a body of flesh and blood. All that he does is to control sufficient of the living physical substance near him to thinly cover a part or all of his spiritual body. This is a common occurrence in the seance room.

It is the understanding and application of these finer forces in Nature which are responsible for the wonderful tales of the Oriental "Masters," "Adepts," "Yogi," "Fakirs," "Jugglers," and in our own western world, of spiritual mediums and hypnotists.

The reader may ask, "What is the difference between Masters, Fakirs, mediums, hypnotists, etc.?" Briefly, however, there are three fundamental differences; first, as to the range of exact knowledge; second, as to actual power over the elements; third and particularly, as to the use made of those powers.

The right use of such knowledge and the right application of such powers determine whether a man is an exponent of "White Magic" or "Black Magic." In other words, they determine whether he is a representative of a true or a false philosophy of life. For illustration, the reader is asked to compare the purposes and the work of a Master such as Christ with the ambitions and exhibitions of the professional fakir, medium and hypnotist.

The Soul Element is the highest life element with which finite science deals. This is true by reason of the fact that this is the highest and most subtle element which enters into the composition of man himself. Water can not rise above its own source when left to itself. Thus far the intelligent soul of man has not risen to an intelligent perception of anything higher than his own highest element.

This, therefore, is the element most difficult of analysis and demonstration. Study of this element and mastery of its principles, therefore, constitute the highest occupation of human intelligence on both sides of life.

"The proper study of mankind is man." This was not intended simply as the study of man anatomically or physiologically. He who really studies man, investigates and analyzes the capacities and powers of the intelligent soul. He does not merely dissect the physical body nor analyze its physical functions.

Knowledge and power on the plane of the Soul element, viz., knowledge of man and worthy dominion over him, are regarded as the goal of finite endeavor.

The several distinct kingdoms of Nature represent the union of physical matter with the several Life Elements in Nature. These several kingdoms are seen to rise successively as one after another the Life Elements are inducted into physical matter.

As already explained, the vitalization of physical matter depends upon the energies which inhere in the universal elements themselves.

For illustration, when electro-magnetism is extracted from iron or steel we have devitalized mineral substance or dead ore. If a bottle of herb extract is left uncorked, the medicine loses its potency or its life. When a man dies physically it means merely that the more permanent living spiritual man has withdrawn from the physical counterpart. It means that the spiritual organism, still animated by the higher life of the soul, survives that event and goes on with existence under new material conditions.

The withdrawal of the Life Element produces the same result in each kingdom of Nature, viz., devitalization or death. The material particles of any physical organism, divested of the controlling Life Element, experience a form of disintegration which we name decay and death.

Physical matter is therefore negative to and subject to the

action of the positive Life Elements. Physical life, as we see it, is therefore nothing more than the manifestation of a certain relation between those animating Life Elements and coarser physical particles.

The principle of polarity or affinity inheres in the universal Life Elements. This means that what physical science calls the law of vibration is, in truth, primarily a spiritual law operating through and upon physical material.

In the Life Elements, therefore, we find the cause of that universal cleavage in physical Nature known as sex, the phenomenon of positive and receptive energy. Each and every one of the Life Elements is dual in its nature and manifests itself as either positive or receptive energy. As a result, the law of polarity, or the law of sex, governs everything known to man, from the chemical atom to the intelligent soul. It will therefore be seen that everything upon this physical plane belongs to either the positive or the receptive department of Nature. This applies to mineral, vegetable and animal substance. It applies also to organized entities, plant, animal and human.

Between the positive and the receptive energies of any one of the Life Elements there exists an inherent attraction and irresistible impulse for union.

As a result, all physical substances, entities and individuals representing those dual energies are impelled to union by the affinities resident in the Life Elements themselves. Thus, each kingdom of Nature is divided into a positive department and a receptive department. Thus, each individual of that kingdom is arrayed on one side or the other, as a positive or receptive factor in evolution.

Between these two departments, between these oppositely polarized atoms, entities and individuals, exists an eternal affinity or impulse for union.

The science of chemistry is based upon those affinities which inhere in Electro-Magnetism and the Vito-Chemical Life Element. The philosophy of life is based upon those affinities which reside in the Spiritual and the Soul Elements of human life.

As already explained, the two lower Life Elements may be

THE GENESIS OF PHYSICAL LIFE 101

assigned to the department of physical science. This is by reason of the fact that they move at so low a rate of vibratory action as to bring them within range of physical experiment. On the other hand, those higher elements which govern animal and human life must be assigned to the realm of purely spiritual forces. This is by reason of the fact that they are so rapid in vibratory action as to entirely elude cognition or analysis by and through physical means. These elements are susceptible to analysis only by the means and methods employed in the higher science.

The process by which physical matter is gradually refined and raised to certain ratios of correspondence with the universal Life Elements constitutes the evolutionary process upon this planet.

The spiritual principle of polarity which inheres in those Life Elements appears to be the agent employed by the Great Intelligence to guide this process.

This principle evolves the four great kingdoms in Nature, successively governed by the four Life Elements.

At this point an attempt will be made to briefly present Nature's working formula of evolution. With this simple outline clearly in mind the evolutionary scheme becomes fairly intelligible. It must be understood, however, that Nature's formula is not presented as a reason for Nature itself. It is, instead, presented as an explanation of the method employed by Nature to generate life and to improve her living products.

In short, it is intended to show how a universal Life Element is inducted into physical matter, which explanation covers the question as to the genesis of physical life.

Careful attention to this formula will enable the reader to understand what is meant by the inseparable phenomena of the refinement of matter and the increase of its vibratory activity. It may enable the thinker to grasp the purpose which Nature is bringing to an earthly culmination by means of one principle fundamental in Nature.

Nature's formula for evolution, if reduced to one fundamental proposition, would stand as follows:

THERE IS A PRINCIPLE IN NATURE WHICH IMPELS EVERY ENTITY

TO SEEK VIBRATORY CORRESPONDENCE WITH ANOTHER LIKE ENTITY OF OPPOSITE POLARITY.

By keeping this one principle in mind the student can not fail to comprehend the entire formula, which is, in truth, but an amplification of this one proposition.

According to all science there is a planetary period antedating all forms of vegetable, animal and human life. During this period, according to Natural Science, the lowest of all the vital elements directly governs physical substance. During this period the cold Electro-Magnetic forces operate directly through and upon all physical matter. These forces operate as positive and receptive energy. In consequence, all cosmic physical substance is either positively or receptively charged with Electro-Magnetism.

As a consequence two important conditions obtain, viz.:

(1.) All physical matter is Electro-Magnetic or, in other words, mineral matter.

(2.) A distinct cleavage exists throughout this mineral kingdom, the one part being positively and the other receptively charged with the governing element of Electro-Magnetism.

The one part, the positive, is in a highly active state. The other part, though equally potent, is simply receptive to the demands of the positive side. Between the two parts there exists a universal attraction or affinity.

This attraction or affinity as between the two parts is, however, maintained through and by the individual attractions of the atoms which compose those parts.

This affinity between individual mineral atoms is governed by THAT PRINCIPLE IN NATURE WHICH IMPELS EVERY ENTITY TO SEEK VIBRATORY CORRESPONDENCE WITH ANOTHER LIKE ENTITY OF OPPOSITE POLARITY.

The result of this attraction between individuals is union. When such union establishes a perfect vibratory correspondence an equilibrium of forces obtains. The result is a permanent cohesion, or an indissoluble union, chemically considered. When such union fails to establish vibratory correspondence repulsion occurs and the divorced atoms seek union elsewhere in pursuance of the law of chemical affinity.

The result of these individual efforts for an individual equilibrium is a ceaseless combining and recombining of mineral substances. By and through these individual efforts for individual adjustment new unions of mineral substances are being continually formed. Gradually these individuals establish permanent unions, after which the same law operates to attract and unite whole groups of individuals. These combinations we term "chemical compounds."

These compounds are the offspring of the Electro-Magnetic energies in Nature.

Thus was our whole earth gradually solidified from its incandescent and gaseous stages.

Moved by this universal principle of affinity, that entire gaseous and liquid mass finally settled into a solid globe in the order of its individual affinities. The law of affinity, however, was not suspended with the crystallization of the mineral gases. On the contrary, it continues to operate through the solid as well as the gaseous and liquid substances. The result of this ceaseless activity of mineral substances is:

(1.) Reduction in the size of the individual atom.

(2.) Increased vibratory action of that atom in the compound.

In consequence, mineral substance is universally (though not simultaneously) refined as a whole and keyed to higher vibrations in Nature. There comes a time when a portion of this mineral substance is raised to certain ratios of correspondence with the vibratory action of a second and higher Life Element in Nature.

This next higher element of Vito-Chemical Life lies universally and coextensively in time and space with the lower element of Electro-Magnetism (as far as science knows).

When the mineral atom has been thus raised to the harmonic relation necessary, it becomes susceptible to the essence and activity of the Vito-Chemical Life Element. Impregnation occurs. The higher Life Element is inducted into mineral substance and the mineral atom becomes a vegetable particle, endowed with Vito-Chemical or vegetable life.

By this process and under this principle is all life generated upon this planet.

The phenomena of the vegetable kingdom are analogous to those of the mineral. Moved by the same principle of polarity or affinity, the Vito-Chemical Life Element operates through and upon vegetable substance as positive and receptive energy. In consequence all vegetable substance is either positively or receptively conditioned. A distinct cleavage obtains throughout the vegetable kingdom. One part represents the highly active or positive energies of the Vito-Chemical Life Element; the other, though equally potent, represents the receptive or absorbing energies of the same element.

Between these two parts, oppositely charged or polarized, there exists a universal and involuntary attraction or affinity. This attraction or affinity of the whole is maintained through and by the individual affinities of entities composing the whole.

The affinity between individual vegetable particles is governed by THAT PRINCIPLE IN NATURE WHICH IMPELS EVERY ENTITY TO SEEK VIBRATORY CORRESPONDENCE WITH ANOTHER LIKE ENTITY OF OPPOSITE POLARITY.

The result of this attraction or affinity between individual particles is union. In the temporary correspondences between individual particles, the powers or generation obtain and reproduction of vegetable cells occurs.

These reproductions represent the positive or male energies of Vito-Chemical Life acting upon the receptive, absorbing, feminine energies of the same element.

The progeny of these combined powers is, therefore, but an incidental result of the effort made by two vegetable particles for vibratory correspondence.

The same natural law of affinity operates to draw individual pairs into definite groups or into definite forms. As a result we have a whole kingdom representing the processes of the Vito-Chemical Life Element, from the simplest lichen to plant, bush, vine and tree.

THE GENESIS OF PHYSICAL LIFE 105

The effect of this ceaseless activity in vegetable substance is to refine it as to particle and to increase the vibratory action of that particle. As a result vegetable substance is universally (though not simultaneously) refined as a whole and keyed to higher vibrations in Nature. There comes a period when a proportion of these vegetable particles is raised to certain ratios of correspondence with the vibratory action of a third and yet higher Life Element.

This third and higher element is known to science as the Spiritual Life Element. It is supposed to pervade all space, as do the lower elements of Electro-Magnetism and Vito-Chemical Life. When the vegetable cell reaches a given ratio of vibratory correspondence with this higher element it becomes susceptible to the essence and activity of that element.

As a result the Spiritual Life Element is inducted into the vegetable cell and converts it into the nucleated animal life cell.

Moved by the same principle of polarity or affinity the Spiritual Life Element operates through and upon animal life cells as positive and receptive energy. This principle operates to produce results clearly analogous to those in the lower kingdoms. In consequence all animal nature is positively or receptively polarized. A distinct cleavage obtains in the animal world, one part being positive or male, the other receptive or female. The one part represents the highly active and aggressive nature of the Spiritual Life Element, the other represents the receptive or absorbing nature of the same element.

Between the two sexes exists a universal and voluntary attraction of affinity. This attraction or affinity between the two sexes, as a whole, is maintained through and by the individual affinities of the male and female animal.

This affinity between male and female animal is governed by THAT PRINCIPLE IN NATURE WHICH IMPELS EVERY ENTITY TO SEEK VIBRATORY CORRESPONDENCE WITH ANOTHER LIKE ENTITY OF OPPOSITE POLARITY.

The result of this affinity between individual animals is union. Where such union establishes vibratory correspondence it becomes a permanent union. Where it fails of Nature's purpose, however, repulsion occurs and the divorced animals seek harmonic relations elsewhere. Reproduction is one of the physical results of this effort for self adjustment. The powers of generation are attained in that supreme effort of the individual pairs.

Reproduction, is therefore, the incidental result of the individual effort for self adjustment between animals.

The principle of affinity operating upon animal life produces:
(1). The positive and receptive activity of protoplasm.
(2). The organization of cells into complex organisms.
(3.) The union of individual animals.

The effect of this ceaseless activity in the animal kingdom produces results analogous to those of the lower kingdoms. Animal substance is gradually (though not simultaneously) refined as a whole, and keyed to higher vibrations in Nature. This refining process in animal life goes on as to physical particle, cell, organism and intelligence. Consequently and correspondingly the vibratory action of particle, cell and organism in animal life is increased.

There comes a period under evolutionary processes when animal substance is so refined and the animal man so raised in vibratory conditions that he comes into a natural and harmonic relation with the fourth and highest Life Element in Nature.

This is the Soul Element, universal in time and space as are the lower elements of Electro-Magnetism, Vito-Chemical and Spiritual Life.

When this period is reached, the most subtle element in Nature is inducted into the animal man and the animal man becomes the human. From this point of progress man is said to possess a soul. From this period man is defined in science as "A Living Soul."*

The birth of the human has been scientifically, though poetically, expressed in Genesis. The poet says, "God breathed into his nostrils the breath of life and man became a living soul." The higher science regards this as merely a poetic version of a

---

*See quotation at the close of this chapter from Alfred Russel Wallace in his "Darwinism."

literal truth. That is to say, when the dual organisms of animal man reach a certain stage of refinement and vibratory action he becomes susceptible to the potencies of the highest Life Element in Nature.

When this evolutionary stage is reached, the infant so prepared breathes it as the breath of a higher life, thus becoming "A Living Soul."

Science holds that the induction of this element confers upon man those distinctive qualities defined as human. From this period man enjoys self consciousness and consciousness of other selves. This means an independent intelligence and a personal identity with the higher powers of reason, intuition and memory. It means also the higher powers of love and a capacity for purely moral development.

These however, are not the limit of the soul's capacities.

The supreme power of the Soul Element is held to be its power of persistence after physical death as an individual and self conscious intelligence. The Soul Element in its positive and receptive energies corresponds with all of the lower Life Elements of Nature. Moved by the same principle of polarity or affinity, the soul Element animates and inspires humanity. It moves in lines of positive and receptive intelligence, the intelligent Ego being essentially masculine or feminine. The positive or masculine soul particularly represents intelligent force of will, while the receptive or feminine soul particularly represents intelligent force of desire.

Thus a distinct cleavage exists in the intelligence as well as in the physical and spiritual organisms of the human family.

It must be understood that each higher kingdom includes all of the energies and functions and powers of the lower Life Elements.

For illustration, a plant includes the energies and powers of Electro-Magnetism as well as those of Vito-Chemical Life. The animal also represents these same two elements though governed by the Spiritual Life Element. Man, however, combines the energies, functions, and capacities of all the elements, though he is dominated by the highest, the self conscious soul. It follows, therefore, that the polar affinities in human life include an attraction in all of the lower elements. The strongest attraction between man and woman, is however, an attraction governed by the demands of the highest element, the soul.

Attraction between intelligent human beings is based upon THAT FUNDAMENTAL PRINCIPLE IN NATURE WHICH IMPELS EVERY ENTITY TO SEEK VIBRATORY CORRESPONDENCE WITH ANOTHER LIKE ENTITY OF OPPOSITE POLARITY.

The result of attraction and affinity between intelligent human beings is union or marriage. Failure to attain the harmonic relation sought results in a natural repulsion, natural divorce and a natural desire to seek self adjustment elsewhere. The physical union of two human beings is an expression of the soul's desire for an individual happiness.

In that physical union, however, the powers of generation are attained and reproduction follows. Reproduction in the human family is, therefore, an incidental result of an effort made by individual intelligence seeking its own satisfactions.

Thus the same principle of affinity which cystallized the fiery cloud ball into our solid earth continues to operate even after it has perfected the four great physical kingdoms and established man as master of them all. The arrest of the physical body occurs with man. Nature has completed the physical instrument of intelligence as far as its mechanism is concerned. Man is now provided with all necessary organs for the uses of physical, spiritual, intellectual and moral living.

NATURE FURNISHES NO HIGHER KNOWN TYPE IN ORGANIC FORM THAN MAN.

The physical organism of man is susceptible to immeasurable refinement and corresponding increase in vibratory action, but not to any known change structurally.

With the arrest of the physical body begins the higher evolution of man.

The Soul Element, though manifesting through physical forms, yet governs the distinct kingdom of intellectual life and therein carries on the higher evolution of man. The greatest kingdom in Nature, the realm of rational intelligence and of ethical development, is the kingdom of the Soul.

Thus, the universal Life Elements appear to have special offices in the economy of Nature. To Electro-Magnetism is assigned the refinement of mineral substance and the solidifying of the planet. To Vito-Chemical Life is given the task of preparing the planet for physical life. The particular office of the Spiritual Life Element appears to be the completion of the physical body. To the intelligent soul is specially assigned the acquisition of knowledge, the acquirement of power, the exercise of love and the attainment of happiness.

From lowest to highest each element has performed its task through the sex principle of positive and receptive energy. Only by comparison is human intelligence able to comprehend this stupendous drama which is being enacted by the positive and receptive powers in Nature. Only by comparing a positive and a negative mineral atom with a man and a woman can the mind grasp the meaning of the word evolution.

Linnæus concisely expressed these evolutionary steps in Nature when he said: "Stones grow, plants grow and live, animals grow, live and feel."

Had he gone one step further and added, "Men grow, live, feel and think," he would have expressed our philosophy.

It requires the greatest effort of intelligence to preceive that all of this marvelous movement, change and progress depend upon that one principle which impels everything that is, to seek correspondence in that which is of opposite polarity. It is, however, these ceaseless efforts of individuals, seeking individual adjustment, which occasion all this mighty movement, variation and improvement. To the effort of the individual must be attributed all evolution, from coarser to finer, from simple to complex and from unconscious and involuntary to conscious and voluntary, to self conscious and independent.

To the efforts of individual intelligence are attributed the rise of consciousness into self consciousness, the addition of rational conceptions to intuitive perceptions, and finally, the evolution from unmoral animal nature to moral human nature. Man, the highest product in Nature, a living soul physically embodied, continues the higher evolution through ceaseless efforts for self adjustment.

Thus, it appears that everything, from the unconscious mineral atom to the self conscious living soul, is seeking an individual relation that is in correspondence or harmony with itself.

Everything that is, atom, cell, organism, animal and human, is in a present state of refining, change and improvement. The physical matter of one kingdom is being perpetually refined for the uses of another. The life of one kingdom is continually preparing the way for the higher life of another kingdom. The intelligence of one kingdom is continually aspiring to intelligence of a higher order.

Thus, the universal principle of polarity or affinity, recognized by physical science as the law of vibration, is, in reality, the fundamental principle of evolution and the generator of all physical life.

Thus, the teaching of the older school comes to fulfill and not to destroy the laws already apprehended by the modern school. It approaches physical science, not to belittle its achievements, but simply to enlarge its scope. It denies none of the facts of physical science. It only refutes certain theories of physical materialism and presents additional facts to support that refutation.

What is denied, and all that is denied, are those deductions which ignore the spiritual principles, elements and forces in Nature and assign the evolution of an intelligent moral being to the blind and mechanical physical forces of digestion. It denies only those deductions which reverse the natural order of cause and effect, and refer that which is spiritual, intellectual and ethical to that which is physical and mechanical.

The higher science does not undertake to explain why the principle of affinity governs the universal Life Elements. It simply accepts both the principle and the elements as it does matter and intelligence, viz., as ultimates in finite science. It does not attempt to explain why evolution is conducted through the sex principle of positive and receptive energy. It merely demonstrates that the affinities and unions of entities thus polarized refine matter, increase its vibratory action, generate life and develop individual intelligence.

It does not pretend to explain how or why Nature improves its products through vibratory correspondences. It merely discovers that the nearer two entities of opposite polarity approach a perfect correspondence in vibratory action, the finer are the effects upon that entity physically, spiritually and ethically. What the higher science has really verified is:

(1) The universality of the principle of polarity in the evolutionary scheme.

(2) The universality of the Life Elements in physical man.

(3) The process by which these polarized elements charge physical matter, combine it, refine it and key it to higher vibrations in Nature.

(4) The effects correlated to such combination, refinement and increase of vibratory action.

These deductions are not presented as explanations of the first cause of either matter, motion, life or intelligence. They may, however, simplify to the mind those principles and processes which are involved in the generation of physical life, the completion of the physical body, and the rise of individual intelligence.

The moralist who claims that "Everything in the moral world has a physical basis" has also declared that "Evolution began in protoplasm and ends in man."

Natural Science, more conservative, would say—evolution begins———? and ends———?

It is thus admitted that finite science has neither comprehended nor demonstrated the beginnings nor the endings of the infinite scheme of evolution. Evolution is regarded as an infinite process. Man is considered as a factor in that process. To actually know the beginnings and the ending of this stupendous drama would be to know God.

That which the higher science does claim to know concerning the genesis of physical life and the evolution of man are:

(1) That the genesis of the animal life cell is the point of

contact between a particle of vegetable substance and the universal Spiritual Life Element.

(2) That point of contact is the stage of evolution at which the vibratory action of the vegetable particle rises to a necessary ratio of correspondence with the Spiritual Life Element.

(3) The genesis of all physical life is governed by the spiritual principle of polarity or affinity.

(4) Polarity, or positive and receptive energy, are powers which inhere in the Life Elements.

(5) The operation of these highly active but opposite energies occasions the phenomena of chemical affinity and of sex in plant, animal and human life.

(6) The office of sex in Nature is to promote and achieve equalization of those positive and receptive energies.

(7) Nature seeks equalization of the whole by and through individual efforts for self equalization of the individual.

(8) The ceaseless attractions, unions, combinations and activities of positive and receptive physical entities entail the refinement of matter, the increase of vibratory action and the genesis of physical life.

Says Alfred Russel Wallace:

The special faculties we have been discussing clearly point to the existence in man of something which he has not derived from his animal progenitors—something which we may best refer to as being of a spiritual essence or nature, capable of progressive development under favorable conditions. On the hypothesis of this spiritual nature, superadded to the animal nature of man, we are able to understand much that is otherwise mysterious or unintelligible in regard to him, especially the enormous influence of ideas, principles, and beliefs over his whole life and actions. Thus alone we can understand the constancy of the martyr, the unselfishness of the philanthropist, the devotion of the patriot, the enthusiasm of the artist, and the resolute and persevering search of the scientific worker after Nature's secrets. Thus we may perceive that the love of truth, the delight in beauty, the passion for justice, and the thrill of exultation with which we hear of any act of courageous self-sacrifice, are the workings within us of a higher nature which has not been developed by means of the struggle for material existence.

It will, no doubt, be urged that the admitted continuity of man's progress from the brute does not admit of the introduction of new causes, and that we have no evidence of the sudden change of nature which such introduction would bring about. The fallacy as to new causes involving any breach of continuity, or any sudden or abrupt change, in the effects, has already been shown; but we will further point out that there are at least three stages in the development of the organic world when some new cause or power must necessarily have come into action.

The first stage is the change from inorganic to organic, when the earliest vegetable cell, or the living protoplasm out of which it arose, first appeared. This is often imputed to a mere increase of complexity of chemical compounds; but increase of complexity, with consequent instability, even if we admit that it may have produced protoplasm as a chemical compound, could certainly not have produced living protoplasm—protoplasm which has the power of growth and of reproduction, and of that continuous process of development which has resulted in the marvelous variety and complex organization of the whole vegetable kingdom. There is in all this something quite beyond and apart from chemical changes, however complex; and it has been well said that the first vegetable cell was a new thing in the world, possessing altogether new powers—that of extracting and fixing carbon from the carbon-dioxide of the atmosphere, that of indefinite reproduction, and, still more marvelous, the power of variation and of reproducing those variations till endless complications of structure and varieties of form have been the result. Here, then, we have indications of a new power at work, which we may term vitality, since it gives to certain forms of matter all those characters and properties which constitute life.

The next stage is still more marvelous, still more completely beyond all possibility of explanation by matter, its laws and forces. It is the introduction of sensation or consciousness, constituting the fundamental distinction between the animal and vegetable kingdoms. Here all idea of mere complication of structure producing the result is out of the question. We feel it to be altogether preposterous to assume that at a certain stage of complexity of atomic constitution, and as a necessary result of that complexity alone, an ego should start into existence, a thing that feels, that is conscious of its own existence. Here we have the certainty that something new has arisen, a being whose nascent consciousness has gone on increasing in power and definiteness till it has culminated in the higher animals. No verbal explanation or attempt at explanation—such as the statement that life is the result of the molecular forces of the protoplasm, or that the whole existing organic universe, from the amœba up to man, was latent in the fire-mist from which the solar system was developed—can afford any mental satisfaction, or help us in any way to a solution of the mystery.

The third stage is, as we have seen, the existence in man of a number of his most characteristic and noblest faculties, those which raise him furthest above the brutes and open up possibilities of almost indefinite advancement. These faculties could not possibly have been developed by means of the same laws which have determined the progressive development of the organic world in general, and also of man's physical organism.

These three distinct stages of progress from the inorganic world of matter and motion up to man, point clearly to an unseen universe —to a world of spirit, to which the world of matter is altogether subordinate. To this spiritual world we may refer the marvelously complex forces which we know as gravitation, cohesion, chemical force, radiant force, and electricity, without which the material universe could not exist for a moment in its present form, and perhaps not at all, since without these forces, and perhaps others which may be termed atomic, it is doubtful whether matter itself could have any existence. And still more surely can we refer to it those progressive manifestations of life in the vegetable, the animal, and man—which we may classify as unconscious, conscious, and intellectual life—and which probably depend upon different degrees of spiritual influx. I have already shown that this involves no necessary infraction of the law of continuity in physical or mental evolution; whence it follows that any difficulty we may find in discriminating the inorganic from the organic, the lower vegetable from the lower animal organisms, or the higher animals from the lower types of man, has no bearing at all upon the question. This is to be decided by showing that a change in essential nature (due, probably, to causes of a higher order than those of the material universe) took place at the several stages of progress which I have indicated; a change which may be none the less real because absolutely imperceptible at its point of origin, as is the change that takes place in the curve in which a body is moving when the application of some new force causes the curve to be slightly altered.

CHAPTER VII.

THE SPIRITUAL BASIS OF EVOLUTION

The word "Evolution" belongs to modern physical science.

The principles and processes, however, which constitute evolution have been known to Natural Science for ages.

Why, then, has Natural Science remained so long silent concerning these known facts of Nature?

This is a question the reader will naturally ask. It is one the writer cheerfully answers.

History and experience show that the instruction of a people must be conducted within the limitations of the average intelligence.

Scientists have been forced to silence, first, by general ignorance concerning Nature, and, second, by the general prejudices against Men who assume to be the superiors or teachers of mankind.

The few who outstrip the many always have suffered and always will suffer embarrassment in their attempts to transmit their knowledge to the world. Between the learned few and the unlearned masses are many barriers. There is, first, the fundamental barrier of ideas; next is the barrier of words, and last, but not least, the universal barrier of prejudice on the side of the unlearned.

Though the world clamors for truth, the history of human development is a long record of persecutions and indignities which the world has heaped upon its teachers of truth. No man nor school of men can teach a science or philosophy except the people have reached a development corresponding to the class of knowledge to be taught.

Not even then can a people be taught except by their own

free choice and desire. Picture the perplexity of a Darwin or an Edison attempting to teach his science to our native American Indians in their own language. Imagine still further the difficulties, should the savage entertain a prejudice against the scientist and condemn his teaching in advance.

The relation between the learned few and the people has, for ages, furnished a parallel to this hypothetical case.

For six hundred years physical science has suffered just such embarrassments. It is a continual struggle between the man who discovers and the masses which are prejudiced against discoveries. Thus has the modern school suffered, although it claims to teach only the visible and tangible facts of physical nature which are demonstrable to any intelligence demanding proof. The older school, however, has a still more difficult task. It undertakes to teach the natural facts of spiritual Nature, which can only be personally demonstrated by a high order of INTELLIGENCE, COURAGE and PERSEVERANCE.

Christ's ministry illustrated the gulf between a "Master" of the higher science and the public mind in Judea nineteen hundred years ago. It will be remembered that the man then addressed as "Master" by the common people continually exhibited his superior powers. For illustration, he clearly read the minds of those about him and those at a distance. He made no attempt, however, to explain the process of mental telepathy. He could not have done so had He desired. At that time there was not sufficient general scientific knowledge. The public mind was a child's mind. It was not the simplicity of the Nazarene but the limitations of His hearers which gave rise to the Parables.

Christ healed the sick by the laying on of hands. He did not, however, discuss either physical or spiritual magnetism. He turned water into wine, yet He did not explain the process which raised the vibratory action of water into that of the wine he imitated. He cast out Devils; that is, He released certain weak-willed persons from the control of vicious disembodied intelligences. He did not, however, explain the law of control and obsession. He did no more than utter a warning against such communication with and submission to "evil spirits."

Had Christ declared at the beginning of His ministry that the world was round or that man evolved through lower forms, it is probable that He would have been crucified several years earlier. He could no more have explained His science to those simple people of Judea than a Roentgen could explain the "X-ray" to native Patagonians.

Nineteen hundred years ago the common intelligence was closed to science. Study and utilization of Nature's forces had no place in the popular mind. Even the better educated had not studied Nature after the rational methods of our modern western nations. So great has been the chasm between the common mind and the trained intelligence of the few that no attempt has heretofore been made to publicly teach the science underlying this philosophy.

The most ever attempted up to this time has been the presentation of ethical codes based upon science.

Even this ethical teaching has appeared mysterious. It has been presented in poetic and figurative speech. The simple facts of Nature have been concealed in symbol and parable, in allegory, proverb and song. A primitive people, like young children, can be taught ethical truth before they have either a craving or capacity for scientific knowledge. Teachers of the primitive mind, like the teachers of young children, have universally adopted a figurative and poetic style.

Only advanced intelligence desires and demands literal and exact knowledge.

Public teaching of any class of knowledge necessitates, first, a common language; next, a general community of ideas and sympathy of purposes, and, finally, a desire for instruction on the part of the unlearned. A common vocabulary and a community of ideas and purposes are, however matters of slow growth. Neither new words nor new ideas can be forced upon the mind that is either unprepared or hostile. A great scientist many coin a few words and inject them into a language. He can not, however, coin an entire vocabulary. A scholar may inject a few new ideas into the common mind. He can not, however, immediately impart an entire science or system of philosophy.

The knowledge gained by the specialists of the higher science is not selfishly withheld from the world. The transmission of that knowledge merely waits upon the preparation and hospitality of the general mind. The dearest ambition of every such specialist is to impart his knowledge. How to impart it has been the problem of the ages. He finds, to his sorrow, that the task of acquirement is infinitely less than the task of teaching.

The past six hundred years have developed an order of intelligence unlike anything in the history of man on this planet. The people of this western world occupy a position that is unique in the intellectual history of the world. Six hundred years along the exact lines of physical science have carried this people to a very high average of intellectual development. During this period of gradual accumulation of knowledge have been evolved a scientific vocabulary and scientific ideas and habits of thought. During this period the common mind has been gradually trained to rational methods of investigation and demonstration.

For the first time in its history the school of Natural Science sees the possibility of explaining its history, its knowledge, its methods and its purposes to the world at large.

As its name implies, physical science is concerned with the physical aspect of Nature. It deals with phenomena visible or tangible to the five physical senses. It deals with man, the earth, the planets and the universe wholly upon the physical side. It confines itself, or attempts to confine itself, entirely, to that which is visible or tangible to the physical sensory organs.

Physical science, however, has not been able to ignore certain universal phenomena, which, it is forced to admit are "superphysical" or "metaphysical." In the first place, it has been forced to consider the phenomenon of intelligence which attaches to all sentient life. Next, it has been forced to concede the existence of certain moral elements which distinguish human beings. It has been forced to consider a certain quality of love and altruism in the human family, neither of which is sex passion. It has noted the fact of pure affection in animals. It has been forced to observe affection and friendship between animals and men.

So universal are these phenomena, so extensive their force, so potent their influence, that physical science has not been able to ignore them.

It even attempts to account for them.

Physical science, by this attempt to explain spiritual and psychical phenomena by physical analysis, betrays its insufficiency in these particular lines.

At the outset these higher phenomena appear so radically unlike physical phenomena that modern physical science is driven into a change of terminology. It is forced to select for this higher phenomena a name which shall distinguish it from physical phenomena. This definition alone furnishes a commentary upon physical materialism which declares as a basic proposition that "All is matter and mechanical force."

Physical science classifies these higher phenomena of life as "psychical." The word "psychical," it will be remembered, is derived from the Greek "psyche," meaning "soul." It is not claimed that physical science acknowledges the existence of the soul by its use of the term "psychical." It does, however, by such definition, confess that Nature embraces phenomena which cannot be classified as physical.

Physical science, after first assuming that everything has a physical basis, sets out to find the physical causes for its "psychical" facts. So long as it seeks only to relate physical cause with physical effect it is admirable. When, however, it attempts to relate psychical phenomena with physical causes it ignominiously fails.

The inevitable logic of such an assumption is to relate intelligence to the digestive organs and to define love as the efflorescence of physical lust. Against such assumptions common intuition, common experience and common sense rebel.

The best intelligence of to-day accepts the physical facts of Nature as collated and classified by physical science. This same intelligence, however, declines to accept the theories advanced by physical science as to the causes of psychical phenomena.

These hypothetical dogmas, based upon only half the facts of Nature, bewildered even so great an intelligence as Mr. Huxley. Having accepted the physical facts of Darwinism, he somehow felt bound to accept Darwin's guesses as to the causes of those facts.

As a result, Huxley repudiated Nature and denounced it as a monster without a single principle that conserved justice or love or altruism. He forced his reason to accept what his intuitions always denied, viz., that all we have been, are, or may become, are merely automatic results of physical feeding, breeding and battle. It is little wonder that this great scientist declared life an unsolvable riddle, intelligence a delusion, love essentially lust and morality without the sanction of Nature. It is little wonder that he said, "I wash my hands of Nature."

Upon just such extraordinary hypotheses as these the modern school of physical science attempts to account for, and to explain a man. These are the theories which stultify intelligence, outrage conscience and violate universal experience.

The best intelligence of to-day declares that such theories and such assumptions explain neither the psychical facts of Nature nor all of the physical facts. As a result, honest and inquiring minds everywhere are seeking more satisfactory interpretations of the higher phenomena. Such investigators are moving forward regardless alike of dogmatic theology and scientific skepticism.

Physical science is also progressive. Within its own recognized boundaries are the few who have outstripped their fellows in both the perception and demonstration of Nature's finer forces. Since Darwin's hopeless pronunciamento fell upon the world, his own students have made discoveries and modified their master's dogmas. The collaborator of Charles Darwin, Mr. Alfred Russel Wallace, after an additional fifty years of investigation, repudiates Darwin's doctrine of physical materialism.

Other and foremost representatives of the modern school are to-day experimenting along the very border line which divides physical material and physical phenomena from spiritual material and spiritual phenomena. Edison, Gray, Tesla and Roentgen are continually extending their knowledge and control over the finest physical element in existence. Keely and others have attempted to do more than this. They have endeavored to utilize the purely spiritual forces in Nature for the practical benefit of society.

Mr. Keely claimed to have passed the limitations of physical matter. He may have done so. He was not, however, able to control and practically utilize those finer and more subtle vibrations of the spiritual elements.

Experiments in hypnotism alone have pushed certain specialists of the school of physical science into the domain of the higher science. Charcot and others of the "Regular School" of medicine, have successfully passed the limitations of physical matter. These men, however, are, as yet, experimenting in total ignorance of the forces and elements employed. The modern hypnotist does not, as yet, really know that the general exercise of hypnotic control would be far more deleterious to man than is physical disease and death.

Physical disease merely destroys the physical body, while hypnotic control, if persisted in, destroys the human powers of will and the faculties of the intelligent soul itself.

Darwinism is now generally accepted as the representative system of modern physical science. "The Descent of Man" is more than a scientific treatise. It is a philosophical doctrine. It is more than a classification of physical facts. It undertakes to explain, not only the causes of those physical facts, but the causes of the "psychical" facts in Nature.

The causes assigned to the "psychical" facts of Nature by physical materialism are wholly inadequate causes. The doctrines of Darwinism are, therefore, misleading and unfortunate.

Darwin thinks he discovers the working formula of evolution. He thinks he discovers, as a fundamental principle in the evolution of man, "A struggle for existence in the midst of a hostile environment." That is to say, Darwin holds that all living sentient entities are primarily engaged in the struggle for physical nutrition. It has been claimed for Darwin that his "Struggle for Existence" had a broader significance than a mere competitive and selfish struggle for physical nutrition. If so, that fact nowhere appears in his own works. From first to last he reduces all phenomena of the higher life of man to this primary struggle for physical nutrition. His own, now famous proposition, confirms this statement.

Darwin, himself, reduces his "struggle for existence" to a mere struggle for physical nutrition when he says: "The struggle for nutrition compels natural selection with reproduction as a consequence, entailing divergence of character and extermination of less improved species."

Darwin may have originally intended this "Struggle for Existence" to include more than the struggle for nutrition. If so, he lost sight of his own intention. The fact remains, that the popular idea of the Darwinian theory is "A Struggle for Nutrition in the midst of a hostile environment." It is, therefore, this popular idea and popular error which command the attention of Natural Science.

Darwinism claims to have discovered two very important facts, viz.:

(1) That the fundamental struggle of sentient life is the struggle for physical nutrition.

(2) That Nature is universally hostile to the life it generates.

Thus, Darwinism fixes upon the digestive organs as the source of all progress, and holds that Nature, so prolific in generation of life, nevertheless provides infinite obstacles to living itself.

By reason of this supposed hostility Darwinism claims that a purely egoistic and competitive struggle for nutrition begins with the dawn of life. This necessity for food forces all living entities to prey upon each other. It establishes life as a perpetual battle in which the physically strong prey upon the physically weak. It establishes a universal slaughter in which only the "fittest physically" survive. It establishes the principle of competition as fundamental in evolution, and sets up a law of self-defense as the primary law of progress.

All of these supposed facts and conditions have, as a result, been taken to mean that man as we know him, a physical, intelligent, moral being, is but an automatic result of food combinations.

It will be seen, therefore, that Darwinism refers all organization, variation and progress of the original life cell to the original hunger of the cell itself. According to Darwin this original hunger of the life cell induces a struggle for nutrition, sets up a competition for food and engenders that species of activity which forces organization, variation and improvement. These activities, be it understood, from life cell to man, are held by him to be essentially mere competitions for food and physical benefit.

Darwinian doctrine, though erroneous in principle and misleading in effect, is both logical and consistent with the purely physical facts on which it rests. Its conclusions are in a direct line with that later premise which declares that all is physical matter, mechanical force and automatic result. While Darwinism is forced to deal with metaphysical phenomena, it denies that there are innate metaphysical principles in the evolution of man. On the contrary, it refers all activity, whether competitive or co-operative, to a demand for physical food, to re-enforcement from without, and to conditions generally hostile to life and living.

Darwinism is, therefore, logical when it determines that all phenomena, including the physical body, the intelligence and the love nature of man, are the combined results of the digestive and generating organs. By "love" Darwinism means a physical passion only. Darwin frankly confesses that he finds no sanction in Nature for either love that transcends physical passion or for altruism which is neither love nor lust. Darwinism, therefore, finds no higher inspiration for human activity than the physical appetites and a self-love engendered by the passions. It finds further, that all this progress, physical, intellectual and moral, has been accomplished in spite of Nature's hostility.

This, in brief, is the dreary doctrine which the best intelligence of the age, while fearing to deny, cannot yet force itself to accept.

This is a doctrine which modern physical science will probably discard in another decade.

Reduced to one fundamental proposition Darwinism would

declare: "All living organisms, together with intelligence and love, are expressions of physical laws and forces."

Thus, the doctrine of pure physical materialism finds Nature to be a mechanical monster without intelligence or moral intent. It finds life to be a godless, loveless and hopeless warfare of the strong against the weak. It finds intelligence an automatic result of food combinations, and love the efflorescence of lust. It finds that the universal activities of matter, life and intelligence have no other meaning than physically improved species.

Darwinism discovers no higher purpose in individual life than a contribution to species. It offers no other reward for conformity to natural law than survival of the "Physically Fittest." It sets no higher ideal before human intelligence than a healthy body and material comfort. To the aspiring soul of man it promises nothing better than total extinction when its contribution to species is accomplished.

Against this unwarranted and demoralizing doctrine, intelligence rebels. It refuses to accept as final those assumptions which degrade life and life's purposes to the level of the physical functions and appetites.

It is easier, however, to protest against the errors of Darwinism than to demonstrate in what particulars they are errors. In truth, no such protest can be successfully offered except one is fortified with exact knowledge of the facts which disprove Darwinism. It is not enough that the common intuitions and the common experiences and the common sense of man himself deny this position of physical science. Proof of its weakness can be conclusively demonstrated only by and through an actual knowledge of the spiritual side of Nature.

This general protest of intelligence has been voiced in a remarkable manner by that popular literary divine, the late Henry Drummond. "The Ascent of Man," his last and most widely read book, deserves more than a passing notice.

This work is remarkable, first, for its beauty of diction; next, for the cleverness of its theory, and finally, for the errors it contains.

The author of "The Ascent of Man" undertakes a superhuman task; that is, he attempts to establish a moral order in Nature based upon the theories, as well as upon the accepted facts of Darwinism. His purpose is to establish love and altruism as the direct results of the physical functions and physical forces. The work fails of its purpose by reason of two conditions: first, the author, in the main, accepts the theories of Darwin along with his facts; and, second, in that *he does not know as a literal scientific fact that there is a spiritual side to Nature.* Because of these conditions Mr. Drummond is not equipped to meet the assaults and assumptions of physical science. Neither is he prepared to formulate, as he attempts to do, a new moral philosophy.

Without a scintilla of absolute knowledge concerning spiritual laws and forces, he attempts to build a spiritualized and moral humanity upon the purely physical appetites and functions. With all its errors, however, the work is valuable, inasmuch as it is the first work which accepts the evolution of man from lower forms and yet postulates a moral order in Nature. In this it is unique, indicating, moreover, the expanding thought of the orthodox church.

Because of the literary merit and the popularity of this work its errors are widespread. For this reason those errors command attention and exposition.

"The Ascent of Man" involves a comparison between the admitted physical facts of evolution and the admitted moral phenomena of human life. This elaborate argument of nearly four hundred pages is intended to prove that all these moral phenomena have their origin in and rest upon the purely physical functions and forces. Drummond's criticism is not that Darwinism is erroneous in theory. On the contrary, Darwinism is accepted in theory as far as it goes. That is to say, the moralist simply asserts that Darwin overlooked certain moral phenomena which are the direct results of a certain physical function which plays an important part even in the Darwinian theory. After this the Darwinian critic proceeds to set forth what appears to him as the true relation between moral phenomena and physical nature.

Now, if it shall appear that neither the skeptic nor the theologian was in a position to know the real factors and causes of either physical or moral phenomena, we must seek further for conclusions.

The initial error underlying "The Ascent of Man" is the author's total ignorance of what the word "Spiritual" means in science. He seems to have no rational conception of the fact that the spiritual world is a material world, that spiritual elements are material elements, and that a spiritual man is as truly a material man as is the physically embodied man. He confounds "spiritual" with "moral," and "spirituality" with "moral regeneration." He speaks of a "spiritual man" when he means a moral man. He speaks of "spiritual principles" when he means ethical principles.

This lack of scientific knowledge of the spiritual side of Nature robs the entire work of ethical as well as scientific value. This ignorance as to the fact of spiritual matter and spiritual elements and forces misleads the great divine when he comes to laying down his major premise.

Mr. Drummond, it will be remembered, introduces his new moral philosophy as a system based upon and supported by the proved facts of evolution.

As a preliminary, therefore, to his own argument, the author states his position with reference to the Darwinian doctrine of evolution. To briefly restate that position is to say that Mr. Drummond accepts not only the physical facts of evolution as laid down by Darwin, but he also largely accepts Darwin's theories concerning those facts. He accepts the basic dogma of Darwinism, viz.: "The struggle for existence in the midst of a hostile environment." He agrees with Darwin that a struggle for nutrition engenders that character of competition which forces industry, commerce and civilization upon the world. He accepts a universal warfare of the physically strong against the physically weak, as a natural mode of progress.

The moralist agrees absolutely with Darwin as to the nature and uses of sex. He defines sex as a "physical device for reproduction." He endorses Darwinism as to the uses of the female in Nature. That is, he holds that the female fulfills her destiny in reproduction.

The moralist, however, accepts these errors with one qualifying clause. By and through this one clause alone are differentiated the two systems of philosophy. This qualifying clause is one which presents morality as a result of natural law.

The specific charge made by the moralist against the materialist is to the effect that Darwinism states only half the case of Nature. He declares that Darwinism discovers but one factor in the evolution of man, when, instead, there are two. He claims that Darwinism postulates but one great physical struggle in Nature, when, in reality, there are two. He finds that Darwinism lays down but one great principle in evolution, when, in truth, there are two. He charges, therefore, that Darwinism considers only the physical and material effects which flow from the one factor, the one struggle and the one principle which he so exclusively studied and analyzed.

That particular factor to which the moralist alludes is "Nutrition." The struggle meant is "The struggle for existence in the midst of a hostile environment." The principle referred to is that of "Competition."

The Darwinian critic, on the contrary, declares that Nature embraces a second factor, a second physical struggle, and a second principle. More than this, he insists that there are moral and ethical, as well as physical and material effects, originating in this second physical factor. The moralist insists that the evolution of man embraces another factor of equal force and another physical struggle equally important and far-reaching in results. He insists that the second struggle governs another principle equally potent with that of competition.

This second factor in evolution is laid down as the physical function of reproduction. The second great struggle growing out of this function is defined as the "Struggle for Reproduction" in the midst of a hostile environment, while the second great principle, dependent upon reproduction, is that of Self-sacrifice.

The moralist charges that Darwinism becomes so absorbed in the function of nutrition that it overlooks the function of reproduction. He claims that such undue concentration upon the physical and material results of nutrition, obscures the moral results which flow from reproduction. While he admits that the struggle for nutrition engenders a universal principle of hostility, he insists, however, that the struggle for reproduction engenders a principle of self-sacrifice, the effects of which must be accepted along with the effects of competition.

Just here the moralist makes his great point.

He skillfully arrays these two great physical struggles, nutrition and reproduction, side by side. He shows the one, nutrition, to be a purely egoistic, selfish and competitive struggle for individual benefit. He shows, on the other hand, that the second great struggle, reproduction, is purely altruistic in its nature, involving self-sacrifice for the benefit of other individuals.

As against that great "Struggle for Self," so exclusively dwelt upon by physical science, the moralist sets forth and graphically delineates that other great struggle which he so well names "The Struggle for Others."

Thus far the criticism is just. Thus far the moralist is in line with Nature. Thus far the discernment of the divine exceeds that of the skeptic. The moralist discovers what Darwin overlooks, viz., the sacrificial struggle for reproduction, its altruistic nature and the ethical phenomena almost universally attaching to the office of maternity.

Thus, the moralist postulates a moral order in Nature, and altruism as a natural phenomenon and not a disease.

In this deduction, and to this extent, Natural Science supports the Darwinian critic. Here, however, the agreement ends.

The purpose of the moralist is to show that morality, love and altruism are based in Nature. While this general assumption is correct, he errs when he seeks to explain the natural causes of these phenomena. The initial error lies in the acceptance of the Darwinian theory, that everything in this world has a physical basis. Darwinism is not disputed as to the physical basis of evolution. The moralist merely criticises it for not having discovered the physical basis of morality and love.

The task set by the moralist for his own ingenuity, is to discover the physical basis of what we know as morality, love and altruism. The point of view from which he goes to this task is distinctly stated when he says: "Everything in the moral world has a physical basis." He further agrees with his adversary when he says: "Life is controlled by its (physical) functions." He goes even further when he touches upon the principle of human evolution, for he says in so many words: "So man, not by any innate tendency to progress, in himself, nor by the energies inherent in the protoplasmic cell from which he first sets out, but by continuous feeding and reinforcing from without, attains the higher altitudes.

Having thus accepted the fundamental doctrine of pure physical materialism, the author sets himself to discover the particular physical causes of this "moral world."

In the search for the physical basis of love he discovers but two possible causes. He finds two universal relationships which exhibit what we know as love and altruistic phenomena. These relationships are sex and maternity. However, having previously defined sex as "A physical device for reproduction," and sex love as "A physical passion miscalled love," the moralist is driven from the consideration of sex as a possible cause. This leaves him but one other physical cause, viz., reproduction. This he accepts and analyzes as follows:

He holds that morality, love and altruism come into the world as a result of the physical pain and the physical sacrifice of the female half of all life. He fixes upon the enforced physical sacrifice of the female in reproduction as the one and only cause in Nature for the evolution of love. He thus conceives the extraordinary idea that Nature embraces an absolutely diabolical plan for forcing love upon the human family, for, to quote directly, the moralist says: "Love is forced upon the world at the point of the sword."

Thus, a great teacher of spiritual truth not only fails to find spiritual principles governing physical evolution, but he insists that the physical functions create the love relationships and the ethical phenomena of human life. Is not this a singular hypothesis for one who is supposed to teach the permanency of that which is spiritual and that which is ethical, and the impermanency of that which is physical?

Drummondism, therefore, does little to enlarge the vision of evolution as laid down by Darwinism.

Having accepted sex as "a physical device for reproduction," it is necessarily treated from that restricted point of view. Sex attraction is everywhere analyzed as physical sex passion. Sex love is everywhere referred to as a biological need for reproduction. The love relation of man and woman is everywhere leveled to the purely physical relation and the physical purpose of that relation. In brief, sex is treated throughout from the viewpoint of physical materialism, viz., as a physical device conserving reproduction.

This interpretation involves another scientific error. The moralist makes no distinction between love which is individual in its nature, and altruism which is general in its nature. He uses interchangeably, and therefore erroneously, the words love and altruism. This new moral philosophy claims that love or altruism is an evolution of feeling based wholly upon the physical function of reproduction. He holds that the enforced sacrifices of the female half of life are the sole causes of all human sympathies, whether those sympathies be expressed as love that is individual, or altruism that is general.

According to this theory any sex relation that rises higher than physical lust must be attributed, first, to the inherited effects of maternity, and, second, to a mutual desire or love for progeny. How far this theory coincides with universal history and universal experience, is left as an open question. How far it contradicts the individual impulses, intuitions and aspirations of the soul, each intelligent reader must determine for himself.

Just how far such a theory contravenes Nature, it has been the effort of the higher science to demonstrate.

The moralist has not added one fact to the general store of human knowledge. Instead, he has merely placed upon the market another theory which the best intelligence and the finest intuitions of men and women reject. He has advanced a theory of love which history and universal experience disprove. He has assigned woman to a place in Nature which woman herself condemns and refuses to occupy. He has so interpreted the love relation of man and woman as to contravene the highest aspirations and ideals of every thinking man and woman.

Carefully reducing this theory to its fundamental proposition discloses its coincidence with that of physical materialism. When so reduced it would simply read: All living organisms, together with love and altruism, are expressions of physical laws and forces.

Thus, "The Ascent of Man," based upon facts of physical Nature, brilliantly written and highly entertaining, is, yet, as erroneous in theory as the doctrine promulgated in "The Descent of Man."

At just this point of debate between scientfic skepticism and orthodox theology, the higher science, with deference to both, desires to be heard. The public is entitled to receive a wider range of fact than that covered by the theorists of either school. The older school presents an array of facts gathered upon two plans of causation. It would, therefore, seem to be the better authority as to the real factors and causes of the evolution of man.

Because of this wider experience and more extended knowledge of facts, Natural Science must not be deemed presumptuous if it contradicts the basic propositions of both the skeptic and the theologian.

As already stated, no effort will be made to explain ultimates in Nature. The higher science does not claim to have discovered the First Cause. It does not claim to have analyzed the Infinite. It does not pretend to explain when or how or why Infinite Intelligence set in motion this evolutionary scheme. It does not claim to have discovered how or why the First Cause selected this particular scheme for the evolution of man. It does not, in short, profess to either know or explain the ultimates of matter, or motion, or life, or intelligence, or love.

Natural Science, like physical science, is forced to deal with Nature as it is. It is also forced to operate with finite intelligence in every department of science. Finite intelligence, dealing with infinite problems, is forced, at least in its earlier reaches, to dismiss the ultimate issues of this stupendous scheme of Nature. The broader science, therefore, deals with the phenomena of two planes of matter, motion, life, intelligence and morality, just as physical science deals with the phenomena of one plane, viz., by study, analysis and demonstration as far as the individual, finite intelligence can penetrate under given conditions.

Physical science finds matter, motion, life, intelligence and love as common phenomena upon this physical plane. Natural Science finds the same phenomena, common also, upon the spiritual plane. It is therefore driven to the general deduction that matter, motion, life, intelligence and love are the correlated properties, elements, principles and activities of both worlds, and that they are universal in time and space. It accepts matter as a universal property of Nature, and motion as a universal mode. It accepts life as a universal element and intelligence as a universal principle. It accepts love as a universal activity of the intelligent soul.

Further than this finite science has not penetrated. It simply finds and accepts these two planes of existence which have correlated properties, modes, elements, principles and activities. It accepts them as ultimates as far as the finite mind is concerned. It holds that these ultimates are unvarying in principle, indestructible in essence, yet infinitely varied in manifestation.

Natural Science deals with man as with any other product of Nature. It studies him upon each plane and explains him in his relation to both. Physical science accepts man as the highest product of this physical plane. Natural Science also discovers that he is the highest manifestation upon the spiritual plane. Therefore, as far as finite science goes, man, in form, intelligence, capacity, individual activity and attainment, is the highest product of Nature in time and space.

Man is a living soul and he has two bodies, the one physical and impermanent, the other spiritual and more enduring. When the student is able to intelligently leave the physical body for investigation upon the spiritual plane, he discovers certain facts that have a bearing as to this "basis of evolution." Among these important facts are:

(1) That he, in common with all men upon this planet, inhabits and operates two material bodies of unlike refinement and vibratory action.

(2) He demonstrates that the physical and spiritual bodies may temporarily separate without causing physical death.

(3) During such separation the intelligent Ego or soul remains with the spiritual body or the "spirit."

(4) It is the intelligent Ego, or soul, who plans for and effects this release of the spirit.

(5) Finite science has not discovered how to separate the intelligent Ego from its ethereal organism which is defined as the "spirit."

(6) By personal contact and acquaintance with ex-human beings, the student finds that men live on indefinitely in the spirit, while their discarded physical bodies disintegrate and return to mother earth.

(7) He thus proves that man is a spiritual being, destined to live indefinitely in the more enduring spirit form. He thus proves that the physical body of man is built upon the permanent spiritual body, or that the earthly man is a spirit about which the coarser physical matter integrates for a given range of years.

(8) What is discovered as to man is true of animals; that is, the student perceives that every living physical animal is modeled upon a superior and more enduring spirit.

(9) As with the animal, so with the tree and the rock. Each of these physical entities is shaped upon a finer and more lasting spiritual model.

What do these facts prove to the investigator? They prove indisputably that every physical entity is also a spiritual entity. They show that what we designate as magnetism, vitality and life, reside primarily in the spiritual models of things. Proof of this last statement is found in the fact that the organic life element passes out of the physical and endures in the "spirit" long after the discarded physical form has disintegrated.

Thus, everything in Nature, from the senseless physical atom to the highest human, is fashioned upon an ethereal or spiritual duplicate. (The duration of these spirits of physical things is not a matter of discussion here and now.) That which is important to man in this connection is the fact that organic life and the intelligence depart with the spirit and continue to operate indefinitely upon the spiritual plane.

These are the facts in Nature which justify science in declaring that THERE IS NO DEATH, that spiritual matter and spiritual forces underlie physical matter and physical forces, and that spiritual principles govern physical manifestations.

These are the facts which justify science in declaring that the struggle for physical nutrition is *not* the fundamental factor in evolution, and that the physical stuggle for reproduction is *not* the fundamental factor in either morality, altruism or love. These are the facts which enable Natural Science to declare without assumption that physical evolution has a spiritual basis and that the ethical phenomena of life are referable to spiritual and psychical principles in Nature.

The student of spiritual phenomena discovers the facts of spiritual nature exactly as the physical scientist discovers the facts of physical nature, viz., by and through the reports which he receives through his several sensory organs. The difference in the nature of the phenomena observed simply demands the exercise of a different set of sensory organs. The advantage of the Natural Scientist in the study of Nature is his ability to use both sets of sensory organs freely, independently and rationally. The physicist sees only the physical side of Nature. The specialist of the higher science sees with equal clearness the spirit of matter as well as its physical manifestation.

For example: The physical scientist sees only the physical man, while the spiritual scientist sees both the physical and the spiritual man. He may see the two bodies alternately or simultaneously. He sees the physical form with the physical eyes and the spiritual form with the spiritual eyes.

How long the "spirits" of men endure after physical death is not under discussion here, nor is the relative duration as between the spirit of a man and the spirit of the animal a part of this work.

Suffice it to say, that man alone represents a principle in Nature which confers self-consciousness, personal identity, and the capacity for persistence as an individualized intelligence. It is not denied, however, that the spiritual body or the spirit of man is subject to change.

That which is alone pertinent to the point at issue, viz., the basis of evolution, is the fact that every physical thing in this physical world is but a manifestation of the potent and more enduring spiritual elements and forces. All we see or touch or know as physical matter is simply coarse physical material which integrates in organic form upon spiritual models.

Sooner or later scientific skepticism and religious speculation will yield to the logic of facts. They will not do so, however, until the limit of physical experiment is reached. They will then, it is believed, accept the principles and adopt the methods of the higher science.

It will be observed that no attempt is made to explain why the "Great Intelligence" sets just the patterns that exist in both worlds. In a general sense it is held that all living organisms are adaptations; that is, that all living entities have, to a certain degree, shaped themselves and been shaped to the uses of the intelligence which animates them. This, however, does not explain why one intelligent entity shapes its body or is shaped to live in a tree, while another shapes itself or is shaped to burrow in the ground.

Darwin in his "Origin of Species" presents an excellent theory from the view-point of a materialist. He finds that the struggle for nutrition in the midst of a hostile environment, forced living entities to move in different channels, and that these enforced selections shaped the physical organism to meet those different conditions.

The specialist of the higher science also theorizes as to the differentiation of species. His theories, however, are not the same as those of the physical materialist. His speculations are based upon a wider range of actual knowledge. They are based upon that already proved spiritual principle which governs evolution, viz., the principle of polarity. We again approach a subject not germane to the question in hand. The theories of Natural Science upon the origin of species are not a part of this work.

It is discovered that Nature embraces certain fundamental principles, properties and elements. By nature, is meant the correlated phenomena of the physical and spiritual planes. It is found, for instance, that matter is a universal property, that magnetism, vitality and life are universal elements. It is found that this universal property and these universal elements are brought into co-operation through that principle of polarity which is universal in time and space.

It is discovered that the four great physical kingdoms, mineral, vegetable, animal and human, represent that universal property defined as matter and those several elements which we define as Life Elements. This co-operation of matter and life is universally brought about through the principle of polarity or the principle of positive and receptive energy. This co-operation of physical matter and the Life Elements is universally conducted through individual processes. That is to say, the evolution of each physical kingdom is the result of the individual activities of entities which go to make up that kingdom. This individual activity of the entity occasions all magnetic change and chemical combination. It occasions all growth, vegetable and animal. It governs the increase of intelligence and the development of reason, morality and love.

It is also discovered that physical change and transmutation of matter are occasioned by the spiritual, rather than the physical part of the individual entity. That is to say, all the operations of the physical individual are occasioned by demands which inhere in the vital spiritual body. This means that the struggle for nutrition, the struggle for reproduction, and all other struggles appearing to be physical, are in truth, prompted by demands which primarily exist in the spirit.

One who studies the spirit of physical matter discovers, first, that all the operations of individuals in the lower kingdom are intelligent. He finds that all activities of animals are governed by conscious intelligence resident in the spirit. He discovers further, that the voluntary acts of men are directly referable to that highest, and apparently indestructible entity, the self-conscious intelligence or soul. Science, therefore, determines in reference to man that this persistent, intelligent Ego operates both bodies in this physical life and departs with the spirit at physical death.

Science discovers vastly more than the fact that there are principles, properties and elements universal in Nature. It goes further than mere analysis of co-operations as between these principles, properties and elements. It goes further than mere demonstration of material processes of evolution, whether those processes be physical or spiritual. It declares, without hesitation and with perfect confidence, that this world of ours is something more than a world of physical forces and functions grinding out automatic results. It declares, instead, that the spiritual forces which underlie physical activities produce results that are purely spiritual results. It declares further, that the intelligent, self-operating principle which animates animal and man is capable of producing another class of effects which are neither physical nor spiritual. These are effects which concern only the intelligence of the animal or of man.

Man, physically embodied, represents all the principles, properties and elements in Nature. He therefore represents all of the energies, capacities and activities of the kingdoms below his own. To these he adds psychical powers, or the energies and capacities of the Soul Element. It is, therefore, declared upon the basis of long investigation and repeated experiment and demonstration, that evolution is the result of spiritual laws and forces. It is also held that the ethical phenomena of human life have their origin in the energies and activities of the Soul. This position, as will be seen, explicitly denies a solely physical basis of evolution, whether the phenomena considered are physical, spiritual or physical. It denies that the evolution of man is the sole result of feeding, breeding and battle. It denies that man, a spiritual being, a living Soul, is the automatic result of the digestive organs or of physical re-enforcement from without.

These conclusions are based upon exact data concerning the "spirit of matter," the nature and the operations of the Life Elements, and the persistence of the Soul after physical death. This is a position which enables science to explain earth's phenomena as representatives of elements and forces which reside in the spirit of matter and endure after physical death.

Thus, science considers the evolution of man and classifies all his activities as results and effects of natural, spiritual and psychical laws and forces. This being true, physical nature is a manifestation of spiritual nature. Morality is an effect of psychical forces and not of physical forces. Love is an activity of the soul and not an efflorescence of the physical functions.

The most careful study and demonstration leave science no choice of conclusions in these matters. It has come to know that the spiritual side of matter is the organic and vital side, that spiritual things are the enduring things, that spiritual forces are the governing forces of physical phenomena, and that psychical forces are the governing forces of ethical phenomena. Familiarity with both planes of existence conclusively shows that everything physical is patterned upon a spiritual model. Acquaintance with man in two worlds conclusively proves that his evolution here rests upon the vital energies and the ethical capacities which inhere in the spirit and the living soul.

These are the facts which constitute propositions new to our modern world, but fundamental in the higher science. These propositions stand out in sharp antithesis to those laid down by physical materialism, whether offered from the view-point of a skeptic or that of a theologian.

Resting upon the carefully proved facts of two correlated worlds of matter, life, intelligence, morality and love, Natural Science declares:

(1) All living physical organisms are expressions of spiritual laws and forces.

(2) Human love and altruism are ethical phenomena governed by the intelligent soul.

Thus it is that the higher science accepts, only in part, the system of evolution as promulgated in "The Descent of Man."

Thus it is that it accepts, only in part, the evolution of love as embodied in "The Ascent of Man."

Without rejecting the minutest physical fact which has, as yet, been discovered by modern physical science, Natural Science hopes to show to the careful investigator that modern intelligence has not as yet correctly interpreted its own array of fact.

On the contrary, the reader is asked to note the fact that the propositions of Natural Science enclose the discoveries and deductions of physical science, as the law of motion and number encloses the principles and propositions of Euclid.

CHAPTER VIII.

THE LAW OF NATURAL SELECTION.

Darwinism is no longer disputed in so far as the logic of actual physical facts is concerned. The physical evolution of man from lower forms is the generally adopted theory—of physical evolution.

While that elaborate system contains philosophic errors, its recorded facts constitute one of the great steps in the higher intellectual evolution of man. Through Darwin's noble life work the common intelligence is familiarized with the idea of evolution, if not with all of its facts and principles.

The world, however, has been unduly impressed and depressed by the Darwinian doctrine itself, by those mere theories by which Darwin accounted for the physical facts he brought to light. The world somehow gives that great scientist credit for furnishing the facts as well as his opinions concerning their origin.

Nature, it must be remembered, furnishes these facts. Mr. Darwin only discovered them to the world.

More than this, Darwin discovered but a limited range of data. He was not in a position to demonstrate all of the facts of the evolution of man. His theories, therefore, as to the factors and causes of evolution, are of no greater value than those of any other intelligent reasoner having the same physical facts before him. It seems inevitable, however, that the public once having accepted the tangible evidences discovered by a man, will accept his opinions also.

Darwinism forcibly illustrates this. For years his opinions filled the world with consternation. It requires considerable reflection to see that a physical fact of Nature is one thing, and a theory explaining that fact is quite another.

It requires a certain amount of independence to declare that Darwinism does not account for man as he really is, an intelligent and moral being, as well as a physical one.

In this connection it will be recalled that Alfred Russel Wallace wholly dissents from Darwin's theories. Mr. Wallace, the co-worker with Mr. Darwin for thirty years, and his collaborator in "The Descent of Man," yet disagrees with the doctrines of physical materialism. During his entire life of investigation as a physical scientist Mr. Wallace was investigating another line of phenomena than those attaching to the physical evolution of man. For fifty-three years he has been investigating modern spiritualism. In this investigation he brought to bear all of the tests and experiments known to physical science. He also brought to bear the cold rational judgments of a man trained in the practical methods of that school.

The result is that Alfred Russel Wallace declares that the phenomena of spiritualism are not demonstrable under physical laws, nor according to physical methods of analysis. He declares, therefore, that Nature has a super-physical or spiritual side which demands the best efforts of science. "The Descent of Man" stands for half the life work of this profound student of Nature. His late work, however, "Miracles and Modern Spiritualism," stands for judgment rendered after a lifetime of investigation.

Thus, it appears that the two men who were co-workers in the same field of science and discovered the same physical facts, have reached diametrically opposite conclusions as to the meanings of those facts. This demonstrates, perhaps, more clearly than any other well-known example, the value of mere opinion.

Mere protest, however, against Darwinism, is no proof of its fallacy or insufficiency. To protest against, or to deny that elaborate array of fact and theory, is quite another thing from disproving either part of it. Mere intuition and faith in super-physical phenomena prove nothing. There is but one class of proof that can be brought to bear in refutation of physical materialism. That proof is simply a personal and exact knowledge of the spiritual side of Nature and the persistence of life after physical death.

While Darwin's doctrine contravenes Nature, its apparently solid basis of fact renders disproof extremely difficult.

This detracts from Mr. Wallace's judgments. Valuable as have been his observations and significant as are his conclusions, they are, nevertheless, only observations and conclusions as far as the public is concerned. Mr. Wallace merely reports those phenomena which are visible and tangible to the physical senses. He is not himself a demonstrator of the spiritual side of Nature. He is not personally able to analyze or to control spiritual elements and forces. He is not in a position to either see or hear the sights and sounds belonging to the spiritual plane. He is not in a position to communicate independently with spiritual people. He is not in position to analyze spiritual material, nor to exercise his psychical powers upon the spiritual plane.

Mr. Wallace is simply a physical scientist reporting the physical effects of unseen, unanalyzed and unconquered spiritual principles, elements and forces. For this reason his carefully collated evidences can have but a limited value. He does not say, "I know." He says, "I believe thus and so." He does not say, "I have traveled in that country; I report the evidences of my own spiritual senses. I know whereof I speak."

Even the higher science cannot furnish, by publication, what is meant by "proof," nor does it undertake such a task. It pretends to do no more than science does everywhere by publication, viz., to report what has been proved by scientific methods. It adopts the usual and only method that any science can pursue by publication, viz.:

(1) It presents an array of scientifically demonstrated facts of Nature.

(2) It refers each inquirer to Nature itself if he demands actual proof of the statements made.

This is all that any student or any specialist or any school of science can do by mere publication.

There is but one known way of proving anything in Nature.

It does not matter whether the thing to be proved belongs to the physical or to the spiritual plane. Actual proof rests upon personal investigation and personal demonstration of each particular fact.

For illustration: No man can prove the science of numbers without acquainting himself with the rules and practice of arithmetic, algebra, etc. He may, through hearsay, believe in mathematics. Proof, however, is self-demonstration.

The world at large has accepted the doctrine of evolution upon its faith in the integrity of Darwin and faith in his report upon Nature. The average Darwinian is no more prepared to prove the facts set forth by Darwin than the theologian is prepared to demonstrate the fact of life after physical death.

After all, public education is conducted mainly by faith. The average mind accepts as truth that which it believes emanates from sane and truthful teachers and investigators. Nor is the world mistaken in its faith. Experience shows that men do not knowingly nor knavishly propound false systems of religion or philosophy or science. The intention of intelligence and the trend of development are ever toward the truths in Nature.

In this respect the higher science looks to public faith in its integrity just as physical science does. It can do no more than to state facts of Nature and challenge the world to an investigation.

Darwinism is something more than a scientific treatise. It is also a system of philosophy. The skillfulness with which the bare, tangible, physical facts of Nature are interwoven with Darwin's theoretical factors and causes, renders criticism difficult. This Darwinian doctrine is nothing more than the best interpretation which Mr. Darwin could put upon the actual, tangible facts which he discovered. Those theories are ingenious. They are not, however, so broad as those of Mr. Wallace. The one propounds a theory which might account for the evolution of the physical body, while the other suggests laws and forces which might account for the higher phenomena of intelligence, morality and love. The one insists that all phenomena are based upon physical laws and forces, while the other holds that spiritual phenomena have super-physical causes.

The Darwinian theory is well summed up in that now famous proposition which reads, "The Struggle for Nutrition compels Natural Selection with Reproduction as a consequence, entailing divergence of character and extermination of less improved Species."

This proposition which is set forth as the unalterable fiat of Nature is, instead, but an assumption of Mr. Darwin himself. It is merely the summing up of his individual opinion as to the factors and causes of evolution.

According to Darwin, this law of natural selection includes that multiplicity of acts and expedients to which all living things are driven by reason of a struggle for nutrition in the midst of a hostile environment. As will be seen, this law rests upon the assumption that the primary cause of all struggle on the part of a living entity is a demand for food. It assumes that there are no other principles involved than the perservation of physical life. According to this law, all that a man is or can become is referable to the original hunger of the original life cell, and the hostile environment which compels a struggle for nutrition.

This theory rests upon two assumptions, viz.:

(1) That of physical hunger as a compelling motive for action.

(2) That Nature, so prolific in the generation of life and so lavish in nutriment of life, is yet hostile to life.

With this construction put upon Nature, the law of natural selection becomes a law of self-defense. The selections dealt with are simply alternatives and expedients to which animals and men are driven in the struggle for nutrition. This wiev of life and the progress of life excludes any principle of a really *natural* selection. It excludes the operation of an individual will and desire on the part of an entity to do thus or so. According to this doctrine progress is effected, not by any innate impelling principle, but is compelled by exterior conditions which force the individual in this or that direction.

This is all that Darwinian doctrine means when one accepts

feeding as the original and only necessity and hostility as a compelling principle. This reading of Nature leaves no place for natural selection, that is, for acts and accomplishments brought about through the uncompelled choice of an organized intelligence. When we say of an organized intelligence that it does thus or so naturally, we mean that it responds voluntarily to a natural and innate impulse. It is meant that an intelligent entity, either animal or man, voluntarily seeks an individual satisfaction in this or that direction. We do not mean that it does thus or so because it is driven into a corner by hostile environment and compelled to make this or that selection in self-defense.

For example, take cattle that are grazing in a meadow which adjoins a field of corn. The cattle, barred from the corn, are feeding upon the grass. In this act of feeding we see the struggle for nutrition. In that the field of corn is fenced, we note the hostile environment which, in this case, is an artificial environment, erected by superior intelligence. If, however, the bars are let down, the cattle at once abandon the grass for the corn. They refuse to feed upon grass.

Why is this? How does Darwin's "law of natural selection" explain this simple act? How does it account for that instant, intelligent and voluntary choice and discrimination as between foods? How is this act explained by a struggle for nutrition in the midst of a hostile environment? In what sense was this preference for corn compelled by the struggle for nutrition?

Does it not rather indicate an innate power of selection in the animal which expresses itself as an individual choice?

A horse would possibly feed upon potatoes should food be scarce, that is, should conditions be hostile. This would be selection in accordance with Darwin's theory of selection, a choice compelled by hostile environment. No one doubts but the horse would return to oats on the first opportunity, with every sign of approval. This last act of voluntary selection is one, however, which Darwinism does not explain.

Indeed, physical science fails to explain the law which governs the selection and discrimination as to foods, on the part of animals as well as men. Attention to this one phenomenon alone would break down the theory of selection by compulsion, or evolution by hostile physical circumstance. It would be discovered that organized intelligence selects its foods in conformity to the law of vibration. Such investigation would disclose another principle of selection, viz., that of affinity, rather than that of hostility.

Darwin's law of natural selection is, in reality, a theory of unnatural selections. It is one that leaves organized intelligence the mere puppet of blind physical forces and of hostile physical conditions. Natural selection under this theory might well be likened to the natural selection a condemned criminal would make if given the choice of death by the rope or the guillotine.

Instead of natural selections Darwin's theory imposes a series of evils from which the victim, animal or man, must make choice or die. His law of natural selection involves simply:

(1) Blind mechanical forces.
(2) Hunger.
(3) Organs of digestion.
(4) Hostile conditions of Nature.
(5) Repeated processes of feeding and breeding.

In the light of such assumptions man is the automatic result of mechanical principles, blind physical demands, hostile conditions and competitive processes. He comes to be what he is solely through physical re-enforcement from without and by compulsory selections brought about by hostile environment. He is simply an effect of the digestive organs. He is a mass of inherited impulses, passions and sensations, acted upon by external physical forces and held together through processes of digestion.

This is a theory which precludes the idea that an organic entity is an individual operated by an individual and intelligent Will and Desire.

The famous Darwinian proposition, already quoted, could well be paraphrased to express yet another view of this "Law of Natural Selection." The struggle for nutrition means, of course, the struggle for physical life, from which may be derived another proposition laying down self-preservation as the first law of life. Darwin would be fairly stated if one were to say: "The struggle for self-preservation compels a battle of the physically strong against the physically weak, with survival of the physically fittest as a consequence, entailing the supremacy of the physically strong and extermination of the physically weak and incompetent."

Physical science, relying upon its "Law of Natural Selection," must also rest upon this correlated proposition which declares that self-preservation is the first law of being. By "self-preservation," physical science always means the self-preservation of physical life. With this as an underlying motive of all struggle, the battle of the physically strong against the physically weak, with survival of the fittest, becomes the natural and only mode of progress. By "fittest" it must be remembered, physical science always means the physically fittest.

There is but one flaw in this otherwise perfect theory, viz., Nature refutes it.

Science shows that this planet of ours passed from a stage of non-life to life. How? By hostility to life, or by hospitality to life? It shows that the simpler forms appear first. It reveals a steady ascent from lower to higher forms, from simple to complex. It reveals the extinction of certain forms of life.

Does it, however, show extermination or even diminution of the lower forms or of the physically weak? On the contrary, does science not discover an increase of life in its simple and weaker forms? The earth, air and water teem with life in its very lowest form. The earth is simply alive with animalculæ. A whole world of this infinitesimal life feeds upon the blood and tissue of the physically strong. Indeed, the physically strong are the store-houses for the physically weak.

If competition instead of co-operation were really the fundamental principle in Nature, this solid earth had not been crystallized, the vegetable kingdom had not appeared, animal life would not have been a possibility, human civilization could not have been accomplished. If life from the beginning were a battle of the physically strong against the physically weak, the weaker insects and animals could not survive and multiply, nor would insect life feed upon the stronger animals and thus breed and multiply.

If evolution really meant a survival of the physically strongest, mastodons would occupy the place of mosquitoes. In truth, however, the mastodon is extinct while the mosquito survives in great abundance. If evolution were a survival of merely the fittest physically, lions and tigers and the strong rapacious animals would have rendered human life an impossibility. If Nature were, indeed, hostile to life, this planet had not progressed from its original state of non-life to this prodigious multiplication of life. When we consider the extinction of the mastodon and the persistence of the mosquito, is there not something absurd in this theory of the battle of the strong against the weak?

In the same way exception must be taken to the theory that self-preservation of physical life is the fundamental law of life. This is another of those "universal laws" which history and experience, as well as the higher science, contravene.

Love of life is a universal passion.

This, however, is not saying that the preservation of physical life is the first law of being. If so, what about the law which impels a widowed bird to refuse to eat, to droop and to die? What of the law which impels a dog to starve upon its master's grave? If preservation of physical life be the first law of life, how shall we account for the fact that men risk death in every conceivable form, too often for causes that are trivial and unnecessary?

The history of man is the history of war, of life risked in battle. If self-preservation of physical life were the fundamental law of life there had never been war, nor deeds of daring and courage, nor heroes, nor sacrifice even down to death. There had never been suicide, nor suicidal neglect of physical life. There had not been a vast army of religious martyrs.

Darwin's law of natural selection covers also what he terms sex selections. Sex, in its physical and functional capacities, receives minute analysis. Sex selections, however, are interpreted to meet the requirements of the general law of natural selection, or selection by compulsion. Primarily, sex is considered purely as a physical desire for reproduction. Sex attraction is interpreted as something compelled by and dependent upon biological need for reproduction. While the value of sex selection is admitted, the sex relation itself is construed to meet the theory of evolution by necessity. Indeed, sex passion and sex love are defined as a "procreative mania," by force of which, organic intelligences are driven into obeying "Nature's first command—reproduction."

Thus, even that profound and voluntary bond between the two great powers in Nature is construed as a relation compelled and forced upon animal and man by a gross physical passion, which amounts to a "mania." Sex, according to Darwin, has no other uses than the physical. Those uses are solely in the interests, not of the individual, but of species. When it has thus analyzed sex it is done with it. Anything higher, by way of intellectual or moral necessity, is undreamed of.

The force and value of physical breeding are exhaustively treated by Darwin. The numerous phases of heredity are noted. He claims, however, that all variety in traits or conditions are but the effects of repeated processes of feeding or of repeated experiments in breeding.

This is a theory of the evolution of man and the uses of sex in Nature, which is fairly logical as long as we consider the physical organism only. When we come, however, to questions relating to increase of intelligence and the origin and development of love and altruism, the theory hopelessly fails. Intelligence is so distinctly super-physical that it refuses to be accounted for by the automatic operations of the digestive organs. Love is so distinctly a super-physical phenomenon that no amount of physical fact nor ingenious theory can explain it as a mere efflorescence of a "procreative mania." Sex passion, considered by itself, scarcely suggests to intelligence what intelligence observes and feels and knows as sex love. Lust without love is everywhere recognized as the most selfish of human passions; whereas, love, including physical passion, is the most unselfish sentiment known in this world. Mere physical sex passion and pure sex love are so unlike in their nature and expression as to appear the very reverse of each other.

While Darwin refers sex love to the physical passions, he

escaped the embarrassment of seeking to account for morality and altruism. He is satisfied that the personal love relations of human life are the outgrowth of the procreative tyranny. When he comes, however, to that class of ethical phenomena which we know as morality and altruism, he avoids the issue by defining them as "abnormal." Finding no logical cause for such phenomena in the battle of the strong against the weak, they are dismissed as unnatural, and therefore unaccountable in science. Having decided, according to the law of compulsory selection, that Nature sought the survival of the physically fittest, Darwin condemns philanthropy. He condemns it first, because it is unnatural, and next because it cares for the unfit and incompetent children of men.

Physical materialism thus sets up standards of life and conduct which are worthy of savagery. It inculcates principles which would destroy civilization.

Nobody believes that Mr. Darwin really wanted to check philanthropy or to exterminate the weak and incompetent. That, however, is his doctrine. The probabilities are that Darwin himself was a philanthropist. His personal life undoubtedly contradicted his own theory. He probably cared as tenderly for his own family, when weak or incompetent, as would the reader. It is likely that he would have been the first to protest had England passed a law to kill all of the weak and unfit subjects of that realm. This barbarous suggestion, which nobody has ever thought of applying to society, simply illustrates how far a man's theories may contradict the common impulses and the common sense of mankind.

This is an interpretation of man and his evolution which fails to satisfy human intelligence. The best intelligence of the age insists that whatever is, is natural. It insists that whatever exists as universal phenomena is susceptible of analysis and demonstration under natural law. It insists that phenomena, universally attaching to organic life under given conditions, must be natural. This includes phenomena which are intellectual and ethical as well as physical and material. It holds that no doctrine of evolution is complete until it accounts for man as he is, an intelligent and moral being, inhabiting a physical body.

Intelligence claims that there can be no "law of natural selection" except at the same time it is a *natural* law of selection, viz., a law of selection whereby the intelligent selector voluntarily chooses to do or not to do a given thing. These are the claims of intelligence. These claims constitute a demand upon science. This is a task for which physical science is not equipped. It is one which the higher science undertakes.

A fundamental error of Darwinism has become a fundamental error of physical science in general, viz., the suppression or obscuration of the individual.

The effect of Darwinism upon the uninformed mind is universally the same. It leaves the individual reader profoundly impressed with the utter insignificance of individual life, aspiration and effort. On the other hand, it advances species to formidable proportions. Nature appears an insatiate monster, engaged in grinding out species at the expense of every individual. There are no forces but mechanical ones, no principles except compulsory ones, no environment that is not hostile. There are no processes except those of battle and competition. There are no motives except selfish ones. There are no rewards for intelligence except physical and material benefits. There is no future for the individual intelligence except as he physically contributes to species.

There is no escape from these conclusions when once the mind accepts the Darwinian theory as to the factors and causes of evolution.

Darwinism is the effort to analyze and account for species. It does not undertake a study of the individual as an individual. To determine the causes which differentiate and improve species is the motive of Darwin's effort. This complete absorption in species almost entirely obscures the individual. The origin and development of species overshadow the life and the purposes of the individual. Absorbed in the study of aggregates, the units lose their value, except as an integral part of the whole. The individual has no place in Nature, except as he conserves or contributes to the whole; that is, to the sum of physical life.

The effect of this doctrine upon the human mind is deplorable. When applied to animals the theory seems rather plausible, for the lives and the employments of animals appear very gross to us. We have little sympathy with the struggles of the individual animal. When we come to man, however, it is very different. Here we protest against any science or any philosophy which wipes out the value of individual life and effort.

History, experience and intuition unite in disclaiming these deductions of physical materialism. They unite in proclaiming the value of the individual, both to himself and to the world.

In still another respect the individual suffers in both Darwinian science and theory. Whatever attention he does receive in the study of species is attention confined to the operation of his physical functions. How and why it feeds, how and why it breeds, are the only questions put to the individual by physical science. Nothing is considered except that which an organism absorbs and digests, or that which it absorbs and reproduces. Living itself is ignored. Neither aminal nor man is considered apart from the physical functions. Life is ignored for that which sustains life, and for that which perpetuates life.

Even in the sex relation the individual plays only a physical part. He is simply the victim of a "procreative mania" which compels sex selection. He is never seen as the voluntary co-operator with another individual for the attainment of an individual object. Physical science discovers nothing in sex but its physical uses and purposes. A particular, intellectual and moral purpose, it has never yet perceived.

The relation of man and woman is held to be the same relation as that which obtains between animals, viz., a physical one for the purposes of procreation. No higher implications as to sex have ever been discovered by physical science. No higher purpose has been seen in that universal cleavage and attraction between Nature's positive and receptive powers.

So completely has this unfortunate doctrine fastened itself

upon modern science that one eminent specialist* publicly deplores our monogamous system of marriage. Instead, he gravely advocates a practically free selection with children reared by the state.

So continually has this erroneous sex doctrine been exploited by physical science that a new moral philosophy is based upon the theory that sex is but a "physical device for reproduction" and that sex love is "essentially lust."

This degenerate view of the sex relation and office is the inevitable result of a science which deals with the physical side of Nature only. This is the inevitable doctrine of a school which subordinates the mental and moral development of the individual to the physical improvement and preservation of species.

Physical science has never yet undertaken to define the sex principle. Indeed, sex has never yet been analyzed, as a principle, to modern intelligence. It has been taught by science in its physical, functional capacity alone. This is an error of science which is reflected back upon society through false literary and social doctrines involving sex questions and relations.

"The Ascent of Man," already quoted as the exponent of a new moral philosophy, falls a victim to these scientific errors of the modern school.

The life and the profession of the author of this new moral philosophy satisfied him that morality, love and altruism are perfectly natural phenomena. His observation and experience among men satisfied him that these are legitimate facts in life, resultant from natural causes. His determination to seek the cause and to explain the process of the moral order in Nature is commendable. His attempt, however, based as it is upon the assumptions of physical materialism, necessarily proves a failure.

In accepting Darwin's "physical basis of evolution," he accepts the "law of natural selection" with scarcely a qualifying clause. That is to say, he agrees that everything that is in the physical world came to be through compulsions or enforced se-

---

*Letorneau.

lections. The moralist goes even further than does physical science. The Darwinian is content to say that everything physical and intellectual was evolved by the operation of a law of self-defense.

The moralist not only admits this assumption, but he goes so far as to say that everything moral was evolved by a law of self-sacrifice imposed upon the female in Nature. Physical science is satisfied to say that physical and intellectual phenomena arise by compulsions. The moralist, however, insists in so many words that love also came into the world "at the point of the sword."

"The Ascent of Man," therefore, practically agrees with the principles laid down in "The Descent of Man." The moralist enlarges the view of Nature only by the introduction of a second class of physical compulsions and a different line of effects which have been overlooked by physical materialism. The moralist simply turns from a consideration of enforced physical competitions to enforced physical sacrifices. He merely insists that the reproductive function equals that of nutrition in power and effect. In both systems the individual is equally the victim of Nature. In both systems the law of selection is a law of compulsion.

The agreement of the moralist with physical materialism is disclosed in the following statement:* "What controls it (life)," says Mr. Drummond, "are its functions. These and only these "determine life; living out these is life. * * * The rationale "of living is revealed for us in protoplasm. Protoplasm sets life "its task. * * * The activities even of the higher life * * * "are determined by these same lines."

The differences indicated here, as between the theories of the moralist and those of the materialist, are differences without distinctions. The materialist simply centers upon nutrition as the cause of physical and intellectual phenomena. The other selects reproduction as the cause of all ethical phenomena. Thus, the moralist admits what the skeptic claims, viz., that the evolution of man has a basis in the physical functions and is governed by hostile environment. The skeptic evolves a physical and intelli-

---

*"The Ascent of Man," pp. 14-15.

gent being out of the universal struggle for nutrition in the midst of a hostile environment, while the theologian evolves a moral being out of the struggle for reproduction as made by the female in the midst of a hostile environment.

Thus, observing those two bright intellectual stars, Darwin and Drummond, from the same point of view, there is no parallax.

It will be remembered that the moralist sets out to show that love comes into the world by reason of the physical sacrifices imposed upon the female half. To do this, it must be proved that sex plays no part in the evolution of life. It must be demonstrated that sex love is a biological need. It must be demonstrated that sex is simply and solely a physical device for reproduction. The author begins his arraignment of sex by saying:*
"It (sex) may be the physical basis of a passion which is fre-"quently miscalled love, but love itself, in its true sense as self-"sacrifice * * * has come down a wholly different line."

However, before the moralist can clear the ground for his theory as to the origin of love, he is forced to meet and dispose of this obtrusive sex question. His review of sex, in this connection, is remarkable as an exhibition of human reason bewildered at every point by spiritual intuitions. His line of reason, based upon physical science, has been carefully marked out. When he began to talk of sex he had intended to show that no natural relationship existed between sex attraction and love.

How well he kept to his carefully prepared line of reasoning may be determined from his own words. Those visions of the sex principle as perceived intuitionally by the learned Doctor, might well stand as the literary effort of some enthusiastic student of Natural Science.

At the very outset the moralist confesses his total inability to grapple with the profound mystery of sex. Astounded by its universal sympathies, he confesses his ignorance of its ultimate meaning and purposes. He does not realize that a few pages

---
*Page 224.

further on he is to forget this confession and drag down his sublime mystery to a physical "device for reproduction."

No more graphic vision of the sex principle was ever recorded than is contained in the following quotations:*

"Realize," says the moralist, "the novelty and originality of "this most highly specialized creation, and it will be seen at once "that something of exceptional moment must lie behind it. Here "is a phenomenon which stands absolutely alone on the field of "Nature. There is not only nothing at all like it in the world, but "while everything else has homologues or analogues somewhere "in the cosmos, this is without parallel. Familiarity has so ac-"customed us to it that we accept the sex separation as a matter "of course; but no words can do justice to the wonder and nov-"elty of this strange line of cleavage which cuts down to the "very root of being in everything that lives. No theme of equal "importance has received less attention than this from evolution-"ary philosophy. * * * How deep down, from the very "dawn of life, this rent between the two sexes yawns is only now "beginning to be seen. Examine one of the humblest water "weeds—the spirogyra. It consists of waving threads or neck-"laces of cells, each plant to the eye the exact duplicate of the "other. Yet, externally alike as they seem, the one has the "physiological value of the male, the other of the female. * * * "When we reach the higher plants the differences of sex become "as marked as among the higher animals. Male and female "flowers grow upon separate trees, or live side by side on the "same branch, yet so unlike one another in form and color that "the untrained eye would never know them to be relatives. * * * "Sex separation, indeed, is not only distinct among flowering "plants, but is kept up by a variety of complicated devices, and "a return to hermaphroditism is prevented by the most elabo-"rate precautions. When we turn to the animal kingdom again, "the same great contrast arrests us. Half a century ago, when "Balbiani described the male and female elements in microscopic "infusorians, his facts were all but rejected by science. But

---
* Page 245.

"further research has placed it beyond all doubt that the begin-
"nings of sex are synchronous almost with those shadowings in
"of life. From a state marked by a mere varying of the nuclear
"elements, a state which might almost be described as one ante-
"cedent to sex, the sex distinction slowly gathers definition, and
"passing through an infinite variety of forms, and with countless
"shades of emphasis, reaches at last the climax of separateness
"which is observed among birds and mammals. * * *
"Through the whole wide field of Nature, then, this gulf is fixed.
"Each page of the million-leaved Book of Species must be, as it
"were, split in two, the one side for the male, and the other for the
"female. Classification naturally takes little note of this distinc-
"tion; but it is fundamental. Unlikenesses between like things
"are more significant than unlikenesses of unlike things. And the
"unlikenesses between male and female are never small, and
"almost always great. * * * What exactly maleness is, and
"what femaleness, has been one of the problems of the world. At
"least five hundred theories of their origin are already in the
"field, but the solution seems to have baffled every approach.
"Sex has remained almost to the present hour an ultimate mys-
"tery of creation, and men seem to know as little what it is as
"whence it came."

It was known to the date growers in Egypt years ago that they must go to the deserts, secure branches of wild palm and wave them over the flowers of the date palm. This had to be done to insure a date crop. Modern science has explained this peculiar custom. Palm trees, like human beings, are male and female.

The garden date bearing palms were females. The wild palm was male. The waving branches transferred the fertilizing pollen from male to female.

Continues Mr. Drummond: "Now consider, in this far away
"province of the vegetable kingdom, the strangeness of this phe-
"nomenon. Here are two trees living wholly different lives,
"separated by miles of desert land; they are unconscious of one
"another's existence; yet so linked that their separation is a
"mere illusion.

"Physiologically they are one tree; they cannot dwell apart. "It is nothing to the point that they are neither dowered with "locomotion nor the power of conscious choice. The point is "that there is that in Nature which unites these seemingly dis- "united things, which effects combinations and co-operations "where one would least believe them possible, which sustains by "arrangements of the most elaborate kind interrelations between "tree and tree.

"By a device the most subtle of all that guard the higher "evolution of the world—the device of Sex—Nature accomplishes "this task of throwing irresistible bonds around widely separate "things, and establishing such sympathies between them that "they must act together or forfeit the very life of this kind. Sex "is a paradox; it is that which separates in order to unite. The "same mysterious mesh which Nature threw over the two sepa- "rate palms, she threw over the few and scattered units which "were to form the nucleus of mankind."

Thus, the moralist, diverted from his rational purpose, practically disposes of his own theory.

Diverted but momentarily from the compulsions and sacrifices of reproduction to the co-operations and fulfillments of sex, he states the love principle in Nature without knowing it.

It will be observed that at one moment he terms sex "the ultimate mystery of creation," and again "the paradox of Nature." Is it not singular that in his very next lesson he defines sex as a "physical device for reproduction?" He nowhere explains how he becomes informed in the interval of those facts which explain sex as simply a physical device for reproduction.

It must, therefore, be clear to the reader that neither scientific skepticism nor theological speculation has discovered the true law of natural selection. Neither has set forth the true factors and causes of the evolution of a physical, intelligent and moral being. Neither has solved the problem of human life nor of human love.

It is almost unbelievable that physical science, familiar with the universal affinities of the positive and receptive energies in Nature, yet fails to discover a principle of co-operation or a law

THE LAW OF NATURAL SELECTION 159

of selection by natural affinity. It is almost unbelievable that a teacher of ethics, conversant with Nature and Revelation, could yet seek a physical basis for love in the enforced sacrifices of the weaker half of all living things.

This is one of the paradoxes of human reason.

There is a science, however, which enjoys a broader acquaintance with Nature than physical science. There is a philosophy which interprets life and love from higher points of vantage than those upon which the churchman stood. This higher science operates upon two planes of matter, life, intelligence and love. This higher philosophy takes into account the spiritual and psychical principles of sex as well as its physical functions. This philosophy also takes into account, not only maternity, but paternity, in both worlds of life.

To this science and to this philosophy we must look for explanation of morality, love and altruism, according to the true law of natural selection. This is a science which explicitly denies the materialist's compulsory law of natural selection. This is a philosophy which clearly refutes the moralist's compulsory law of love. On the contrary, our science demonstrates a really natural principle of selection, and our philosophy elucidates a voluntary principle of love.

To this higher science the world must look for discovery of a struggle in Nature which is infinitely greater than either the struggle for nutrition or the struggle for reproduction. To the higher philosophy it must look for illustration of a principle that is far mare potent in character and effect than either competition or sacrifice.

In short, the higher science and the higher philosophy, founded upon Nature, physical, spiritual and psychical, refute all theories of "Natural Selection," governed by compulsions.

They declare, instead, that there is a natural law of selection which is neither compulsory nor competitive. They declare that there is a natural law of love which is neither compulsory nor sacrificial.

CHAPTER IX.

THE NATURAL LAW OF SELECTION.

Darwin's law of natural selection is a theory and not a law.

This correction is offered by a science which has extended its investigation beyond the point arrived at by Mr. Darwin.

An exposition of the true factors and causes of evolution rests, primarily, upon the proved fact of life after physical death. Secondarily, it rests upon the fact of a correlated development of individual life upon two planes of existence. The exposition of these true factors and causes is, at the same time, an exposition of the Natural Law of Selection.

Darwinism is a masterpiece of cold intelligence focused upon the physical plane. Spiritual intuition plays no part in his work. The profound effect of his discoveries was inevitable. Average intelligence looks to exceptional intelligence for leadership. It was inevitable that the lesser scientist and the less acute reasoner should defer to his masterly logic as well as to his indisputable facts. There was none among his own schoolmen to dispute him. The people were simply confounded. Many men, both learned and unlearned, rejected his conclusions upon their own intuitions as to a spiritual side of life. None, however, had the facts at command nor the rational evidence to support those spiritual intuitions and convictions.

All false systems of philosophy rest upon misinterpretations of Nature.

Nature, both in fact and in principle, is changeless.

For example: Our solar system has been changeless since man has inhabited this planet. The theories, however, which have been projected in relation to that system have been numerous and contradictory. Even the opinions relating to the origin of our planet would make a considerable volume.

Mr. Darwin discovered nothing but bald physical facts. He based his theories upon these partial data alone. He did not, apparently, entertain a single intuition of super-physical matter or super-physical forces. It was, therefore, natural that Darwin should see in man merely an improved ape.

The singular fact remains that neither Darwin nor the school of modern physical science has actually bridged the gulf between man and animal. They have come down to it on both sides, but the link is still missing. Nature does not furnish hybrid types. There are animals resembling humans. There are humans resembling animals. Nobody, however, has discovered a normal type that it can not easily assign to the animal or the human kingdom. The fact remains that monkeys and men, though unquestionably related, are yet distinct products of Nature. If the evolution from the ape to man were a mere matter of degree, the chasm which now exists between them would be filled with hybrid types, with crossings and recrossings of that which could be classified as neither ape nor man.

That element which differentiates man from animal is as clearly disinguishable as that which differentiates plant from animal.

The physical facts of evolution, as reported by Darwin, have been known for ages to certain special students. These few, however, arrived at their knowledge through the study of spiritual, rather than physical, processes. Hence the extreme difficulty of presenting that knowledge to the world. Indeed, such presentation were impossible now except for the preparatory work accomplished through the development of the physical sciences.

For ages the higher science has recognized all struggles of intelligent life upon the physical plane. Neither the struggle for nutrition nor the struggle for reproduction has been overlooked. In reality, science demonstrates that all of the achievements of intelligent life, physical, mental and moral, are the results of effort or struggle. By reason of its extended research upon two planes of existence and over long periods of time, science is able to say:

(1) Nature is not hostile in principle nor competitive in operation.

(2) Neither the struggle for nutrition nor the struggle for reproduction is the cause of the evolution of man.

(3) Evolution is not a battle of the physically strong against the physically weak.

(4) Survival depends upon other laws and conditions than physical.

(5) Sex is more than a physical function.

(6) Individual destiny is something more than a contribution to species.

Familiarity with evolutionary processes during long periods of time enables the higher science to say:

(1) The intelligent spiritual principle of polarity in operation constitutes the true and Natural Law of Selection.

(2) Evolution is progress by intelligent affinities and not by blind physical compulsions.

(3) Evolution is a race as between the intelligently strong and the intelligently weak.

(4) The Natural Law of Selection conserves the survival of the intelligently fittest.

Science holds further, that the government of inorganic matter and of organic life is conducted under two aspects, viz.:

(1) Under a universal principle of intelligence which conserves the refinement of matter, the increase of vibratory action and the improvement of physical organisms.

(2) Under an individual principle of intelligence which conserves the increase and persistence of individual intelligence and governs the individual and ethical relations of intelligent beings.

The universal principle governing the general operations of all material substances may, therefore, be said to demonstrate the mathematics of evolution.

For example: The involuntary unions and combinations of particles of frozen moisture into the perfect snow crystal illustrate the mathematical intelligence of Nature.

On the contrary, the individual principle governing the particular operations and the ethical relations of organic life, may be said to demonstrate the Harmonics of Nature.

For example: The love relations of man and woman illustrate the voluntary and individual operations of intelligence.

By general operations of Nature is meant the involuntary response of mineral, vegetable and animal substance, to the intelligent demands of the universal principle of polarity.

By particular operations of organized life are meant the voluntary relations of individual intelligences; relations impelled by an innate independent and self-operated Will and Desire.

For example: A male animal (and somtimes a man) will attack a rival and force him to relinquish his suit for the favor of some particular female animal (or woman, as the case may be). Here is the direct operation of an individualized and intelligent will which directly serves the individual interests of that particular male animal (or man). Or, again, the same principle is illustrated when a female animal, or woman, bestows her favor upon the victor. Here is the direct operation of an individual and intelligent desire which directly serves the individual interests of that particular female animal or woman.

While it is the general intelligence of Nature that impels male and female to unite, it is the individual intelligence of organic life which establishes the law of Individual Preference and equips the particular entity to seek and attain relations which conserve the individual Will and Desire.

The attempt to define intelligence, other than the governing principle in Nature, would amount to an absurdity. The finite mind can neither grasp nor define ultimates. Neither on this side of life nor on the other has the finite thus far comprehended the Infinite, either as to matter, motion, life, intelligence or love. Science is forced to rest upon the principle itself. It can do no more than occupy itself with demonstration and classification of phenomena which go to prove the principle. The finite mind thus far is forced to content itself with the fact that all operations of Nature are intelligent, whether guided by a general principle of intelligence or by an individual intelligent Will and Desire.

This being true, science holds that no adequate theory of evolution can ignore intelligence, either in its general or its individual aspect.

This position directly contravenes physical science. Not Darwin alone, but almost the entire line of authorities in that school, ignores these most patent of all facts in Nature, viz., the manifestation of a general intelligence in every operation of inorganic nature, and the phenomenon of an individualized intelligence rising out of those general processes of inorganic nature.

This complete overlooking of so self-evident a principle is the one error of modern science which, to the trained student of Nature, appears as "inexplicable." To account for the universe as the outcome of blind mechanical energies, and to account for human intelligence as a result of blind physical forces, do not necessarily indicate lack of intelligence. They do, however, indicate intelligence restricted and confined to very narrow limits. Pessimistic philosophy is the inevitable result of science which, ignoring the fundamental phenomena of intelligence, confines itself to the study of bald physical fact.

In this particular the higher science marks a new era in scientific declaration. It puts aside those unsatisfactory theories of mechanical principles and blind forces, and gives to intelligence a more rational explanation of itself.

It declares, instead, upon carefully demonstrated facts of Nature, that evolution is progress by and through intelligent principles, universal and individual.

This being true, science accepts intelligence as the primary cause of all that is, and not as a recent and incidental result of a digestive apparatus.

Finite science does not hold itself competent to account for intelligence. It does claim, however, that science is able to point out the general evidences of intelligence. It is also able to trace the processes of individualizing intelligence by means of improved physical organisms. The evolution of intelligence is a more marvelous process than the evolution of the physical body. In the eyes of the true scientist (one who knows the persistence of intelligence after physical death) the physical body can never appear to be more than the instrument which persistent intelligence slowly fashions and perfects for its uses on the physical plane.

Some one has said: "Intelligence sleeps in the stone, dreams in the animal, wakens in man." The case of Nature had been more fully stated had it been said: "Intelligence sleeps in the stone, dreams in the plant, wakens in the animal and acts in man."

Intelligence, however, whether manifested in the involuntary activities of material substance, or in the independent activities of intelligent entities, remains as the ultimate mystery in Nature. This is the principle which rationalism names "law." This is the principle which intuition calls "God."

This, indeed, is the "Great Unknown," both to finite science and to finite faith.

Physical science, it will be remembered, bases the evolution of man, physically and mentally, upon the struggle for nutrition in the midst of a hostile environment.

It will be recalled that a new moral philosophy, based upon this theory, claims that man, as a moral being, is the result of the struggle of the female for reproduction in the midst of a hostile environment.

Upon such assumptions physical science declares its natural law of selection.

The student of Nature, however, who analyzes life and encounters intelligence in two worlds, is in a position to reject the theory of evolution by non-intelligence. He discovers and demonstrates those facts and principles which disprove such assumptions.

Nature, it is found, embraces other principles than mechanical ones, other forces than blind physical forces, other methods than compulsions, and other purposes than the preservation of species or the creation of a family. Evolution is something more than a battle of the strong against the weak. There is another struggle than the struggle for nutrition or the struggle for reproduction. Sex plays a higher rôle than the mere perpetuation of species. Living is more than feeding and breeding, and the intelligent individual has a nobler destiny than "rearing the largest number of healthy progeny."

On the contrary, it is demonstrated that those other and highest principles in Nature are intelligent principles, those other forces are spiritual and psychical forces, those other methods are co-operative, those other and higher purposes are intelligent and beneficent in character. Demonstration of these highest principles, forces, methods and purposes, enables the higher science to law down the Natural Law of Selection and to point to the survival of the intelligently fittest as the final outcome of evolution.

The fundamental principle in Nature is an intelligent principle of fulfillment, viz., a principle of equalization of forces through vibratory correspondences. The phenomenon of polarity, or the co-operation of the positive and receptive powers in Nature, constitutes the expression of this universal principle.

For illustration: That principle which impels one physical particle to seek vibratory correspondence in another particle of opposite polarity, must be recognized in science as an intelligent principle. The act of the atoms is an intelligent act. In this case, however, the intelligence is not an individual intelligence residing in the atom. It is, instead, that general intelligence which physical science is content to name "Natural Law."

The phenomenon of love between two rational beings, man and woman, constitutes an expression of the individual principle of intelligence as well as the involuntary effects of the universal principle.

For illustration: That principle which impels one rational being to voluntarily seek correspondence in another such being of opposite polarity, must be recognized as an act of independent, individual intelligence. In this case the impelling principle resides in the individual, not merely in the universal principle which governs unconscious physical substance.

The dawn of organic sex marks that stage of evolution where general intelligence enters upon the process of individualizing intelligence. Animal life, from its lowest to its highest expression, stands as indisputable proof of the struggle of general intelligence for an individual expression.

Human life illustrates the highest achievement of that purpose. The universal intelligent principle of polarity first raises the individual product to the point of an individual, self-operating intelligence. After this, individual intelligence raises itself by individual efforts which directly assist Nature in the general purpose it has in view for the individual.

Thus, universal intelligence and individual intelligence are working out the universal purpose of Nature and the particular purpose of the individual. Any other interpretation of Nature, of life and of man, libels Nature and stultifies intelligence.

Physical science demonstrates that this planet of ours was once a shapeless mass of fiery gases revolving in space.

It is the triumph of the spectroscope, under the light of scientific intelligence, to have demonstrated to an absolute certainty that there are in existence at this moment other immense sources of gaseous substance from which planets are evolved.

When the human mind looks out upon this world alone, recalling the history which science unfolds, it becomes an absolute impossibility to define this stupendous generation of physical, spiritual and psychical phenomena as an efflorescence of blind and non-intelligent forces.

The evolution of man rests upon that principle in Nature which impels intelligent entities to select their vibratory affinities. The individual, be it understood, makes such selections, not that he may further Nature's general purposes, but because he is so impelled through the expectation of and desire for an individual and ethical self-content. In this intelligent principle of co-operation and fulfillment the higher science finds its Natural Law of Selection.

This being true, organic sex represents this principle in operation. It follows, therefore, that the individual co-operations of male and female and of man and woman constitute the true principle of natural selection in the kingdom of individual intelligences.

From what has already been said concerning this intelligent

principle of polarity by way of the refinement of matter, the increase of vibratory action and the ascent of life, the reader is prepared to hear that evolution means progress by intelligent affinities, instead of progress by blind physical compulsions. The question may be asked, What are the evidences which show that the evolution of man is conducted by a law of selection through affinities rather than of selection by compulsion? It may also be asked, What are the evidences that sex selection is voluntary selection by intelligent affinities, rather than compulsory selections in answer to a biological need?

The discussion of the law of polarity has already covered the intelligent operations of the lower kingdoms of mineral and plant. It has already been explained how evolution is maintained in the lower kingdoms by the impulse of the individual entity seeking vibratory correspondence in another entity of opposite polarity. It has already been explained how these operations, conducted by the general principle of affinity, yet intelligently conserve the evolution of an organized and self-operating individual intelligence.

What we now have to deal with is the particular principle which directly governs the evolution of man from the early reaches of organic life. Science is now called to explain, not merely the involuntary affinities of matter governed by the universal principle of intelligence, but the voluntary affinities of organic life governed by the principle of individual intelligence.

With the generation of physical life and the dawn of organic and individual intelligence, there arises a distinctly new class and order of phenomena. When Nature, under its general principle, succeeds in evolving a sufficiently delicate physical organism, the phenomenon of sensation is evoked. This is the first distinct evidence of an individual intelligence. The capacity for sensation is evidence of a certain character of intelligence which does not reside in a stone or a plant. This is, as it were, a point of intelligence, individual to that particular living entity, which experiences sensation.

The cruder the physical organism the poorer the instrument for the uses of intelligence. In consequence, the duller are the sensations and the fainter the perceptions. Evolution appears under one general aspect of intelligence until the generation of individual intelligence takes place. Until individual sensation, perception and volition obtain, there is no apparent distinction as between the operations of the general principle of polarity and the response of the individual to that principle. That is to say, all the operations of individuals below the point of a self-operating intelligence appear as purely automatic. They appear as activities controlled by intelligence. When, however, Nature succeeds in evolving an organic intelligence, of even the meanest capacities, evolution proceeds by a distinctly double process, viz.:

(1) Under the general guidance of the universal and intelligent principle of affinity.

(2) By the direct impulse of the individual and intelligent Will and Desire of an organic entity.

Until this stage is reached the phenomena of Nature appear simply as mathematical and involuntary results worked out under intelligent direction. After this the higher phenomena of life consist in ethical effects, to which individual intelligences voluntarily give rise. From this stage science must deal with processes and effects which are individual and ethical, as well as those which are universal and mathematical.

From this period, we may very properly say, begins the evolution of man, an individualized intelligence, inhabiting and operating two material bodies.

The Natural Law of Selection rests upon that principle which impels this individual intelligence to exercise its individual will and desire to secure its own ethical content. The individual intelligence, like the individual atom, seeks self-adjustment, primarily, through affinities with another individual. The individual intelligence, therefore, is impelled to such union by an individual will and desire, as well as by the involuntary impulses due to the general law of correspondence in material substances. While the general law is taking care of the chemical affinities, the individual intelligence is seeking those ethical conditions which satisfy the individual needs and requirements of intelligence.

Nature demonstrates that the higher activities and relations of intelligent life are governed by an independent Will and Desire, rather than by the general mathematical principle. The powers of sensation and volition raise the entity from automatic to independent action. Everything in Nature that has attained to this point moves and acts thereafter upon an innate principle of selection which conserves that particular entity alone.

This, therefore, becomes the true and Natural Law of Selection, in that it is the one and only principle which *impels* but does not *compel* intelligence to action.

The most unfortunate effects of Darwinian doctrine are due to his repeated declaration of a compulsory struggle for nutrition, and therefore, a compulsory law of selection. He says, it will be remembered, that "the struggle for nutrition compels natural selection," etc. That very statement, the basis of his doctrine, embraces a contradiction, that is, if we interpret the words "compel" and "natural'" in their usual and accepted sense. To be consistent with his fundamental theory, that everything improves and progresses under physical compulsion, Darwin includes sex selection in the category of "compulsions."

Darwin's error is due to his inability to discriminate between general necessities, imposed by general environment, and the particular powers of intelligent entities operating within that environment. He fails to observe that whenever and wherever the really voluntary, and therefore Natural Selection of intelligence is interfered with, degeneracy occurs. He fails to perceive that the voluntary will and desire of an individual intelligence governs acts or selections which are voluntary. He fails to perceive that anything and everything done in accordance with an innate will and desire is an act *impelled* and not *compelled*. Each act is, therefore, a voluntary act and not a compulsion.

Darwin rightly conceives that improvements and higher types may be obtained by cross-breeding. To his mind, however, that improvement rests wholly upon the selected physical conditions. In reality, such improvement is due, primarily, to the judicious crossing or assimilation of intelligences. This principle, difficult to prove in animal life, is clearly demonstrable in human society. Any arbitrary system of sex selection which ignores the natural spiritual law of selection means degeneracy.

For example: It is the commonest fact of history that the arbitrary marriages made in the interests of royalty invite deterioration and degeneracy to that particular line. Scientifically and philosophically, such marriages are the lowest types known to human society, viz., marriage in the interest of fictitious position which ignores physical, spiritual and psychical fitness. Marriage brought about by purely physical passion is even less ignoble, since here, at least, is a temporary physical affinity.

If Darwin's idea of marriage, based upon physical fitness alone, could be enforced, the moral degeneracy of the race would result. We would have, instead, fine animals in human bodies. Physical passion would increase, but love would wane. The marriage system, or the individual marriage, which ignores either the general principle of affinity in Nature or the individual will and desire of the intelligence, means moral degeneration to a nation or to an individual.

China and Japan are among the pitiful examples of an arbitrary interference with the intelligent and Natural Law of Selection.

It is admitted that the marriage of the diseased and delicate does entail weakness and suffering upon offspring. That misfortune, however, could not be compared with the moral degeneracy that would flow from the attempt to enforce marriage among the physically strong by arbitrary selection of individual pairs. It is man alone, in his individual pride of intelligence, that has ever conceived or executed a really compulsory law of selection. It is man alone who introduces hostilities into evolutionary processes.

From all that has been said, it must now be understood that the co-operations of sex constitute the Natural Law of Selection, under which individual intelligence seeks self-satisfaction. The voluntary will and voluntary desire of a masculine and a feminine entity for each other, constitute the primary factors and causes of the evolution of man.

Physical science lays down a struggle for nutrition and a principle of hostility as the basis of evolution. The higher science discovers in Nature another and a higher struggle, as well as another and a higher principle. Natural Science discovers and demonstrates a struggle of organic intelligence for an individual persistence and an individual and ethical satisfaction. This supreme struggle of intelligence for individual expression, satisfaction and persistence, is conducted, primarily, through sex co-operations, or through those natural selections induced by vibratory correspondences with that which is of an opposite polarity. Universally the normal individual seeks self-adjustment, primarily, in another individual of opposite polarity. Universally, therefore, the development of the individual rests, primarily, upon the relations it effects with its polar affinities. All of this means that individual development is, primarily, a question of sex and sex selection.

The deductions of physical science as to sex rest entirely upon the study of physical nature, from protoplasm to man. The deductions of the higher science rest upon the study of physical matter, spiritual matter and intelligence, in their combined operations, from protoplasm to man, physically embodied and physically disembodied.

These are deductions which universally consider the correlation of forces upon both planes. They rest upon investigation and analysis of physical functions, spiritual principles and psychical activities. By means and methods, not as yet generally recognized as scientific, it is found that everything, from protoplasm to man, has its spiritual duplicate or pattern. It is discovered further, that the principle of intelligence resides, primarily, in the spiritual part. It does not matter whether that intelligence appears as general or individual intelligence.

As already explained, the Life Elements are dual in character, being positively and receptively conditioned in their relation to each other. That is to say, the dual energies of any one of the Life Elements are positive and receptive to each other, or mutually dependent in their operations.

When the Spiritual Life Element is inducted into matter it does something more than generate life. It causes living proto-

THE NATURAL LAW OF SELECTION

plasm to separate into two departments which diverge more and more from the initial state. The one department is charged with positive, generative, male energy. The other is charged with reproductive, or receptive, feminine energy. Organic sex thus represents the inherent energies of the Spiritual Life Element, and is, therefore, a spiritual principle instead of a physical function.

The general principle of intelligence in Nature causes protoplasm to co-operate until by such co-operation is evolved a physical organism that admits of an individual expression. Sex is, therefore, an intelligent spiritual relation manifesting through physical co-operations, instead of a moral phenomenon evolved by the automatic operations of the digestive and generative organs. The gradual individualizing of intelligence through lower forms would admit of volumes in illustration. The facts, however, germane to this subject, relate, first, to Nature's struggle to improve the physical instrument of intelligence, and second, to the struggle of individual intelligence to satisfy itself through that organism.

This intelligent struggle of Nature and this struggle of individual intelligence are as well defined, from oyster intelligence to ape intelligence, as are the improvements in the form, functions and capacities of the physical body.

Thus, evolution by intelligence and the self-evolution of intelligence rest, primarily, upon the sex principle. Physical sex union thus represents the struggle of intelligent entities for spiritual as well as for physical union. This being true, human marriage represents spiritual relations with spiritual consequences, as well as physical relations with physical consequences. Physical paternity and maternity, therefore, represent an effort for individual adjustment, rather than an effort for reproduction. Thus, the Natural Law of Selection is disclosed as a series of intelligent co-operations, rather than a series of blind compulsions. It is seen as a system of fulfillments, instead of a system of sacrifices. It is survival of the intelligently fittest, rather than of the physically fittest. It is, in brief, the evolution of intelligence by way of vibratory correspondences and ethical satisfactions, rather than by physical compulsions and physical sacrifices.

The advanced student, familiar with the spirit of matter, knows that both the vital and the intelligent powers reside, primarily, in the spirit. He knows that the forces which integrate physical material are, primarily, spiritual forces; that physical functions express spiritual energies, and that physical relations are spiritual relations, conserving the development of intelligent life.

Even the demands for physical nutrition are, primarily, spiritual demands. Indeed, all appetites and passions called physical, are, primarily, spiritual requirements of intelligence seeking self-satisfaction through physical conditions.

This reading of Nature throws new light upon the demand for nutrition and the capacity for reproduction which inhere in protoplasm. It throws a new light upon sex which appears with the dawn of life itself. Under this reading of Nature the physical functions and the physical activities of intelligent life are relegated to their proper place in the evolutionary scheme. In the light of this struggle of intelligence, the struggle for nutrition and the struggle for reproduction fall into line, the one as a duty incidental to the body, the other as a duty incidental to the race. Both of these minor functions and activities are thus seen as mere conservers of intelligent life during its operations upon the physical plane.

With this understanding of the Natural Law of Selection, progress appears as a system of Harmonics. Nature is revealed as an overshadowing power, intelligent in operation and beneficent in intent. With this interpretation of sex, the union of two animals or of two humans represents, not merely a biological need, but an innate necessity of organic intelligence for an ethical self-content. The physical organs of sex thus become the mere instruments for the accomplishment of certain functions incident to physical life and earthly development. The male organs of generation represent the positive energies of the vital elements, while the female organs of reproduction represent the receptive energies of the vital elements.

Physical science lays down its doctrine of sex honestly. Nevertheless, that doctrine is demoralizing in effect. Society can no more escape the penalty of a false philosophy than the individual can escape the penalties of his own ignorance. Ignorance is the root of all social evil, just as partial knowledge is the basis of intellectual controversy. Disputation lessens in an exact ratio with the definite knowledge brought to bear. Every new proved fact in Nature settles some sharp debate. Perhaps the oldest and fiercest controversy ever waged by human intelligence covers the question of predestination and free will. To-day science declares that each combatant represents a true principle in Nature. The predestinarian stands for the principle of universal intelligence, while his adversary represents the principle of an individual free will and desire. Science demonstrates that the evolution of man is a result brought about by the selections of free-willed entities moving under immutable principles.

By free will is not meant free execution, nor even the immediate power of execution. What is meant is the fact of an internal, free choice or motive, or a natural, independent impulse to do or not to do a certain thing. A man desires, wills and universally attempts to execute his individual desires and designs. This innate power of individual will constitutes the motor power in the self-evolution of intelligent beings. At every stage of being, individual will and desire and individual efforts and executions move upon and modify environment.

The desire to do and the will to execute are as characteristic of the worm, the fly, or the dog, as they are of man. It is the universal possession and operation of these intelligent capacities which refute the theory of evolution by physical feeding and breeding alone. Nature, on the contrary, shows that evolution is the gradual rise of a free-willed entity to a self-conscious and rational compliance with the demands of universal intelligence. The degree of development attained by an individual, a family or a race, depends upon the degree of intelligent compliance exercised as to these universal principles in Nature.

Physical science, as far as evolution is concerned, is committed to the theory of predestination. That is to say, it presents Nature as a series of fixed mechanical principles and hostile conditions which remorselessly move, first upon protoplasm, and lastly upon man. Engrossed with the physical evolution of species through the digestive organs, physical science continually ignores individual intelligence and the individual impulses and objects of individual life. It fails to perceive that while Nature is working out a general result of individualizing intelligence, the individual is striving for a particular result individual to itself. Engrossed with the study of physical processes of nutrition and reproduction, it ignores the intelligent operations and the ethical consequences which attach to the life of the individual consumer and reproducer.

The predestined extinction of individual intelligence is the fiat of a science which has not, as yet, discovered the spiritual side of Nature, nor perceived the evolution of intelligence.

Physical science fails to discriminate between the general operation of a universal principle of intelligence and the particular selections of an individual intelligence. It perceives a general principle which apparently compels material substance to its activities. It fails, however, to perceive that it is an individual will and desire of intelligent beings which shape and control the higher ethical activities of organic life.

For illustration: Physical science perceives that the primitive man is compelled to find nutrition if he would preserve his life. It does not, however, explain why the savage prefers one kind of food to another, nor how and why he deliberately sets out to secure it. By reason of a scarcity he may be temporarily compelled to eat less palatable food than he desires. This unpalatable food, however, sustains life and thus fulfills the demands of nutrition. But what course does this primitive man pursue? Does he continue to eat that which he does not individually desire? Does he submit to environment or does he seek to overcome it?

The history of human development answers this question.

That history is not simply a record of external conditions beating and buffeting a non-intelligent stomach into a civilized man. Instead, it is the history of crude intelligences pushing out, striving for, and achieving individual satisfactions through an individual force of will and desire.

Physical materialism entirely ignores that supreme capacity of individualized intelligence, viz., the power to contravene, defy and pervert those very principles intended by Nature to govern the physical, intellectual and moral development of the individual himself.

The higher science fully considers the universal principles in Nature. It considers the conditions and relations which arise by force of these principles and of external circumstance. It recognizes selections that are compulsory. It takes into account the facts of competition. It recognizes the sacrifices in human life. It does more than this, it recognizes also the universal principle of fulfillment through the general effects which flow from vibratory correspondences. It recognizes also the far-reaching effects of individual, intelligent forces which go to modify and shape and change environment.

Science that has gone far enough to discriminate as to the forces in evolution, declares that the principle of harmony outruns conditions of hostility. It declares that individual intelligence co-operates with general intelligence and that natural selections far outweigh, in force and effect, all unnatural selections or selections by compulsion. It finds, in brief, that enforced selections, made in the interests of nutrition and reproduction, are of little value in the higher evolution of man, as compared with the voluntary selections by affinity.

On the contrary, Nature discloses a struggle of intelligent life which far overshadows these lesser struggles of physical life. It discovers a purpose in Nature and life far more inspiring to the individual soul of man than the preservation of species or the creation of a family. In this supreme struggle of Nature science reads the past, analyzes the present and forecasts the future of man in the body by way of physical refinement, intellectual power and moral achievement.

It follows, therefore, that Natural Science presents to the world a new proposition. Darwin has said: "The (physical) struggle for nutrition compels natural selection with reproduction as a consequence, entailing divergence of character and extermination of less improved species."

Nature, on the contrary, declare that *the intelligent Spiritual Struggle for an Individual Self-Completion induces Natural Selection, with Reproduction as a consequence, entailing Divergence of Character and appearance of more improved Individuals and Species.*

Familiarity with Nature over long periods of time enables the higher science to declare:

(1) Evolution is an intelligent scheme.

(2) Evolution is conducted under two principles of intelligence, the universal and the individual.

(3) The spiritual principle of polarity represents universal intelligence.

(4) The individual Will and Desire of organic beings represent the principle of individual intelligence.

(5) The universal principle of intelligence evolves by its operations, individualized, or self-operating intelligences.

(6) The self-operating intelligence conducts its own evolution along the lines laid out by Nature, viz., through vibratory correspondences between the positive and the receptive individuals.

(7) The intelligent spiritual principle of polarity foreshadows the true and Natural Law of Selection, viz., Selection by Affinity.

(8) The individual struggle of organized intelligences for self-adjustment, constitutes the Natural Law of Selection in the evolution of man, viz., Selection by Intelligent Affinities.

(9) Evolution is a race between the intelligently strong and the intelligently weak.

(10) Evolution means Survival of the Intelligently Fittest.

(11) The surpreme struggle in Nature is the Struggle for Completion.

(12) The object of universal intelligence is the Completion of an Individual.

(13) The purpose of the individual intelligence is Self-Completion.

CHAPTER X.

A QUESTION IN SCIENCE

Does Nature embrace a purpose?

Here is a question of greater moment to the well-being and happiness of intelligent man than is a mere knowledge of Nature's scientific processes.

Upon this question the findings of physical science and those of Natural Science radically disagree.

Physical science concerns itself with the evolution of the physical man. It seeks only to analyze the physical processes by which the physical body is evolved. It seeks only to trace the physical causes which give rise to the phenomena of sensation and intelligence.

It does not concern itself to discover any ultimate purpose of Nature in these processes. It does not seek to know why Nature has finally produced this complicated physical, intellectual and moral being—man.

It finds no other uses for man in Nature than his operation of those functions by which the physical body is sustained and the species propagated and improved. The entire range of physical science leaves no other impression upon the mind than that Nature exists for the sole purpose of physically improving species.

This, indeed, is the entire argument, and intent of physical materialism.

Natural Science, on the contrary, does something more than to enumerate facts and analyze processes, whether those facts and processes be physical or spiritual. It is not content to investigate material phenomena alone. It is not satisfied to simply discover the physical and spiritual processes involved in the building of the body and the individualizing of intelligence. The higher science seeks to know the why as well as the how. It aims to know why man exists in two worlds as well as how he exists. Natural Science, like physical science, is concerned with the study of natural phenomena. Unlike physical science, however, it is even more deeply engrossed with the study of the higher ethical phenomena attaching to intelligent life.

Physical science has one motive. The higher science has two. Where physical science ceases its inquiries Natural Science goes forward. Where physical science presents only the past physical history of man upon this planet, Natural Science forecasts his spiritual, intellectual and moral possibilities in two correlated worlds of life.

Darwinism declares that the seemingly purposeful in Nature is merely a series of adaptations forced upon species in the struggle for nutrition in the midst of a hostile environment. It perceives nothing in evolution which indicates anything that could be properly called an intelligent purpose. It foreshadows a result, truly, but simply a result affecting physical life. It does not contain one hint of any natural purpose which meets the natural intuitions and aspirations of human life and intelligence.

"Natural Selection," as laid down by Darwin, foreshadows simply and only a "physically improved species" brought about through the "survival of the fittest" in that universal battle of the physically strong against the physically weak. This "physically improved species" is held to be the "fittest" under Nature's fundamental principle of hostility. A physically improved species is, therefore, accepted as the noblest result possible under Nature's working formula.

Darwinism finds in Nature no more subtle principle than physical appetite. It finds no higher struggle than a physical one. It conceives no higher standards than physical improvements. It forecasts no higher evolution than a physically strong and healthy race. With one-half of Nature obscured from his senses, it is little wonder that the physical scientist promulgates a doctrine which is, at once, a stultification of intelligence and the death knell of human ambition, hope and aspiration.

Absorbed in the examination of purely physical facts and

functions, physical science ignores, as far as possible, the universal accompaniments of physical life, viz., spiritual and psychical phenomena. Accepting the organs of digestion as the primary cause of the evolution of man, it becomes necessary to refer all super-physical phenomena attaching to the evolution of man to the same cause. As a result, physical science passes over the spiritual and psychical implications attaching to human life, as minor issues, or as a mere efflorescence of physical feeding, breeding and battle.

Darwinism considers the individual solely as an agent for the perpetuation of species. He defines the individual as a mere result of past condition incident to the struggle for nutrition. His destiny is completed in what he may contribute to the physical improvement of species.

This is what a man counts, and all that he counts in Darwinian doctrine.

The value of individual life under this theory is summed up in the general assumption that the sole intent of Nature is the improvement and preservation of species. Upon this fallacious premise Darwinism argues, first, that philanthropy which cares for the weak and unfit children of men, is a violation of natural law. Next, it declares that the highest duty of the individual man is "the rearing of the greatest number of improved progeny."

Thus, in a physically improved species we find the ultimate object of Nature. In the rearing of the largest number of improved progeny, we discover individual destiny—according to Darwin.

Darwinian doctrine, reduced to a decalogue, declares:

(1) Life is a struggle for nutrition and physical benefit.

(2) The business of life is the struggle for nutrition and a struggle for reproduction.

(3) The purpose of evolution is the physical improvement and preservation of species.

(4) The individual exists for species.

(5) There is nothing to live for except physical posterity.

(6) Intelligence is an emanation of food combinations.

(7) Love is essentially lust.

(8) Philanthropy is unnatural and therefore is a disease.

(9) The expectation of life after death is a superstition.

(10) Individual ambitions, hopes and aspirations which transcend the requirements of nutrition and reproduction, are delusions and dreams based in superstitions or indigestion.

These are deductions and dogmatisms which are rejected as insufficient by the common intuitions and the common experiences of human intelligence. These are doctrines which the intelligent soul of man denies. These denials of the soul are supported, analyzed and explained by a broader science. The limitations of physical science are responsible for the theory that evolution is based in digestion and conducted by competition. Those limitations are responsible for the idea that intelligence is merely an emanation of physical food, that love is an efflorescence of lust and that morality and philanthropy are abnormal.

Inevitably this out-of-focus view levels man to the needs and requirements of his physical body. Inevitably such deductions end in gross materialism. Scientific skepticism does not contain the merest shadow of a purpose in Nature that appeals to either an intelligent or a moral being.

The protest against Darwinism has never been on account of the facts set forth. It was the appalling theories which accompanied those facts that shocked the spiritual intelligence of the world. Even average intelligence has not found it so difficult to accept the physical facts which show the gradual evolution of the physical body. It is the intelligent soul which refuses to accept the explanation which Darwin makes in connection with the physical facts. The man of keen spiritual intuitions does not reject Darwinism because it allies man, structurally, to the ape. He rejects it because it reduces man to the kingdom of the ape, makes him the plaything of blind physical forces and limits his destiny to improvement of species. He rejects it because it levels life, intelligence and love to the gross needs of the body, and passes the death sentence upon the living soul.

After the first shock to its spiritual faith, the church began to consider Darwinism in the light of cold reason. To-day an enlightened clergy, for the most part, have laid the terrors of the theory and accepted the facts of Nature as recorded in "The Descent of Man." The enlightened Christian world now agrees that the evolution of the physical body from lower forms does not necessarily mean that man is merely an improved animal, that life has no other purpose than feeding and breeding, that competition is the principle of progress and that death ends all.

That new moral philosophy (already referred to), based alike upon the facts and theories of Darwinism, finds what it believes to be a purpose in Nature. The reading of Nature by the moralist embraces certain improvements upon Darwinism. It does not, however, touch upon the fundamental errors of physical materialism.

"The Descent of Man" postulates a physically improved species as the best result obtainable under evolutionary law. "The Ascent of Man" declares that the great purpose involved in evolution is the creation of a family.

The moralist who seeks to both support and criticise Darwinism has a difficult task. He is right when he declares that love, and not warfare, is the greatest thing in the world. He is wrong, however, in the pathway he selects for love. He is wrong when he introduces love into the world "at the point of the sword." He is mistaken when he formulates a principle of sacrifice as the true principle of love. He is wrong when he declares that the creation of a family is the purpose of evolution.

He finds that the great moral task of Nature, as revealed in evolution, consists in forcing love upon the world "at the point of the sword." This end, he claims, is effected by and through the endless pain and sacrifices of maternity. That is to say, love is forced upon the world through the physical disabilities of the female half. To achieve this moral result it is claimed that Nature is, therefore, principally engaged in the manufacture of mothers. This manufacture of mothers is explained as the necessary process in achieving the final, moral purpose of human life and earthly development, viz., the creation of a family.

The moralist, along with Darwin, accepts Nature as a series of compulsions. He agrees with his adversary, that physical competitions accomplish all physical and intellectual results. He insists, however, that the enforced physical sacrifices of woman accomplish the moral results. He finds, as it were, a double purpose in Nature, viz., the manufacture of mothers and the creation of an improved family. It is not always clear which he regards as the more important, since at different points he assigns each to the leading rôle. The preponderance of argument, however, indicates that he regards the family as the ultimate purpose in the evolution of man.

The place assigned to woman in Nature is frequently explained by the moralist through analogies. One of these interesting analogies is presented when seeking to show the moral intent of Nature, even in the lower kingdoms.

"For reproduction alone is a flower created; when that process is over it returns to the dust." This is what the moralist says when considering the endless sacrifices of maternity and the universal office of the female in Nature. A natural corollary to this would read: For reproduction alone is a woman created. When that office is discharged her usefulness to society is ended. The moralist does, in reality, say the same in effect when he declares that a woman fulfills her destiny "in paying the eternal debt of motherhood."

Again the moralist discusses the natural office of the female, by analogy, when he says:* "No one * * * reverences a "flower like a biologist. He sees in its bloom the blush of the "young mother; in its fading the eternal sacrifice of maternity. "A yellow primrose is not to him a yellow primrose. It is an "exquisite and complex structure added on to the primrose plant "for the purpose of producing other primrose plants."

Logically applying this analogy to human life, it would read: No one reverences a woman like a sociologist. He sees in her bloom but the blush of the young mother, in her fading the sign of past usefulness. A woman, to him, is not a woman, but a group of complex female organs added on to the woman for the purpose of producing other human beings. Or again, this means

---

*"The Ascent of Man," p. 248.

about what it would to say: No one reverences man like the anatomist before the dissecting table. To him man is not a man, but a highly specialized complex organism of bone, tissue, muscle and nerve, which was previously occupied by an intelligence for the purpose of feeding and breeding and operating that mass of bone, tissue, muscle and nerve.

When the moralist leaves the vegetable kingdom and reaches man, his vision as to the purposes of Nature is not enlarged. In the midst of this nineteenth century development, in close acquaintance with women of individual genius and of individual aspirations, the propounder of this new philosophy found it possible to say:* "In as real a sense as a factory is intended to turn "out locomotives or clocks, the machinery of Nature is designed "in the last resort to turn out Mothers. You will find Mothers "in lower Nature at every stage of imperfection; you will see "attempts being made to get at better types; you will find old "ideas abandoned and higher models coming to the front. And "when you get to the top you find the last great act was but to "present to the world a physiologically perfect type. It is a fact "which no human mother can regard without awe, which no "man can realize without a new reverence for woman and a new "belief in the higher meaning of Nature, that the goal of the "whole plant and animal kingdoms seems to have been the cre-"ation of a family."

Had the writer of this curious theory really possessed any exact or definite knowledge concerning the principles of evolution, he had possibly been able to perceive that part of the intent of Nature is the evolution of a *woman*. He had then perceived that this intent necessarily involves not merely mothers, but also wives, sisters, aunts and other female relatives in every stage of imperfection.

With such a reading of Nature it was inevitable that the author should say in so many words: "A woman completes her destiny in her children." The moralist, of course, in this instance, refers to earthly destiny. But even in this limited sense,

---

*"The Ascent of Man," p. 268.

such a deduction offends against the intelligence of woman herself.

It is a common error of the masculine mind, however, to assign woman to her particular place in the order of Nature without consulting woman herself. The assumption of the moralist upon the question of woman's place in Nature, is one which woman—generally speaking—is alone qualified to endorse or contravene.

This new moral philosophy, if reduced to its basic propositions, would read something as follows, viz.:

(1) Life is a struggle for physical and moral benefit.

(2) The struggle for physical benefit is egoistic and selfish; the struggle for moral benefit is altruistic and sacrificial.

(3) The business of evolution is the manufacture of mothers.

(4) The object of Nature is an improved family.

(5) Life is controlled by its functions, and the destiny of the individual is fulfilled in following lines laid out by nutrition and reproduction.

(6) The female is created for reproduction.

(7) Love is forced upon the world through the physical disabilities of the female.

(8) Sex is the physical device for reproduction.

The difference between Darwinism and Drummondism, thus appears to be a difference only in degree. The one subordinates both man and woman to the struggle for nutrition, while the other subordinates man to nutrition and woman to reproduction. The moralist plainly says:* "Man's life, on the whole, is de-"termined chiefly by the function of nutrition; woman's by the "function of reproduction. Man satisfies the one by going out "into the world, and in the rivalries of war and the ardors of the "chase, in conflict with Nature, and amid the stress of industrial "pursuits, fulfilling the law of Self-Preservation; woman com-"pletes her destiny by occupying herself with the industries and "sanctities of home and paying the eternal debt of Motherhood."

If these words mean what they say, they do not remotely suggest the realization of those ideals and aspirations which are

---

*"The Ascent of Man," p. 257.

individual to the intelligent soul itself. Were this, indeed, the case of Nature, then the individual man must accept as fulfillment of his destiny the most successful struggle for nutrition and physical benefit which he can make. Woman, on the other hand, should cease to look to a more personal and individual destiny than that of paying the debt of motherhood.

Here we have graphically presented two great struggles said to be taking place in Nature, viz.: "The Struggle for Life" and "The Struggle for the Life of Others." The first is a purely egoistic struggle for physical benefit. The other is an enforced physical sacrifice for posterity, which process the moralist defines as altruistic.

These are held to be the two great sruggles of all living Nature, the two main activities of intelligent life, the two great motives of action, the two sources of inspiration to human endeavor. Here, in brief, are shown the two factors of evolution, the physical causes of all we know as life, of all we admire as intelligence and of all we reverence as love.

The first theory, as will be seen, conceives Nature as working out its physical and material benefits under a law of self-defense. The second theory conceives Nature as working out its moral purposes and benefits under a law of self-suppression and self-sacrifice. Neither of these theories finds a purpose in Nature which justifies these physical competitions for life, nor these physical sacrifices for the life of others.

Both of these doctrines agree that Nature is a monster. The one theory sees Nature improving species at the expense of the individual man and woman. The other holds that Nature is improving the family at the expense of the individual woman.

Neither materialistic science nor materialistic theology perceives a principle in Nature that impels but does not compel. In neither struggle depicted is there a hint of that higher struggle which sustains intelligence during its struggle for nutrition and its struggle for reproduction. Nowhere in these doctrines is there any recognition of that universal motive which inspires every created thing to action, from atom to man. Nowhere is there any recognition that intelligent human nature embraces a principle of life, of progress and of love, which is neither competition nor sacrifice. Nowhere is there recognition of the principle of co-operation and fulfillment, which is the principle of individual content.

The failure to recognize the universal principle of affinity in human life is especially surprising, since the moralist so clearly observes it in the under-world of unconscious substance. Not until he closes his work does he really discover the principle upon which his argument should have been based. Here he catches a glimpse of that universal law which governs the evolution of love from its faint foreshadowings, in the mere rest or equilibrium of two unconscious atoms, to the self-conscious happiness of two intelligent souls.

In his closing chapter the moralist says:* "The earliest con-"dition in which science allows us to picture this globe is that of "a fiery mass of nebulous matter. At the second stage it con-"sists of countless myriads of similar atoms, roughly outlined into "a ragged cloud-ball, glowing with heat, and rotating in space "with inconceivable velocity. By what means can this mass be "broken up, or broken down, or made into a solid world? By "two things—mutual attraction and chemical affinity. The mo-"ment when within this cloud-ball the conditions of cooling "temperature are such that two atoms could combine together "the cause of the Evolution of the earth is won. For this pair "of atoms are chemically 'stronger' than any of the atoms im-"mediately surrounding them. Gradually, by attraction or affin-"ity, the primitive pair of atoms—like the first pair of savages— "absorb a third atom and a fourth and a fifth, until a 'Family' of "atoms is raised up which possesses properties and powers alto-"gether new, and in virtue of which it holds within its grasp the "conquest and servitude of all surrounding units. From this "growing center attraction radiates on every side, until a larger "aggregate, a family group—a Tribe—arises and starts a more "powerful center of its own. With every additional atom added, "the power as well as the complexity of the combination in-

---
*"The Ascent of Man," p. 337.

"creases. As the process goes on, after endless vicissitudes, re-
"pulsions, and readjustments, the changes become fewer and
"fewer, the conflict between mass and mass dies down, the ele-
"ments passing through various stages of liquidity finally com-
"bine in the order of their affinities, arrange themselves in the
"order of their densities, and the solid earth is finished.

"Now recall the names of the leading actors in this stupendous
"reformation. They are two in number, mutual attraction and
"chemical affinity. Notice these words—Attraction, Affinity.
"Notice that the great formative forces of physical evolution have
"psychical names. * * * In reality, neither here nor any-
"where have we any knowledge whatever of what is actually
"meant by Attraction. * * * To Newton himself the very
"conception of one atom or one mass, attracting through empty
"space another atom or mass, put his mental powers to con-
"fusion. And as to the term Affinity, the most recent chemistry,
"finding it utterly unfathomable in itself, confines its research at
"present to the investigation of its modes of action. * * *
"Here, as in every deep recess of physical nature, we are in the
"presence of that which is metaphysical, that which bars the way
"imperiously at every turn to a materialistic interpretation of the
"world."

Thus, the philosophy of the moralist recognizes a metaphysical principle in unconscious Nature which he has previously denied to conscious Nature. He recognizes the affinity of atoms as the impelling power in the lowest kingdom, yet postulates competition and sacrifice as the compelling forces in the highest kingdom. The author of "The Ascent of Man" was evidently inspired by the vision of a solid earth, slowly evolved by individual atoms settling in the order of their affinities. At the same time he turns with disdain from the suggestion that a spiritualized and moral humanity was evolved by individual intelligences uniting in the order of their affinities. He recognizes a "psychical affinity" between mineral atoms, but he finds only physical passion as the bond uniting man and woman.

Here, in brief, are set forth two popular theories which rest wholly upon the physical functions of Nature. The first theorist, absorbed in the struggle for nutrition, declares that a physically fittest species is the highest result attainable through evolution. The other theorist, concentrating upon reproduction, declares that a morally improved family is the object sought in evolution.

It will be observed that both theories ignore Nature in one important particular. Neither considers that which forms the very basis of the physically improved species and the morally improved family, viz., the Individual, through whom Nature must improve species and perfect the family.

This, indeed, is the fundamental error of science and philosophy which seek to interpret man as a result of the physical functions. It is just here, and in this particular, that human intelligence rebels. It is here that the highly developed man finds himself unable to accept the reading of physical materialism and of so-called rationalism. He admits the physical facts of Nature as reported by physical science. He accepts the doctrine of a physical evolution from lower to higher forms. He perceives in Nature a struggle for nutrition. He perceives, also, a struggle for reproduction. He may even attempt to force reason to accept the theories which attempt to explain how those facts came to be.

The common intuitions of man, however, revolt at the final conclusions of scientific skepticism. His intelligence refuses to accept those interpretations of himself which level him to the needs and requirements of his physical appetites and passions. He refuses to abide by those decisions which leave him a mere contribution to species or the mere progenitor of a family.

Against such interpretations of human life and destiny, the common intuitions, the common experiences and the common sense of man rebel. The self-conscious intelligent Ego knows itself to be an individual. Man feels and knows that every impulse of his nature, every concept of the brain, every act of his life, every aspiration of the soul, emanates from himself and has its effects upon himself as an individual.

Theories to the contrary, there is something in man which seeks explanation of himself as an individual, which seeks to wrest from Nature the cause of that individuality and the final purpose and destiny of his own being. Physical science and moral philosophy to the contrary, no theory of evolution of either the body or the intelligence or of life will be accepted if such theory obscures the individual as the mere contribution to species or to family. The individual intelligence will finally reject every theory that limits the destiny of man to the struggle for nutrition, and the destiny of woman to the struggle for reproduction. Individual intelligence demands of science some rational explanation for man, as he is, in his present stage of development. It demands of science that it shall analyze and intelligibly explain a being who possesses individual motives, impulses, aspirations and powers; one who believes that he has an individual place in Nature and an individual destiny to fulfill.

To this demand of intelligence physical science has not responded.

Neither the theories of physical materialism, nor philosophy built upon those theories, has thus far recognized the individual principle in Nature. Neither has thus far suggested the true factors and causes which create an individual. Neither do these theories contain a hint as to the true destiny of the individual. The intelligence of man will no more accept a "physically improved species" as the highest result obtainable in Nature, than it will accept a "morally improved family" as such. These are theories which simply bewilder intelligence and leave man a greater puzzle to himself than he was before science and philosophy undertook to account for him.

Men and women, demand of science that it shall furnish the key to the existence, the office, and the destiny of men and women as individuals. Science is asked to trace the path of individual evolution, to discover the purposes of individual life, and to forecast the possibilities of individual powers. Highy developed men and women find it impossible to accept themselves, with all of their individual impulses, ambition, aspiration and acquirements, as either the automatic results of nutrition or as incidental promoters of reproduction. Man refuses to believe that his individual destiny, as a man, is fulfilled and completed in the activities of nutrition and by contribution to species. Woman also refuses to believe that her individual destiny, as woman, is completed in "paying the eternal debt of motherhood."

Such, indeed, is the attitude of thinking men and women of to-day. Evolution has so far refined a large proportion of our western people that they will not accept physical materialism, even though it is clothed in the garb and authority of science. The natural processes of physical refinement have so far spiritualized and sensitized the average man and woman as to develop spiritual intuitions of other forces and conditions and purposes than the physical.

Such as these, having only intuitions to guide them, nevertheless, refuse to accept theories which reduce life to the level of its physical functions, even when such theories appear to have a substantial basis of fact. Such as these who have no actual or rational proof of the spiritual side of Nature, who have only intuitions to support them, are still withholding judgment. They are looking eagerly to science for confirmation of those intuitions which declare that individual life is more than feeding and breeding and that individual destiny is not fulfilled in species nor in family.

Such as these are looking to the self-declared students of Nature for more rational answers to those universal questions: "Why I am *I*?" and "For what am I created?"

CHAPTER XI.

THE QUESTION ANSWERED.

The Completion of the Individual.

Natural Science lays down an intelligent principle of co-operation as fundamental in evolution.

It does more than this. It demonstrates that every operation of Nature conserves a definite purpose. These purposes, on the one side general, and on the other individual, are demonstrated as intelligent purposes.

The processes by which these purposes are wrought out furnish science the key to this marvelous upward movement in Nature which we term Evolution.

Evolution represents a stupendous mathematical design. It discloses an intelligent mode of operation. It foreshadows a sublime ethical purpose.

Universal intelligence is employed in working out the mathematical designs of Nature. Individual intelligence is employed in working out its own ethical purposes, as well as in discharging its spiritual and physical functions.

This primary purpose of Nature and this primary purpose of the individual govern the greatest known struggle in the universe, viz., the struggle of intelligence in the midst of a seemingly hostile environment. This greatest struggle in Nature displays itself:

(1) As the struggle of universal intelligence to evolve and complete an Individual Intelligence.

(2) As the independent struggle of the intelligent Individual to complete himself.

Physical materialism sets forth, as the goal of evolution, a physically improved species by reason of the struggle for nutrition. Theological materialism sets forth as the purpose of Infinite Intelligence, an improved family, by reason of the physical sacrifices of the female in Nature.

Nature, however, refutes these assumptions. Without contravening the most insignificant physical fact discovered by modern science, the higher science is still able to declare:

(1) Evolution embraces a general purpose for the Completion of an Individual.

(2) Individual life embraces the particular purpose of Self-Completion.

(3) The effort made by both universal and individual intelligence, to accomplish these purposes, constitutes the universal Struggle for Self-Completion.

Thus are briefly presented the final deductions of a science which considers evolution as a correlated process upon two planes of existence. Here are the basic principles of a philosophy which considers man in his correlated lives in two worlds of intelligent activity.

The evolution of man rests upon co-operations as between a universal principle of intelligence and a particular principle of individual intelligence. Though universal intelligence is working along lines which appear as purely mathematical, and individual intelligence is working along lines which appear to be purely ethical, both, in fact, are working out the same result. Science demonstrates that universal nature, upon its inorganic side, is co-operating to individualize and improve intelligence. It demonstrates further, that individual intelligences are co-operating to serve the very purpose which universal intelligence has in view, viz., the individualizing, persistence and completion of an Intelligent Being.

From this it must appear that the higher science postulates, first, an evolution by general intelligence, and second, an individual intelligence as working out definite purposes which are both general and individual in character. This position diametrically opposes that of physical science which declares, first, an evolution by non-intelligent principles, and second, physical and mental improvement as the outcome of such blind forces.

Nature, carefully investigated from both planes of matter, life and intelligence, disproves these assumptions.

Physical science is not responsible for its errors. Neither a man nor a school of men is to be condemned for that which has not been discovered. Physical science, however, throughout its entire investigations, strangely overlooks, or underlooks, the most marvelous fact of Nature, viz., the omnipresence of intelligence. This oversight, or undersight, includes, of course, the individualized intelligence, as well as the general intelligence in Nature. Here we have the most singular error of physical science; this complete ignoring of the intelligent individual and this complete absorption in the physical functions of that same individual.

This obscuration of the intelligent individual in Nature takes from science its true motive. It leaves philosophy without proper foundation. It robs even religion of its highest inspiration.

In physical science the individual is lost sight of in the preservation of species. In materialistic philosophy the claims of the individual are swallowed up by the demands of society. Even religions hold up to the individual a life of self-suppression and resignation rather than one of fulfillment and happiness, as the highest life attainable in this physical world. The higher science, on the contrary, insists upon due consideration of that upon which Nature bestows all of her energies, viz., the individual. Nature accepts, and therefore true science and true philosophy accept the individual as the starting point. Indeed, for all natural, scientific and philosophical purposes, the intelligent individual is the center of the universe. From this center true science and true philosophy must radiate.

Beginning thus, it is not only possible, but clearly within the province of every individual to become a true scientist and a true philosopher. With such a beginning, the student of Nature is ever building outward from the true center. In an ever widening circle he will thereafter extend his investigations to family, to species, and to races.

The modern exponents of the higher science hold in reserve, however, the public consideration of family, species and races, until satisfied that they have explained somewhat of the functions, capacities and purposes of the individual intelligence. Having mastered this primary lesson, the student may proceed to larger subjects with some degree of confidence.

Resting upon this peculiar method, Natural Science declares that the intelligent individual is the first object in Nature, and has a life, a place, a purpose and a destiny distinctly his own. These are accounted as individual to himself, as are his face and form, his habits and tastes, his physical and spiritual powers, his intellectual ambitions and moral aspirations.

Nature demonstrates that the individual man comes to be what he is by reason of the struggle in which his own intelligence is forever engaged.

The discovery, analysis and demonstration of this greatest struggle in Nature, and the principle involved in that struggle, enable science to determine both the general purpose of universal intelligence and the particular purpose of individual intelligence. Such discovery, analysis and demonstration enable philosophy to declare:

(1) Universal intelligence is engaged in a general Struggle for Completion.

(2) The intelligent individual is engaged in the particular Struggle for Self-Completion.

(3) The intelligent soul achieves its purpose and fulfills its destiny, primarily, in that individual completion.

Thus, it appears that the general purpose of Nature and the particular purpose of the individual are the same purpose, being wrought out under different principles of intelligence. It follows, therefore, that Nature is beneficent, rather than hostile, in its attitude toward organic intelligence. It appears, indeed, that Nature has expended unmeasured energies during countless ages to improve and complete this individual. It appears also that for uncounted ages individual intelligences have assisted Nature by their individual struggles for self-completion.

There are, it is true, apparent hostilities and conflicts between Nature and the individual. These, however, must be ascribed to the undeveloped perceptions and conceptions of individual intelligence. They must be set down merely as errors of individual judgment, concerning those things which the individual seeks as a benefit or a pleasure to himself. The triumph of modern science is exposition of the fact that human intelligence is slowly but surely utilizing hitherto "hostile forces" of Nature, and is converting them into material benefits to the individual and the world.

In principle, in execution and in purpose, universal Nature is both hospitable and intelligent. It moves toward its beneficent purposes in spite of the errors of individual intelligence. The individual, in spite of his ignorance, is indirectly guided and admonished and educated by the immutable principles which he finds he cannot alter.

Thus, guided, warned and instructed by the great silent Teacher of Laws, man gradually advances to an independent, rational and cheerful compliance with those laws.

How does science demonstrate that the purpose of evolution is the Completion of an Intelligent Individual, physically, spiritually and psychically? How does it demonstrate that Self-Completion is the purpose of individual intelligence?

As already explained, the actual demonstration of the higher science cannot be conveyed through published statements. No chemist can demonstrate the properties, powers and processes of nitro-glycerine in an essay. It is admitted that no material tangible proof of the intelligent Struggle for Completion exists except the portentous fact of evolution itself. Nothing which physical science would call "proof" is obtainable in the laboratory. It is not discoverable under the dissecting knife. It is not revealed to the vivisectionist. It defies the microscope.

If, however, the reader ask how such purposes are discovered and demonstrated, he may receive at least an intellectual conception of the science involved. If he seek corroborative evidences, these are easily obtainable. The evidences of the Struggle for Completion crowd upon each other in every field of human activity and investigation. For such evidences the inquirer may turn to history. He may consult physical science. He may investigate Nature. He may study his neighbor, and analyze himself.

Everywhere, at every turn, in every record, in every activity and circumstance of life, in every individual motive, he will find himself confronted with that one principle in Nature which governs the Struggle for Completion.

He will find this principle under many names. He will find it as the law of polarity, the law of motion and number, the law of vibration, the law of affinity. Under whatever name he finds it, however, he will perceive it as a fundamental principle of positive and receptive energy, forever seeking equilibrium through individual entities. When he has ranged in his studies from chemical affinity to human love he will be familiar, theoretically at least, with the universal Struggle for Completion.

When the student finally investigates the operations of the individual will and the individual desire of intelligent human beings, he will find that he confronts only another phase of that many-sided principle which refines matter, increases the vibratory action of matter, generates physical life, perfects the physical body, and individualizes intelligence.

A study of the Struggle for Completion means something more than observation of the phenomena of physical life.

Such study properly begins with a consideration of the intelligent operations of unconscious substance, mineral, vegetable and animal. It next includes a knowledge of that individualized intelligence which inspires and operates the physical organism of both animal and man.

To study unconscious Nature through the affinity of atoms and the conjugation of cells is to discover the universal Struggle for Completion in its lowest manifestation. To study man through the operations of the intelligent soul is to disclose the Struggle for Self-Completion in its highest known form. Inquiry, therefore, as to the purposes of evolution must include a recognition of intelligence as well as of matter. It must include all of the operations of Nature as far as intelligence itself can penetrate. The attempt to solve the problem of evolution through the physical functions of digestion and reproduction is like trying to understand the law of harmonics by studying the fiber, mechanism and office of a piano case.

For knowledge of this greatest struggle which directly bears upon individual destiny, the student is recommended to the study of human life. By the study of life is not meant the study of the physical functions, but rather the play of that intelligence which inspires man to daily action. Such study, of course, would include a knowledge of the physical functions, appetites and passions.

The student is asked to carefully review the history of civilization and the development of science, art, philosophy and philanthropy, and then to decide whether such accomplishments are referable to the powers of digestion or the powers of the brain. He is asked to study the motives which underlie the ethical phenomena of love, and determine whether such development is compelled by a physical struggle, or impelled by the individual, intelligent struggle for self-satisfaction. He is asked to study individual character and its effects upon the world. He is asked to analyze the secret impulses and ideals of his own soul. He is asked to determine the motives which inspire his own life to daily action.

Such study and observation and such self-analysis will convince him that there is something underlying human activity infinitely more subtle and more potent than the blind demands of the physical functions. He will convince himself that individualized intelligence is engaged in a struggle that is of infinitely greater importance to himself than his physical feeding or breeding.

This Struggle for Self-Completion eluding the physical senses of man is observable and tangible to intelligence only. The higher phenomena of intelligent life and the higher struggles and the higher activities of intelligent beings have no explanation in physical science. It follows, therefore, that exact proof of the Struggle for Completion is obtainable only by such means and such methods as are employed in the higher science.

In the universal impulses, ambitions, activities and relations of human intelligence, however, the investigator will find ample evidences of the Struggle for Self-Completion. As a particular point in evidence the reader is referred to the life and the works of Charles Darwin. They furnish an interesting commentary upon his own theories. Here is a life spent, not in competitions for physical benefit, but in the accumulation of knowledge. Here is a life of not only laborious research, but research accompanied with a compilation of knowledge which, for accuracy and wealth of detail, astonished the world. According to Mr. Darwin's theory, his own life is contrary to Nature's purpose, viz., the preservation of species by way of the most successful struggle for nutrition, and the "largest possible number of healthy progeny" on the part of the individual. The life and the works of Darwin, however, prove exactly the reverse of what he lays down as the working formula of human life. The motive which impelled Darwin to his gigantic task may have been, at the beginning, merely a love of knowledge. It certainly was not the "struggle for nutrition" nor "the struggle for progeny." However that may have been, the motive which prompted the publication of that knowledge was the desire to transmit it to his fellowmen. By that publication and transmission of knowledge he proved his desire to benefit humanity. He became a philanthropist. Thus, in the very face of his own theory of a selfish struggle for nutrition and of selfish competitions, he lived a life of unselfish devotion to purely intellectual pursuits, and illustrated the highest altruism in his efforts to serve truth for the good of mankind.

For such a man and for such a life Mr. Darwin finds no sanction in Nature. This illustrates the ease with which a learned man may theorize in defiance of his own life and motives, as well as in contradiction of the universal impulses and experiences of mankind. The case of Mr. Darwin is not unique in itself. The individual struggles and achievements of intelligent beings constitute the history of the evolution of man. The final triumph of individual intelligence over unknown and apparently hostile environment, is the commonest fact of human progress. This power of individual intelligence to modify, conquer and create environment testifies to the supremacy of individual intelligence in the evolution of man.

In a broader sense than scientific skepticism conceives, there is a struggle for life. In a far nobler sense than speculative theology dreams, there is a struggle for love. From the lowest to the highest expression of Individual Intelligence there is a struggle for life, or for self-persistence as an individual. From the lowest to the highest sentient and conscious entity there is an individual struggle for an individual satisfaction. This struggle for individual persistence and this struggle for individual satisfaction go to make up the individual Struggle for Self-Completion.

Thus, the struggle for individual persistence is something more than a struggle for physical life. The struggle for individual satisfaction is something more than a struggle to reproduce. Nutrition sustains life but it is not life. Reproduction conserves love but it does not create love. Above and beyond the involuntary operations of the physical functions are the voluntary Will and Desire of an intelligent entity. Especially is this true of human life. Above and beyond the functional struggles of nutrition and reproduction extends the intelligent struggle of the soul for its own ethical satisfaction and happiness.

As already explained, the universal principle of intelligence governs the mathematical progress of matter and life. It seeks this end through and by vibratory correspondences between individual particles of material substance. As already shown, this general purpose is maintained by the action of individual particles seeking vibratory correspondence in other particles of opposite polarity.

Thus, in Nature's lowest kingdom, viz., that of unconscious, inorganic mineral substance, originates that tremendous struggle whose later developments enrich the kingdom of man. How well that struggle is rewarded appears in the refinement of matter, the generation of physical life, the individualizing of intelligence and the development of Love.

With the dawn of this higher sentient life and self-operating intelligence arises a new class of phenomena which attach to organic intelligence alone. From this period the Struggle for Completion displays two distinct phases, viz.:

(1) The coldly mathematical results which are involved in

the refinement of matter and the increase of vibratory action, thus assisting intelligence to an individual expression.

This is the purpose for which Nature struggles.

(2) The purely ethical effects which accrue to individual intelligence through those mathematical correspondences and refinements.

This is the purpose for which the individual struggles.

Thus, on the side of Nature is a universal principle seeking, primarily, to complete an individual, while on the side of the individual are the individual affinities of male and female intelligence, which impel them to complete themselves through an individual satisfaction or content.

From the dawn of individual intelligence this greatest struggle in Nature thus proceeds, on the one side exhibiting the mathematical law of vibration, and on the other the ethical effects enjoyed by individual intelligences.

This mathematical principle and these ethical effects of the struggle for completion are illustrated when two birds mate for life. On the mathematical side we have simply a certain ratio of vibratory correspondence in the physical and spiritual organisms of the two birds. On the individual side, however, there obtains that recognition of such adjustment which is, in effect, content to the intelligence of the birds. Birds, so mated, do not separate. They have fulfilled the universal principle of vibratory correspondence. They have attained those individual results which satisfy their intelligence. They have, indeed, reached completion in bird life.

The Struggle for Self-Completion is conducted, primarily, as a co-operative struggle of masculine and feminine intelligence.

Back of the minor functional struggles of both animal and human life stand the eternal co-operations of positive and receptive individuals, male and female, and man and woman. Back of the competitions of nutrition and the sacrifices of reproduction stand the affinities and fulfillments which are mutually enjoyed by individual intelligence. Back of all lesser struggles, competitions, compulsions and sacrifices exist the eternal co-operations of Nature's divided forces, male and female, man and woman.

The completion of individual life, therefore, rests, primarily, upon a harmonic relation between two intelligent beings of opposite polarity.

The mathematics of evolution are represented in Nature's effort to accomplish this result through vibratory correspondences in the physical and spiritual organisms of intelligent beings. The ethics of life are represented in the efforts of intelligent individuals to effect this purpose through harmonic relations which satisfy the intelligence. The word "ethical," therefore, applies only to those experiences which accrue to individual organic intelligence during its Struggle for Self-Completion.

In lower Nature, that is, in the inorganic world, the Struggle for Completion is rich in effects. They are not, however, ethical effects.

For example, chemical particles seek and obtain equilibrium, or vibratory correspondence, in their electro-magnetic energies. That equilibrium, or correspondence, or chemical affinity, cannot be defined as ethical, since it includes neither individual sensation, perception nor enjoyment. Such unions and such effects are but the faint foreshadowings of those which obtain in the highest kingdom under the same principle. The values and effects of chemical life and chemical activities are but the suggestion of those infinitely richer values and effects of human life and human intelligence. While the principle of affinity operates universally the same, the effects of that principle, in value and variety, increase almost infinitely with the induction of each higher life element.

For example, perfect affinity, or completion, in chemical substances merely amounts to a permanent cohesion of individual particles. That perfect cohesion simply constitutes physical solidity and durability. It may also induce color or transparency and brilliancy, as in the diamond, simply physical effects, as far as observation goes. Neither here nor in vegetable affinities do we find effects which may be termed ethical. The dawn of ethical life appears with the operations of organic intelligences. With the induction of the Spiritual Life Element, the generation of sentient life, and the appearance of conscious intelligence, the ethics of life begin. When Nature has guided the individual to the point of a conscious participation in the scheme of Nature, ethical phenomena are evoked. The animal is endowed, not merely with an individual intelligence and impulse to seek its affinities, but it has also the intelligence and the will to repulse that which is not harmonic with itself. As a result, the Struggle for Completion, even in the animal kingdom, is conducted by Nature along the mathematical lines of vibratory correspondences, and by the individual along the ethical lines of an individual satisfaction.

Animal affinities and unions range, in their individual values and effects, from a temporary union and temporary satisfaction, to permanent union and permanent content.

For example, is cited the conjugal habits of the jackal, which mates and separates with little intelligence and with but a fleeting suggestion of an ethical content. On the other hand, the enduring conjugal relations of two lions represents a complete and enduring ethical satisfaction on the part of the lions. Such union represents the Completion of the Individual—in the animal kingdom.

It is not, however, until the soul element has been added to the powers of individual intelligence, that the ethical values of life overshadow all other considerations. The physical body is now completed. The individual is no longer governed by the Spiritual Life Element. The intelligence is now inspired by a higher element. This higher self-conscious soul makes new demands. New energies, new requirements and new capacities have been added to lower energies, requirements and capacities. The operator of a physically perfected organism has a new and a higher line of achievement than the operator of the incomplete animal organism. Man is debtor to the animal for the physical body, the perfect instrument for the uses of intelligence upon the physical plane. Man, however, by reason of this physically perfected body and the inspiration of a higher element, moves into higher activities and higher necessities. Animal activities and satisfactions do not meet the requirements of the living soul.

Equipped with the physical instrument which readily responds to his will and his desire, the living soul enters upon the Struggle for Self-Completion though ignorant of the infinite possibilities beyond the purely animal plane. This being true, he demands, primarily, correspondences and satisfactions of the soul, rather than correspondences and satisfactions of the physical body. However, as man represents all of the life elements in Nature, he demands correspondence in the energies of each and all of these life elements. Thus, the demands of the intelligent soul, for a harmonic relation, include the necessity for correspondence also in the Electro-Magnetic, the Vito-Chemical, and the Spiritual Life Elements, which go to make up the lower man.

From this it must appear that human life is far richer in ethical effects than is the life of the animal below him. That which so largely increases the ethical satisfactions of man over those of the animal are the affinities which reside in the soul. Man enjoys all the lower satisfactions of the animal. He is capable of a purely animal content. The demands and capacities of the soul, however, create a higher line of activities and a higher class of necessities which represent the ethical satisfactions of human intelligence. The ethical content of the animal rests upon perfect correspondence in those lower life elements of which he is made up. With man, however, the struggle is to satisfy the soul, the highest element, as well as the lower elements inferior to the soul. Nothing less will satisfy the soul of man. The lower elements have no power to permanently satisfy the higher. The energies and offices of the lower nature do not satisfy human intelligence. Those energies and offices influence but they do not govern the higher evolution of man.

While it is true that feeding and breeding consume a large part of human life, they are nevertheless but incidents to intelligent development, just as the feeding of coal into a locomotive is but an incident to the purpose sought. Impulses born of the lower elements are continually mistaken for the necessities of the highest element. Sooner or later, however, the intelligence perceives its error and renews its search for permanent satisfaction.

A perfect harmonic between soul and soul is the one and only relation or condition which will confer upon man the consciousness of individual completion and a permanent individual happiness. This is the relation which science declares must obtain before man and woman can be prepared to take up still higher lines of personal development. This relation, once established, results in a voluntary and indissoluble union, here and hereafter.

In the establishment of this relation Nature accomplishes its primary purpose, viz., the Completion of the Individual. On the other hand, man and woman accomplish what appears to them as an ultimate purpose, viz., Self-Completion and Happiness.

The reader must clearly understand just what is meant by Individual Completion. The word "Completion" must not be confused with the idea of "perfection," nor must it be taken to mean the completion of individual development. Science has nowhere discovered a state or condition of perfection in man. Neither has it discovered any individual relation or attainment which means the end of individual effort and achievement.

Individual completion means merely that state of vibratory correspondence, equilibrium of energies, and ethical satisfaction which may be obtained in the union of two intelligent individuals of opposite polarity. It means that individual relation which mathematically and harmonically fulfills the design of Nature and the needs of individual intelligence.

Nature works out its general purpose when it establishes a perfect vibratory correspondence between the physical and spiritual organisms of a man and a woman. Man and woman, however, work out the particular purpose of human life when they arrive at a reciprocal relation which satisfies every requirement of body, spirit and soul. Nature's effort to thus complete the individual is attended by steadily increasing mathematical results. The individual Struggle for Self-Completion, at the same time, moves forward with ever increasing ethical gains.

Thus, we have in intelligent life a fundamental Struggle for Completion which is co-operative in principle, vibratory in action, mathematical in design, and ethical in effect. Science holds that this struggle of intelligence involves the physical, spiritual and moral completion of the individual man and woman in earthly life. In this supreme struggle and in this supreme purpose science finds explanation, justification and compensation, for all lesser struggles, compulsions and sacrifices.

The individual Struggle for Self-Completion, therefore, looks neither to competitions, compulsions nor sacrifices. It looks only to co-operations and fulfillments which satisfy the individual life. The physical functions are now seen as activities which conserve completion, but which do not complete the individual. They now appear merely as duties incidental to life and to the development of intelligence, and the enjoyment of love. Physical nutrition supplies material to sustain life, but it does not supply life. Reproduction furnishes objects to love, but it does not create the love nature. Nutrition and reproduction involve intelligent service, but they do not generate intelligence. In nutrition man pays the physical debt to his physical nature, but he does not satisfy either the spiritual nature or the psychical. In reproduction woman pays a physical and moral debt to the race. She does not, however, discharge the debt to her own spiritual nature nor to her own soul.

Nutrition is a contribution to the body. Reproduction is a contribution to the world. The one is purely egoistic and individual in its nature; the other, by necessity, is altruistic and impersonal. Neither activity develops nor satisfies the intelligent soul. In neither struggle is the destiny of individual intelligence fulfilled. Man is not merely a function of nutrition. Woman is not merely a function of reproduction. On the contrary, men and women are living souls. They are intelligent individuals who perform these physical functions during their earthly struggle for Self-Completion.

This is the one and only reading of Nature which explains man as he is, and gives dignity and value to individual life.

Up to this present time our popular science has not discovered the individual man and the individual woman in Nature. It deals with man only as he is related to species. It has not discovered nor analyzed nor explained those highest and inseparable intelligences, the individual man and woman. Up to this date we have in science "male" and "female" only. The real work of physical science, in this connection, ends when it has set forth the physical functions and the physical relationships of "male" and "female."

A new moral philosophy, based upon physical science, has gone one step farther. It discovers a "mother." It perceives a scientific and an ethical value in the reproductive capacity of the female. Neither physical materialism nor theological materialism, however, has recognized man and woman as individual intelligences, having a scientific and ethical value as such in Nature.

Man in his relation to nutrition, and woman in her relation to reproduction, are the only points of view from which modern science has studied the highest known products in Nature. Physical materialism is concerned with the preservation of species. Moral philosophy based upon physical science is concerned with the creation of a family.

Nature, on the contrary, is primarily concerned with the Completion of an Individual.

Darwin discovered the "male" and "female" in Nature. Drummond discovered the "mother."

It is left for higher science to discover "man" and "woman." To it has been assigned the task of analyzing these intelligent beings through the highest as well as the lowest elements they represent.

The position of physical science is particularly unfortunate in its treatment of the "female" in Nature. Woman is persistently analyzed by physical science through her reproductive capacity alone. It is true that the individual man is subordinated to the physical improvement of species. At the same time masculine intelligence and masculine energy are given some measure of credit in the evolution of society. Man, as the main factor in the struggle for nutrition, is the accredited mental force in civilization.

No recognized work of physical science has, as yet, traced the moral influence of feminine intelligence in the evolution of man.

Science, however, that recognizes the intelligent soul, as well as the physical organs of repduction in woman, declares that she is an intelligent being, having a place in Nature, a part to perform in life, and a destiny to fulfill, that are individual to herself here and hereafter. It takes into account the spiritual and psychical as well as the physical nature of woman. It recognizes the intelligent activities of the soul, as well as the physical functions of the body. It analyzes her intellectual and moral capacities, as well as her physical capacities for reproduction. Taking these into account, the higher science analyzes the feminine Struggle for Completion, by and through the study of those Life Elements which are operated by the feminine principle of intelligence.

Primarily, everything that represents the receptive, absorbing and pacific powers of the Life Elements is termed "feminine." Evolution upon the feminine side of Nature, therefore, includes what are commonly termed in physical science the negative properties in Nature. The word "negative," however, is inadequate as a definition of the "feminine" in Nature. Though feminine nature may be called negative to the masculine, it is not a negation in Nature. It is, instead, a very definite power, viz., the power to receive and absorb. The feminine nature represents the powers of absorption and non-resistance.

Evolution, under the feminine principle in Nature, includes the receptive, absorbing, nourishing and non-resistant entity of each kingdom in Nature, viz.:

(1) The mineral atom which is receptive to a positive atom, or to the atom positively charged with electro-magnetism.

(2) The receptive and absorbing particles and parts of vegetable substance which receive and reproduce that which is generated by the positive male particles or parts.

(3) The female animal which is receptive, absorbing and nourishing in its nature. It is also non-resistant to the positive and dominating energy and will of the intelligent male animal.

(4) Woman who is physically, spiritually and psychically receptive to the positive, physical, spiritual and psychical forces of man.

Nature, history, religion and common experience support these deductions.

The physical receptivity of woman to man is proved in maternity, which is the most patent fact in Nature.

The spiritual and psychical receptivity of woman to man is revealed in the history of civilization of government, of art, of literature and of science. All of these reveal man as the aggressive, organizing and creative factor in human life. Even the history of religions shows where women find their systems of faith and ethical codes of life. Woman accepts her religion from a Buddha, a Moses, a Christ, and even from a Mohammed and a Pope.

These are facts to which a certain type of feminine pride may not be ready to yield. They are facts, however, which, when fully recognized, will serve to better explain woman's real power, viz., the powers of the receptive, absorbing and pacific elements in human life.

In the lower ranges of human society this non-resistant nature of woman subjects her to deplorable injustice. Her weaker body; spirit and will are enslaved, dominated and coerced by man, who has not learned the uses and the purposes of strength in Nature.

Woman is first physically enslaved by man, and made the victim of his stronger appetites and passions. She is coerced by his stronger will in the individual relation. Socially she is restricted and restrained by his arbitrary codes of law. It is the masculine mind that projects educational systems which exclude woman, thus retarding her intellectual development. It is man who gives to woman even her systems of faith, enacting canons that bind both her reason and her conscience. All this man does in ignorance of the mutual relation and office of the sexes. These are the penalties which he inflicts upon himself through woman. These are the crimes of his ignorance. These are the burdens woman bears until she rises to a rational conception of her own place in Nature, and develops the strength to maintain that place.

Masculinity is one thing, femininity is another. Every product in Nature must, of necessity, represent one or the other of Nature's dual powers, viz., positive or receptive energy. That cleavage which obtains among unconscious mineral atoms and vegetable particles, which appears with the dawn of organic life and which characterizes the highest product of Nature, the human, is an eternal and unchangeable cleavage as far as finite science has discovered. As far as science can determine the co-operative relation of positive and receptive, of male and female, and of man and woman, is unchangeable. The same division of mankind which obtains in this world exists in the next. The same attractions and co-operations persist upon both planes of life.

That which is masculine and that which is feminine never lose their essential qualities of positive and receptive energy. Man and woman may come into a better adjustment in their mutual relation. They cannot, however, exchange places, nor become the same thing. There is assimilation, but there is never exchange nor transformation in the basic elements of masculine and feminine nature. There may be an increasing intellectual comprehension of each other's qualities. There are only closer vibratory correspondences upon the material plane and closer harmonics upon the psychical plane. There is never a stage of development, so far as is known, where the soul loses sex, where man ceases to be man, where woman ceases to be woman, in those essential qualities of positive and receptive power.

The highest purposes in human life are bound up in this true relation of man and woman. To arrive at this true relation is the fulfillment and the end of the Struggle for Self-Completion in its purely individual sense.

Unmeasured time has been consumed by the human race in arriving at our present stage of physical, spiritual and psychical sex development. Unmeasured time has been consumed in arriving at our present intellectual comprehension of the laws of sex, and the reciprocal relations, powers and possibilities of man and woman. We have now reached that stage of development which admits of a rational conception of the true and scientific relationship of the masculine and the feminine. We have risen to a degree of intellectual development which admits of a practical effort to meet the requirements of the law of sex.

The initial phase of sex, as seen in the positive and receptive character of mineral atoms, is accentuated in each higher kingdom. Each additional life element gives rise to new capacities and new variations. They are, however, only variations of the same positive and receptive energies of the masculine and feminine elements. Each added life element reinforces masculinity in its positive, aggressive, generative, and organizing powers. Each added life element also reinforces feminity in its receptive, absorbing, nourishing and pacific powers.

The atom and man represent the extremes of evolution upon the physical plane. Compare, for example, the positive and receptive powers of two mineral atoms with the same essential powers of a highly developed man and woman. Consider the addition to the original values in form, nature, capacities and achievements. Compare the strength and tenacity of the attracting forces which bind two mineral atoms and those which bind two intelligent human beings of opposite polarity. Separate the two mineral atoms by the fraction of an inch, and the force of their attraction is overcome. Each particle is ready to unite with another particle of opposite polarity. This is the limit of the attracting force of electro-magnetism.

How is it with man and woman who have once experienced that superior attraction which obtains only in the realm of the intelligent soul?

Such as these never separate in that realm. Such attraction as this includes also an attraction in all lower and lesser elements. Let these two be separated by the limits of earth and that bond exists. Let them be separated by every bar which custom, law, convention or circumstance may erect, that tie remains unbroken. Let time and death intervene, even then these two are bound by the natural law of affinity.

This, therefore, is union which is superior to space, time and circumstance. This represents the attracting powers of the soul, as compared with those of the mineral atom. When so compared that feeble attraction and cohesion between mineral atoms of opposite polarity is scarcely a suggestion of the attraction and union beween the self-conscious, positive will of the intelligent masculine soul, and the self-conscious, absorbing desire of the intelligent feminine soul.

Man and woman, however, endowed as they are with such superior capacities, represent the same primitive powers of the positive and receptive mineral atoms. In their ultimate completed individual relation man and woman are but variations upon the primary affinities of mineral atoms. This fundamental difference and this fundamental attraction and co-operation of masculine and feminine energies manifest themselves throughout the infinite gradations and variations of inorganic substance, of organic life, and of individual intelligence. The power of attraction and the basis of co-operation between the lowest and the highest representatives of masculine and feminine nature are essentially the same, viz., aggressive force on one side and absorbing power on the other.

Universally man is attracted to the receptive qualities in woman. His aggressive intelligence, as well as his stronger physical nature, seeks its polar opposite; that is, he seeks those absorbing and pacific powers of intelligence, rather than those which are aggressively forceful. Universally woman is attracted to the strength, the courage and the power of man. It does not matter whether those qualities appear as physical, spiritual or psychical.

Force of will on one side and power of desire on the other constitute the principle of affinity between two intelligent souls of opposite polarity.

These facts and these principles do not, however, argue an inferiority of woman to man, nor of man to woman. To thus interpret the masculine and feminine in Nature were as foolish as to debate the relative merits of heat and light in the economy of Nature. Man and woman simply represent the two indispensable and vital factors in the evolution of man. While they differ in Nature, in offices, in capacities and in attainments, they are, however, equal in their uses and one in their purposes.

These facts of Nature merely go to show that man stands for completion by aggressive activities while woman represents completion by pacific activities. One represents progress by force of an individualized intelligent will. The other represents progress by the power of an individualized intelligent desire.

Physical materialism is repeatedly compelled to observe these inherent qualities of masculine and feminine nature. It does not pretend to explain them. It simply infers that masculine force is a result of the struggle for nutrition. It merely infers that the pacific qualities of feminine nature are the result of the struggle for reproduction in the midst of a hostile environment. Physical science has entirely overlooked the psychical powers of individual will and individual desire which underlie the activities of intelligent human beings. It analyzes man and woman as mere functions for feeding and breeding. It ignores the individual will and the individual desire of self-conscious independent beings. It fails to perceive that nutrition and reproduction in reality depend upon this individual will and this individual desire which are forever seeking a purely ethical self-satisfaction. It fails to perceive that human intelligence is never permanently satisfied in the exercise or the enjoyment of the physical powers, functions and passions.

The struggle for nutrition furnishes a modicum of employment to masculine intelligence. It does not, however, furnish him with satisfaction or content. Indeed, it absorbs energies which he would gladly employ in other lines of activity. Maternity develops the love nature of woman. It does not, however, develop her intelligence. In reality, woman's purely rational development proceeds in spite of and not by reason of maternity.

"A flower is created for reproduction. When its usefulness is over it returns to the dust." Thus declares the moralist when he would illustrate the uses of the female in Nature.

Not so, declares science, turning to Nature for corroboration.

A flower, like a woman, is an entity, an individual created for life. While it lives its bloom, its beauty and its fragrance are its own. While it lives it absorbs life from its natural elements, earth, air and water. While it lives it reaches toward the sun, its source of being. While it lives its individual charms are a contribution and a blessing to the world. A flower, like a woman, is simply an individual representative of universal elements and principles having a capacity for reproduction.

If reproduction were the sole purpose of the female, Nature

has certainly wasted time. The normal reproductive period of a woman's life is about half of her natural lifetime. If this function covered the purpose of the female in Nature, then Nature has unduly wasted time and energy. If this is the vital issue of woman's life, then the individual woman cheats Nature. It does not need science to prove that woman persists very definitely and individually long after her reproductive usefulness is past. Not only this, but she persists under the conviction that she has not outlived her usefulness, that she has a place in Nature and in society. She continues to desire life. Her intelligence still occupies itself with plans and purposes that are individual. Her nature is still susceptible to an individual love. Her soul still yearns for an individual completion—happiness.

While it is true that the woman nature impels union with man, it also binds woman to her children. This, however, is not saying that the intelligent needs and requirements of woman's nature are fulfilled and satisfied in those children.

No normal woman will deny, and no man is in the position to deny, one universal desire that lives in the soul of a woman. This is a desire which wakens in childhood and persists to the end. It is a desire that survives all conditions of a woman's life. It is one that lives on, too often unfulfilled, in the midst of a growing family. That one desire which never dies, however skillfully or conscientiously concealed from the world, is the desire for an exclusive, individual, love relation. It is the longing for an exclusive companionship and a personal happiness in that companionship.

The higher science, giving heed to this voice of Nature, delivers to woman a more hopeful message than physical science or moral philosophy or orthodox religion has ever brought to her.

Nature declares that woman is created for an intelligent self-completion which primarily includes that exclusive love relation, that individual companionship and that personal happiness which her soul forever craves. It declares that woman is not created to conserve merely the family. On the contrary, it finds that she is created for an individual life, love, knowledge, attainment and influence, here and hereafter.

Science has determined that this long-sought self-completion of the individual man and woman depends primarily, upon the relation they establish between themselves. In the attainment and preservation of this perfect mutual relation is laid the substantial foundation of all further powers, activities and achievements, here and hereafter.

In such a relation the primary purpose of intelligent human life is achieved.

Such completion involves consequences to the individual which are both material and ethical. On the material side are both the physical and spiritual organisms of a man and a woman keyed to the same vibratory conditions. On the ethical side are two intelligent souls who respond to each other in all of those activities, governed by the positive masculine will on one side, and by the absorbing feminine desire on the other.

Such is the completion that Nature demands, even upon this physical plane. Such is the completion that science has demonstrated. In such completion, and in this alone, philosophy claims that man and woman are fitted to properly discharge the physical functions and altruistic obligations to the race. In such completion, and in this alone, is human intelligence equipped to exercise its highest powers and to achieve its noblest possibilities.

Thus moves this marvelous scheme of Nature toward its completion of the Individual. In every department of intelligent life are the same principle and the same purpose demonstrated. Guided by the universal law of affinity, and inspired by the subtle demands of the intelligent soul, man moves to his own completion through an infinite series of harmonics, physical, spiritual and psychical.

Here are explained the rise, and the purpose of that rise, from an involuntary to a voluntary activity on the part of Nature's individual products. Here are explained the method and the purpose of evolution which display, first, an unconscious response to a general intelligence, second, a conscious impulse, and finally, the rational will and the rational desire to comply with that general law. Here are explained that Struggle for Completion, and the purpose of that struggle which engages all of the energies of every entity of every kingdom.

Thus, the seemingly purposeful is the actually purposeful. The adaptations which occur in the Struggle for Completion are seen to serve the ethical purposes of individual intelligence, as well as the requirements of the physical body. Under this reading of the law Nature becomes the guide, guardian and benefactor of mankind. Life has a purpose. Individual intelligence has an individual destiny. Love has a natural evolution. Happiness is a scientific possibility.

Scientifically and mathematically stated, the *universal* Struggle for Completion looks only to the Completion of an Individual through vibratory correspondences. Scientifically and ethically stated, the *individual* Struggle for Self-Completion looks only to individual happiness through its individual harmonic relations.

The key to all of these processes and all of these purposes is found in that overshadowing and immutable principle of Affinity between the Positive and the Receptive energies in Nature. The attraction between man and woman is the key to all other attractions. The union of one man and one woman represents the principle of all other unions. A harmonic relation between one man and one woman rests upon the conditions which govern all harmonics. The individual completion and happiness of one man and one woman furnish the key to the universal harmonics of Nature and to the future completion and happiness of mankind.

Upon man and woman, as the mathematical center of all vibrations and the ethical center of all influences, the harmonics of human life depend.

This being true, Natural Science, and philosophy founded upon that science, present to the world certain new propositions.

Physical science postulates a struggle for existence in the midst of a hostile environment, as the basis of the evolution of man.

Nature, on the contrary, demonstrates such basis to be a struggle of individual intelligence for an individual adjustment and completion in the midst of environment only seemingly hostile.

Physical science fixes upon the physical functions of nutrition and reproduction as the compulsory causes of all progress; whereas Nature demonstrates that the spiritual principle of Polarity, or Affinity, is the impelling cause of all we perceive as physical evolution and as material refinement.

Physical science declares that a physically improved species is the highest result obtainable under physical compulsions.

Nature, on the contrary, demonstrates that a completed individual is a scientific possibility under universal law of attraction.

Physical science interprets the highest duty of the individual as the "rearing of the largest number of healthy progeny."

Nature, on the contrary, demonstrates that the first duty of the individual man and woman is Self-Completion and Individual Happiness.

Physical science is, therefore, the science of evolution through feeding, breeding and battle; whereas, Nature demonstrates the evolution of intelligence through living, learning and loving. The doctrines and dogmas of physical materialism restrict the purposes of human life to the needs and requirements of the physical body. The philosophy of life, however, based upon Natural Science, extends the purposes of life to the needs and requirements of the intelligent soul.

Thus, the school of physical science, familiar with physical fact alone, determines that the whole purpose of intelligent human life is the preservation of species. This is a science of natural hostility, of selfish competition and of cruel sacrifice. This is the doctrine of the suppression of the individual, the doctrine of doubt, despair and annihilation.

On the other hand, the higher science, familiar with physical, spiritual and psychical facts of Nature, demonstrates, first, that Nature is intelligent in design, co-operative in principle, harmonic in operation, hospitable in purpose, beneficent in results. It demonstrates that the primary purpose of intelligent human life is an Individual Completion and Happiness.

This is the science of natural hospitality, co-operation by affinity, fulfillment through harmonics, the science of individual development and of individual love. This is the philosophy of faith, hope and happiness, and of the persistence and progress of the Completed Individual in another and higher world than this.

CHAPTER XII.

MASCULINE WILL AND FEMININE DESIRE.

The ancient wisdom religion acknowledges what it defines as the "Father-Mother" principle, as the governing principle in Nature.

Translated to Anglo-Saxon, this means a recognition of the intelligent spiritual principle of sex which displays itself as positive, generative and masculine energy on one side, and as receptive, nourishing and reproductive feminine energy on the other.

The ancient "Father-Mother" principle is a far more scientific definition of the creative principles in Nature than the "Father" principle which is alone recognized in Christian philosophy.

Man represents the intelligent Will Force in Nature. Woman represents the intelligent powers of Desire. In these essential natures the male animal and man correspond, as do the female animal and woman. Man and woman, however, as representatives of the psychic or soul element, are almost infinitely stronger in both the powers of Will and the powers of Desire. When the intelligent animal will and desire have been reinforced by the highest element in Nature, they display powers and capacities which may well be termed "God-like."

When animal intelligence has been rationalized by the induction of the highest element in Nature, we have then an individual, rational Will and an individual rational Desire, as the motor powers in the evolution of man.

Every normal, physical entity, risen to the point of an individual intelligence, represents either the positive principle of intelligence which we define as "Will," or the receptive principle which we define as "Desire." Not, however, until the powers of the soul are added to the powers of intelligence are there what we define as the "Rational Will" and "Rational Desire." Universally the positive male entity is the generator of life and the organizer of forces. Universally the receptive feminine entity is the mother of life and the conserver of forces. When the stage of man arrives, we find man as the generator of physical life and the organizer of physical, spiritual and psychical forces. We find woman the mother of life and the conserver of physical, spiritual and psychical forces. Universally the feminine entity is the willing co-operator with the masculine entity in all of his functions, occupations and activities.

Thus, the sex relation is determined by the innate principles of positive Will and receptive Desire. It is, therefore, not a relation which has been evolved by the functions of the body nor by man-made customs codes, and conventions. Man establishes himself as warrior, ruler and leader, by innate force of body and Will. Woman, on the contrary, establishes herself by the power of Desire, as the most potent influence in the life of the warrior, ruler and leader.

Man seeks achievement by force. Woman seeks accomplishment through self-surrender. The one seeks to control, the other to influence.

Physical science dimly perceives these fundamental differences of masculine and feminine nature. It deals with them so confusedly, however, as to increase rather than to dispel the mystery. When it explains the physical differences which obtain in sex, as results of the physical functions, that explanation appears plausible. When it comes, however, to these marked distinctions in super-physical nature, it is forced to admit that such distinctions are "inexplicable."

Thus far physical science has speculated in vain as to the causes of sex distinction. It publishes volumes of theory and statistical data. It has studied the habits, analyzed the blood, weighed the brain and followed out in detail the smallest clue which might lead to a solution. It has, however, analyzed, weighed and measured physical properties only. The means and methods thus far employed have failed. The subtle distinctions in sex nature have not been revealed at the dissecting table, in the laboratory, nor under the microscope. All that physical science has thus far discovered as to sex, may be summarized as follows:

(1) Sex inheres in all living matter.

(2) The generative entity is defined as male.

(3) The reproductive entity is defined as female.

(4) The male half of Nature is characterized by greater physical strength and greater force of Will.

(5) The feminine entity is characterized by a more delicate physical organism and the more pacific temperament.

(6) The sexes are differentiated in appearance, in organic structure, in physical function, and in their intellectual activities and psychical temperament.

(7) In human life, the mental as well as the physical activities of man are the more aggressive; those of woman are the more pacific.

(8) Man is distinguished by the more forceful application of the rational powers, woman by the larger employment of the intuitional faculties.

(9) The activities of masculine intelligence more particularly conserve the acquirement of knowledge and the exercise of power. The activities of feminine intelligence more particularly conserve the development of the love relations and the preservation of established order.

Darwin alludes to these general distinctions as phenomena incidental to the struggle for nutrition, rather than evidences of fundamentally differing sex principles. Mr. Drummond, struck by what he terms the "mysterious bias of sex," says:*
"The tasks which demand a powerful development of muscle "and bone, and the resulting capacity for intermittent spurts of "energy, involving corresponding periods of rest, fall to man; the "care of children and all the various industries which radiate "from the hearth and which call for an expenditure of energy "more continuous, but at a lower tension, fall to woman."
"Whether this or any theory of the origin of sex be proved or "unproved, the fact remains, and is everywhere emphasized in

---
*P. 256.

"Nature, that a certain constitutional difference exists between "male and female, a difference inclining the one to a robuster life "and implanting in the other a certain mysterious bias in the "direction of what one can only call the womanly disposition."

This "mysterious bias," when properly analyzed, is discovered to be merely the logical effects which flow from the inherent energies of the feminine life elements. Science further discovers that all of the infinite variations arising in the masculine and feminine departments of Nature are, after all, only variations growing out of the positive energies which govern the masculine and the receptive energies which govern the feminine half of Nature.

There is one fundamental sex characteristic in both the animal and the human kingdoms which has received but an incidental attention from physical science. This peculiar characteristic, dismissed as "inexplicable" by physical science, is, in fact, the very key to the higher evolution of man through sex selection. The law referred to is that of *individual preference,* which phenomenon obtains almost as universally in the animal kingdom as in the human.

That principle in individual nature which impels one bird to seek another particular bird as its mate, or which impels a man to seek a particular woman as his wife, is the principle which refutes selection by blind and automatic processes. In the animal kingdom this law of individual preference is governed by the intelligent Will and intelligent Desire of the male and female. In human life this law of individual preference is governed by the intelligent, rational Will and intelligent, rational Desire of man and woman. When the individual Will and Desire of the animal have risen to the stage of rationality, the law of individual preference is immeasurably strengthened. Here we have something more than an individual Will and an individual Desire moving in a certain direction, bent upon a particular object. We now find instead, Will and Desire reinforced and put into execution by rational calculation and by a rational control of, and triumph over, environment. To just the extent that man brings rational intelligence to bear upon sex selection, to that extent he rises above the animal and demonstrates a higher evolution.

Science that claims to have solved the problem of evolution should be able to explain why both birds and men have died of grief when separated from the particular object of their desires. How will it explain this character of individual preference as the result of a "blind procreative passion?" Thus far, physical science has not tried to explain this phenomenon. It contests itself with defining such manifestations as "inexplicable." So long as physical materialism fails to explain this general law of individual preference in sex selection, it should not rest in its search for the factors and causes of evolution.

This phenomenon of individual preference alone clearly refutes the theory that sex is no more than a physical device for reproduction, that it has no other motive, impulse nor meaning, than procreation. Instead, here is the phenomenon of physical life which clearly demonstrates the fact of spiritual affinities in Nature, as well as a blind biological need.

Darwinism, and indeed, all modern materialism, rationalism and theology, find in sex but one meaning and one purpose, viz., reproduction. Anything higher or finer, by way of an individual, intelligent and ethical necessity, has not yet been conceived by modern science and philosophy. Thus far, it has been left to the poets and to the common intuitions of mankind to preserve the true ideal of sex love.

Neither scientific materialism nor theological materialism has disclosed the principle which impels the aggressive and pacific energies in Nature to combine. Neither has portrayed that greater struggle in which aggressive male intelligence and receptive feminine intelligence are engaged. Neither has conceived of the ethical purposes which the rational masculine Will and the rational feminine Desire are slowly working out.

To the findings of physical science we owe the rise of pessimistic philosophy. To these we must credit those late doctrines of sex and of love, against which developed intelligence protests. Because of such findings, cause and effect are reversed. The highest and most subtle ethical values in life are made to appear as the mere efflorescence of the grossest functions of Nature. Because of this out-of-focus view, the true spiritual sex relation has been reduced to a "biological need."

Physical science thus substitutes mechanical for intelligent principles, blind physical forces for intelligent spiritual and psychical forces. It substitutes physical functions for spiritual relations, and universal laws for individual Will and Desire. It universally substitutes the needs and requirements of the physical body for the needs and requirements of the intelligence which operates the body. Modern science demonstrates chemical affinity. Is it not singular that it denies to the intelligent soul as subtle a principle as that which governs an atom of dust? Is it not singular that the same intelligence which grasps the metaphysical principle of chemical affinity is offended when asked to recognize the metaphysical principle which draws intelligent beings together?

Indeed, it has come to be almost unsafe to use the word "affinity" in connection with human relations. The word has been prejudiced by reason of its misuse and misapplication by the ignorant, the designing and the vicious. The word has been misused to cover up immoralities induced through "spiritual mediumship." It has been misapplied by would-be reformers who, with but a mere glimpse of the law, have sought to regulate society. This, together with the fact that numerous men and women excuse their weaknesses upon the ground of "affinity," has brought reproach upon the word itself. It has lost caste in society. As a result, the conventional fear it. Both science and theology have come to regard it with suspicion. It has come to be generally recognized as a synonym for "free-love."

All of this, however, does not abrogate the principle nor rob the word of its value. It merely illustrates the difficulties in the way of establishing a fact or a truth in Nature.

It is not especially surprising that physical science overlooks the higher implications of sex. Restricted as it is to the purely physical uses of that relation, it is natural that it should interpret the highest ethical relations of the soul as an effect of the grossest physical function. It is not surprising that physical science has overlooked the universal spiritual principle of affinity. This total ignorance, however, as to the spiritual side of Nature, brings confusion and humiliation into all modern discussion of sex. Scientific ignorance as to the true principles and purposes involved has debased the name. It has leveled the intelligent affinities of the soul to the unconscious affinities of physical substance.

If the individual intelligence of man did not continually dispute physical science and so-called rationalism by its own intuitions and impulses, the youth of this age would be hopelessly demoralized. If intuition did not continually contravene scientific skepticism, no man could preserve the true ideal of love. No man could exalt his own love relations above their reproductive office. A general acceptance of the deductions of physical materialism, as to sex, would be the end of that which is most sublime in the relation of man and woman. These deductions as to sex, marriage and reproduction constitute, at once, the most unscientific, and the most debasing doctrine that ever emanated from recognized authority.

These so-called scientific theories, together with the widespread immoralities of sex, have prejudiced the question. A discussion of sex, the most vital problem of the age, is practically tabooed.

Human ignorance brings confusion into the sex question, as into every other vital issue of life. Here, as elsewhere, the free Will and Desire of ignorance pervert Nature's laws. In this relation, as in every other, man is a free moral agent. He may occupy himself in the lower elements and offices of sex, thus arresting his own development. He is left perfectly free to invite the degeneracy and extinction of his own family or line. Nature, however, in the end, safeguards its higher meanings and purposes. Nature, in the end, defeats perverted and degenerate sex practices. It wipes out the offending family or community. The natural law of affinity and the needs and requirements of the intelligent soul are the forces continually at work to overcome false theories and abnormal practices.

There is nothing prejudicial to sex in scientific analysis of its physical functions. On the contrary, it is as important as are the analyses of the spiritual and psychical powers and activities. Nor is there anything gross in the true physical sex relation. On the contrary, that relation should be accepted as a legitimate office of human life. Properly viewed, it is merely one of love's expressions. Rightly guarded, it conserves the highest interests of the race. The physical sex relation is a proper function of life. It is not, however, *the* purpose of life. Nor is it a function properly discharged until man knows the meaning and purpose of his own life.

That which alone is prejudiced to sex is that *theory of sex,* promulgated by physical science, which is likely to be adopted as the rule and guide of marriage. That which alone is detrimental to the sex relation is the deduction that "a biological need is the prime cause of marriage," and the basis of love between man and woman.

Many who secretly revolt at such doctrines have not the courage to contradict them. Others of gross inclinations adopt this position as an excuse for their own naturally low estimates of the sex relation. What is said is not meant to impeach the motives of physical science. The scientific skeptic is undoubtedly sincere. He desires only to uncover Nature to the intelligence of man. His position is due to his limitations and not to a deliberate intention to mislead. Both Huxley and Haekle accepted theories of evolution and of love which shocked their own spiritual natures. Indeed, the writings of most of our great specialists along these lines indicate their reluctance to level their own ideals of life and of love to the theories of their own school.

Physical science tells us that the evolution of man rests upon blind physical forces. Nature, however, demonstrates that it rests, first, upon the spiritual affinities, and next, upon the Will and Desire of the individual. Nature demonstrates that the higher evolution of man is a process governed by his rational Will and rational Desire. While the unconscious affinities of the lower elements influence, they do not control the life of the individual man or woman. The supremacy of the Will and Desire of the soul over all of the lower elements is the triumph of human nature.

The ascetic who cheats nutrition and defeats reproduction illustrates the supremacy of the individual Will over the demands made by the physical functions. The martyr who dies, rather than recant, and the soldier who goes to death in battle, represent forces that override the demands of physical nature. They prove that there is something in man even stronger than the instinct for self-preservation. The woman who dies of disappointed love demonstrates that a desire of the soul may outweigh the involuntary demands of either nutrition or reproduction.

The higher intelligent and moral development of man presents the record of an individual Will and an individual Desire, brought to bear, either in support of or to defeat the involuntary demands and impulses of the lower nature. The higher life of man is a perpetual struggle between his rational judgments and the involuntary impulses of the lower life elements. Our present stage of evolution demonstrates the triumph of rational intelligence over the unconscious forces in Nature.

When analyzed, this higher evolution reveals another and a far greater struggle in human nature than the struggle for nutrition or the struggle for reproduction. It reveals a struggle conducted by intelligence through the mediumship of the brain, as well as a struggle conducted automatically by physical matter through the digestive organs. Nature reveals the fact that the rational Will and Desire of mankind are engaged in a co-operative struggle for an object peculiar to intelligence alone. Man, as Will, and woman, as Desire, are seeking another class of affinities and another class of results than those which pertain to nutrition and reproduction.

This fundamental struggle in human life has for its final object neither species nor family, but instead, a state or condition of intelligence which we can define as ethical only. This ethical condition, so eagerly sought by mankind, is translated to human intelligence as an individual content or an individual happiness. These are attainments which accrue to individual intelligence and have to do with individual life only.

It follows, therefore, that the higher evolution of man rests, primarily, upon the Will and Desire of rational intelligence. This higher evolution is an expression of the Will and Desire of intelligent entities, and not of blind physical forces. The evolution of man, physically, spiritually and psychically, involves:

(1) The universal spiritual principle of polarity or affinity.

(2) An individual free Will and Desire.

(3) The individual struggle of rational intelligence to execute that Will and attain that Desire.

Evolutionary processes are, therefore, expressions of a progressive struggle through and by which individual Will and individual Desire reach adjustment under the immutable principle of affinity.

Evolution accomplished, means the final bringing of the rational Will of man and the rational Desire of woman into full, free, and intelligent compliance with the unchangeable laws of the universe.

The efforts made by the individual man and woman to reach this state of completion is defined as The Struggle For Happiness.

The vibratory affinities which characterize this progressive struggle constitute the Harmonics of Evolution.

CHAPTER XIII.

THE STRUGGLE FOR HAPPINESS.

The secret springs of human action are neither the organs of digestion nor those of reproduction.

This is the ultimatum of a science that has demonstrated the fact of life after physical death, and the progress of development in another world than this.

To know what moves an individual is to know what moves humanity. To know what improves an individual is to know what improves the race. To learn the secret impulse, motive and desire which inspire the individual man to action, is to discover the secret spring of universal human activity.

It will be remembered that each kingdom of Nature is directly controlled by the highest element which goes to energize and vitalize the products of that kingdom. Mankind, therefore, while combining the energies and potencies of all lower elements, are yet dominated by the highest which enters into human nature Though moved by the impulses, and susceptible to the involuntary affinities of all lower elements, man's life is nevertheless directly controlled by his highest nature, the intelligent soul.

It is, therefore, to this essential and intelligent soul, or Ego, that we must look for the springs of human action. In this alone can we hope to find a rational explanation of man as he has been, is, and aspires to become.

This statement rests upon certain knowledge of man's psychical activities in two correlated worlds of intelligent life. A long and careful study of the intellectual and moral activities of man in two worlds demonstrates that he is moved to such activities by motives which are ethical and individual to himself. This is the fact, whether the intelligent Ego be physically embodied or physically disembodied.

Nature's general struggle for the completion of an individual is accompanied, therefore, by a purely ethical motive or struggle on the part of the individual intelligence. This greatest struggle in Nature dawns in animal life, but not until the stage of man does it rise into paramount importance. The difference in the ethical possibilities of animal and human nature is the difference between the potencies and capacities of the spiritual life element and the potencies and capacities of the intelligent self-conscious soul.

Though the individual struggle for an individual completion is the commonest activity of human intelligence, it is nevertheless a struggle wholly obscured, as such, from the consciousness of the individual. There are good reasons why this greatest struggle in existence is unknown to the popular intelligence. It has never heretofore been disclosed as a scientific principle. Neither physical science nor popular philosophy has, up to this time, apprehended this struggle as a natural and distinct activity of intelligence. It has never been observed as a legitimate process of evolution.

It was not until Darwin's exposition that the world discovered the universal struggle for nutrition, and in the same way the tremendous moral significance of reproduction was overlooked until it was laid bare by Mr. Drummond.

Thus, while the struggle for self-completion is the universal struggle of human intelligence, the average individual lives, struggles, and dies, without having recognized the real motives of his own activities. It is safe to say that not one out of every hundred ever formulates the ultimate motives of his own daily activities. It is true that the poets and singers have dimly grasped this great law of completion, just as they foreshadowed the evolutionary ascent of physical man. These, however, we classify as prophets or dreamers, whose visions simply furnish us literary entertainment.

In the great active world of human affairs this greatest struggle passes under countless disguises. It is known under an infinite variety of names. Every man is engaged in it, yet none appears to realize that he is so engaged. Men appear to themselves and to others to be doing an infinite variety of things. This, in a physical sense, and even in a purely intellectual sense, is quite true. In an ethical sense, however, all men are in reality doing the same thing. In reality, all men are struggling for the same ultimate ethical state of the soul.

Though the struggle for self-completion has hitherto found no expression through science, though it is nameless to the popular mind, it is a struggle too well known to the soul of every individual. Be the man high or low, civilized or savage, wise or simple, he is none the less conscious of one impulse that moves him, one motive that sustains him, and one desire that inspires him.

If we substitute another term for the scientific struggle for self-completion, the ethical struggle which accompanies the visible, physical activities of life shall be made clearer to the reader. It is possible that by a mere change of terms we shall disclose that which is the most familiar state of our own being.

When we declare that the main activity of human intelligence is The Struggle for Happiness, we have only stated the struggle for completion in another way. We have only stated the commonest fact of daily life, and the most familiar experience of our own souls.

This, then, a state of individual, ethical content and happiness, is the inspiration and the goal of every human activity. Here we finally come to that struggle in human nature which must take the place of Darwin's "Struggle for Nutrition," and Drummond's "Struggle for Reproduction." It takes the place of those struggles, but not in the sense of denying their existence, their value or their influence. The Struggle for Happiness merely assumes the leading rôle in human life, in place of those involuntary activities which sustain the physical processes of nutrition and reproduction.

This universal struggle of the intelligent soul is foreshadowed in every activity of every lower entity. In each human being is repeated all of the demands of all the lower elements. That is to say, the physical body represents the play of the electro-magnetic and vito-chemical life elements, while in the physical appetites and passions are displayed the energies of the spiritual life element. In man, however, there are demands which transcend both the chemical requirements of the body and the physical satisfactions of the appetites and passions. These other higher demands have their source in the soul, and require satisfaction in the nature and quality of the soul element.

These are demands which govern the rational and moral development of human nature and have to do with the evolution of happiness.

For that which the soul demands, human language has found but one definition. There is but one word for it, and this is an ethical word. No other word nor elaboration of words could better explain or define that certain ethical desire which universally exists in the intelligent soul. When the higher science declares that the main activity of human intelligence is The Struggle for Happiness, it has explained the nature of the condition desired, as explicitly as human language permits.

Up to this time the world has never seriously considered happiness as either a natural phenomenon, as a normal state of being, or as the scientific basis of the philosophy of life. Physical science, in its engrossment with the involuntary operations of food combinations, with the instincts of reproduction, and the influences of environment, has absolutely ignored the operations of individual intelligence which govern the higher phenomena of life. The intellectual and moral energies and capacities and motives of the individual man have been obscured by a method of study which excludes everything but the physical improvement of species.

A science, therefore, which is prepared to sponsor a natural philosophy of human happiness must be prepared to meet the criticism of scholars who have reached other conclusions.

Physical science has no room in its method for the study of intellectual and moral phenomena. On the other hand, pessimistic philosophies, contaminated by continual association with disease, crime and abnormality, come to regard human happiness as a delusion of the mind, the dream of dreams, without basis in fact, a hollow echo of physical appetites and passions. Even religion, which recognizes a struggle of the soul for happiness, has not located that struggle as directly bearing upon this mortal life. On the contrary, it has almost universally regarded this struggle as one whose aim and fruition belong to another world than this.

Sacred literature nowhere contains a nobler inspiration of human intelligence than that embodied in our own Declaration of Independence. When it affirmed that among the inalienable rights of man are "life, liberty and the pursuit of happiness," the United States Government took its stand upon the most exalted interpretation of Nature.

Nobody, be he scientist, or pessimist, or theologian, will deny that happiness is the most desirable state of being. If honest with himself, he will admit that a longing for this condition is the secret spring of his own higher activities. Indeed, if one studies, not the operations of physical functions, but the lives of his neighbors, he will discover that the hope and expectation of happiness sustain the individual man during the struggle for nutrition and the individual woman in her reproductive sacrifices.

How to secure this individual content is the first question which human reason puts to Nature. To this search of immature minds for an individual and ethical content we owe the steady development of intelligence and morality.

To discover the most direct route to this desirable estate has absorbed the energies of individual intelligence for ages. The knowledge acquired throughout the process constitutes our civilization with all of its science, art, literature and learning, together with all of its religions, philosophies, and philanthropies.

Ideals as to what constitutes happiness have, therefore, varied as much as individuality itself. They have differed as much as do men in point of time, race, country and development.

The important question to science is not what manner of ideals men have set up to mark the goal of human will and desire. It is not what men seek and *name* as happiness. That which is important to science is the fact that the individual man cherishes such an ideal and seeks a certain ethical condition as the end and aim of all his acts and accomplishments.

Ignorant of the vibratory principle in Nature, and ignorant of evolutionary processes and possibilities, man works out his happiness under seemingly hostile conditions. Those conditions, however, are only seemingly hostile, for to one familiar with the correlation of forces they are seen as beneficent conditions only. Only just such conditions could have developed a rational and moral human being. At the beginning the individual has no remote conception of the laws which govern individual happiness. He does not even speculate as to why he desires this state of being. He has no theories as to a law of harmony in life. Only after ages of experiment and penalty does he discover that there is an immutable principle in Nature which governs physical equilibrium and ethical harmony or content.

The primitive man has only his crude impulses and ambitions to guide him. His rational intelligence is not equal to the consideration of general laws. All he feels are his impulses of attraction and repulsion. He is conscious only of his necessities and desires. He is conscious only of love or hate, or sorrow or satisfaction. He only knows that he is content in the possession of that which he craves, and discontented or unhappy when deprived of his desires.

Because of this undeveloped reason, the Struggle for Happiness has been subject to every character of experiment which the imperious will and strong passions of man, and the weakness and stupidity of woman could suggest. It has been subject to gluttony and lust, to fierce ambition, avarice and vanity. It has displayed every variation that animalism, cruelty and folly could suggest to ignorant men and women as means to an end.

The history of human development proves that happiness— the goal of human desires—is a state of consciousness that does not depend upon the physical appetites and passions, nor upon the acquisition of material wealth. It proves further, that not even power or position or fame or honor is the guaranty of this coveted estate. Because of this universal fact, painfully demonstrated throughout the ages, there should be little wonder that the scientist and philosopher become skeptical on the subject of human happiness.

It is nevertheless true that the repeated failures of the past have not served to wipe out the hope and expectation of happiness from the minds of succeeding generations. This is a fact which can not be accounted for by the laws of heredity, as laid down by physical science. Does it not rather suggest that Nature, here as elsewhere, has really furnished adequate laws for what appears to be a universal necessity?

If happiness were the outcome of physical satisfactions, man had never progressed beyond the limit of Nature's sufficiency. If the organs of digestion and reproduction were the real inspirations of life, then human intelligence had never risen beyond appreciation of the pleasures of appetite and of lust.

Physical nature is, in fact, easily satisfied. So easily, indeed, that without the higher intelligent ambitions and desires, life had never passed the stage of savagery. If material possession (surpassing physical necessity) created happiness, then the words "wealth" and "happiness" were synonymous. Is this the fact? On the contrary, what more abject being exists than the miser, the soul who has bent its energies along lines below its natural level? Is it not also true that the sorrows of the rich are as commonly exploited as the miseries of the poor?

Nor is happiness the result or outcome of a biological need. On the contrary, most of the disease and much of the crime and sorrow of human life are the direct results of perversion in the physical sex relation. The most unhappy and the most degraded of mortals are those who seek happiness through the physical appetites and passions. The personal life history of the glutton, the libertine and the prostitute would carry its own proof of this statement. The attempt to satisfy the soul through the body always has been and always will be the most lamentable mistake that a rational being can make.

Nor is happiness bound up in maternity. The universal sacrifice of woman in this relation needs no comment. The unhappiness imposed upon woman by masculine lust and by compulsory child bearing, is a matter well known, at least to woman herself.

Children do bring love. They may prove to be a pleasure and a consolation, but a woman never yet realized the sum of her desires in "paying the eternal debt of motherhood."

Human happiness results from neither a competitive struggle for nutrition nor a compulsory struggle for reproduction. On the contrary, the sources of happiness as far transcend the physical activities of life as the demands of the soul transcend the requirements of the body.

The history of intellectual and moral development is the history of self-conscious intelligences seeking satisfactions which appeal to intelligence alone. The greatest struggle in Nature, therefore, rests upon ambitions and ideals which have nothing to do with the involuntary operations of physical nature. To relate this ethical struggle of the soul with conditions growing out of physical nature is to set intelligence an impossible task. By no trick of imagination can we logically relate the universal ambitions of intelligence with feeding and breeding. By no process of reason can we confuse the purely ethical rewards which intelligence seeks with the purely physical satisfactions which the body demands.

Indeed, the physical deprivations which intelligence suffers in the pursuit of its higher necessities are the commonest facts of human history and individual experience. All that man has accomplished above and beyond animal accomplishment represents the ambitions of an intelligent soul, seeking to gratify itself through possession, power, fame, honor, knowledge and love.

The motives and ambitions which move men to their daring accomplishments reside in the soul and not in the body. The inspirations which have evolved both the word and the ideal of "heroism" have their seat in the brain and not in the stomach.

Take, for illustration, the history of exploration and discovery alone. Before this record of voluntary physical sacrifice, deprivation and suffering, how insufficient appears the theory of a "struggle for nutrition in the midst of a hostile environment." Among these historical adventurers have been men of strong physical nature and strong physical desires. There have been men of wealth and comfortable condition; men who literally abandoned certain physical comfort for certain physical discomfort, for probable disease, for possible death. What relation, however remote, exists between the demands of nutrition, the instinct of self-preservation, and that spirit of daring which braves the unknown dangers and deprivations of the burning tropics or the frozen north? Do any or all of the laws of physical heredity account for those intelligent processes which override the demands of physical nature, mapping out a path of achievement perilous to physical health and even to life itself?

What other possible motive than a purely intellectual ambition could lie at the basis of such adventures?

The history of exploration and discovery is but the history of all higher human achievement. Such achievement is everywhere the record of individual intelligence, aggressive, ambitious and masterful, moving out in new lines, seeking those rewards which only intelligence appreciates. It does not matter whether the particular ambition which moves a certain individual to action, be the desire for wealth, power, knowledge, fame or even love itself. That which is important is the fact that the higher development of man rests upon psychical ambitions and not upon physical functions.

Whatever may have been the particular motive which inspired a Nero, a Napoleon, a Washington or a Lincoln, that motive cannot be even remotely associated with "the struggle for nutrition in the midst of a hostile environment." The thirst for power and the love of liberty are qualities of the ambitious and intelligent soul. They are not even remotely expressions of physical functions or physical compulsions.

What is there in the struggle for nutrition to account for missionary zeal and adventure? Here we have to deal with rational beings who voluntarily resign the physical comforts of life and voluntarily imperil health and even life. For what? Merely that they may educate the ignorant in what is regarded as truth, thereby satisfying an internal ethical need. In this case we have neither ambition, vanity, scientific zeal, nor the love of wealth or power to account for this phenomenon.

Indeed, here is renunciation of not merely physical comfort, but of most of the things which intelligence regards as sources of happiness. The missionary, nevertheless, has done that which, according to his soul's necessities, promises him the greatest measure of ethical content and happiness.

Again, where or how shall we find any relation between the struggle for nutrition and the act of the millionaire who piles up another million? Through what processes of digestion shall we account for the miser, that unhappy being who denies the demands of nutrition to satisfy a purely intellectual greed for treasure? Or, again, how does the Darwinian theory account for the soldier who rushes to certain death in the midst of battle? How does it account for the deprivations to which the scholar will subject himself in the mere acquisition of knowledge?

To what character of food combinations and to what class of physical competitions must we look for explanation of the inindividual achievements of scientists, inventors, poets, painters and singers, throughout the world's history?

Almost universally the history of intellectual development is the history of physical deprivation. It is the record of individual intelligences choosing between comfortable physical conditions and their intellectual ambitions. It is almost universally a history in which the demands of digestion are made secondary to the demands of intelligence. The enforced struggle for nutrition may be a spur to indolence and stupidity, but it is the perpetual stumbling-block to the intellectually ambitious. The absurdities into which the Darwinian theory leads, are not so clear until we attempt to connect the highest expression of pure intelligence and moral heroism with the involuntary operations of chemical substances. The attempt to account for a Christ or a Buddha, or for a Shakespeare or a Plato, by the findings of physical science, is an effort that confuses reason and belies every intuition.

If the theories of evolution seem inadequate to account for intellectual ambition and achievements, how much less does the procreative tyranny account for the love story of the world. How, according to this theory, shall we account for even that law of individual preference which has rescued the animal world as well as the human, from a promiscuous sex relation? With countless opportunities for gratifying purely physical passions, what is there in intelligent Nature which limits individual choice and erects the barrier of a "natural repulsion"? What is there in the theory of a "tyrannic need of procreation," which explains the death of a human being when disappointed in love, or when separated from the chosen lover by circumstance or death? What is there in human nature that impels a man or woman to suicide when deprived of the love and association with—shall we say the body, or the soul, of some particular individual?

By what perversion of reason can a "biological need" be made to account for such love and such a relation as existed between Petrarch and Laura, Abelard and Heloise, Dante and Beatrice? To claim that such love, even remotely, depends upon the lusts of the flesh, is to debase reason and to desecrate the noblest ideals of our own souls.

Indeed, the study of the intelligent side of animal life goes to show that the individual animal is given up to the business of living, and is seeking individual satisfactions as eagerly as man himself. The man or woman who owns and loves a dog, knows that the dog has capacities for love and friendship which are wholly unrelated to procreative instincts or even to the affinity of species. The dog that starves and dies upon his master's grave is a singular commentary upon the theory of evolution by digestion and a procreative tyranny.

The common sense of the world has come to acknowledge that we must look elsewhere for explanation of a higher evolution of man than in physical feeding, breeding and battle. When science discovered the universal law of completion in Nature, and the universal struggle for self-completion in the individual that explanation was found. When science recognizes the demands of the soul, as well as those of the body, when it perceives a "Struggle for Happiness," as well as a struggle for nutrition and a struggle for reproduction, it is then prepared to analyze man as a physical, intellectual and moral entity.

Intellectual progress was made possible only by the fact that intelligence refuses to be satisfied with those necessities, pleasures and powers which belong to the physical plane. The very fact of an intellectual and moral evolution is witness of this higher struggle of the soul to achieve satisfaction in terms of its own essential nature.

The higher philosophy insists that the universal is the natural. It realizes that the desire and the struggle for happiness constitute a universal impulse and activity. Men toil, not merely for nutrition and physical necessities, but that life may be sustained to achieve results. They toil, hoping to win from toil the conditions for happiness. Men toil, not merely for bread, but for the further satisfaction of that higher nature which is not satisfied by bread alone. They toil that they may achieve knowledge or power or possession, or that they may form those ties which shall guarantee content or happiness to the soul.

Happiness and the ideal of happiness are no more the outcome of the struggle for nutrition, than love and the ideal of love are the outcome of the tyrannies of lust or the sacrifices of maternity.

If human development and happiness were indeed the outcome of the struggle for existence in the midst of a hostile environment, then the Laplander and the African should represent the standards of enlightment and happiness. The truth is, however, that the happiest and the most highly developed human beings are found in the prolific belts of the temperate zones. Nature, therefore, develops through giving rather than through withholding, through hospitality rather than through hostility. It develops the body through generous nutrition. It stimulates intelligence by furnishing it intellectual ideals. It confers happiness through a natural principle of fulfillment.

Even in savagery the struggle for nutrition is subordinated to the struggle of ambitious intelligence for satisfactions appealing to intelligence only. The savage delights far more in the scalplocks at his belt, than he does in the spoils of battle. The scalplocks are the visible sign and symbol of his individual strength and courage and victory over his enemies.

So the millionaire rejoices in his surplus, the result of his individual genius, the passport to public admiration, the sign of his intelligence and power among men.

Thus also the miser enjoys what appeals to intelligence only.

He cheats physical nature for a benefit that is solely and only an ethical benefit, a state of consciousness.

It is probable that the Prince of Wales hopes to succeed to the crown of England. If so, what is the basis of that desire? The struggle for nutrition or the Struggle for Happiness? He may look to that succession with mixed motives. His anticipations may be the egoistic pleasures of sovereignty, or altruistic designs for his subjects. In either case he anticipates satisfactions which flow from the honors of the one or the privileges of the other. In no remote sense can this occasion contribute to his physical necessities or physical pleasures. On the contrary, here is a man whose entire life represents the complete gratification of every physical necessity, appetite and impulse, but it is safe to say that the Prince of Wales has made as vigorous a struggle for individual happiness as the lowliest of his subjects.

The entire history of art is the history of intelligence seeking satisfaction, not because of, but in spite of the exactions of physical nature. It is the history of individuals who have almost universally sacrificed purely physical demands for purely intellectual ambitions. Indeed, the struggle for nutrition has been the obstacle of genius instead of its inspiration, the world over.

Edgar Allen Poe had better served the struggle for nutrition by following the plow instead of the pen. He would have been far more comfortable physically, but who shall say he would have been better satisfied or happier? The life of Poe is but a type of life in Bohemia, where genius suffers and starves that it may pursue its intellectual ambitions and immortalize the ideals of the soul.

Who that considers the voluntary physical deprivations of intelligence the world over, but feels intuitively that the sources of art are as far removed from purely physical functions as the ambitions of the soul are removed from the cravings of the stomach.

Thus also with the motives which inspire men to patriotic and religious martyrdom. The one satisfies conscience in battle for the (to him) ethical principles of right. He battles for life and liberty that he may be free to pursue the struggle—not for nutrition—but for happiness according to his conscience. The other dies by torture that he may not forfeit his soul's happiness by denial of his God.

Thus moves the world of intelligent life. Thus moves the intelligent individul in search of those objects, relations and conditions which shall afford a purely ethical content and happiness to the soul.

The ideals of, and the capacities for happiness, are as infinite as individuality itself.

The Hottentot and the Christian gentleman entertain very different ideals of happiness. So, too, their individual enjoyments are as widely separated as are their ideals. The difference in their ideals and in their ethical satisfactions is measured by the physical, spiritual and psychical differences in their individualities. Happiness, to the highly developed, not only appears more, but is vastly more, by way of effects, than it is to the man of low degree. It is more by way of effects, in that the physical body, the spirit and the soul of the highly developed are keyed to the higher vibrations and the higher harmonies in Nature.

A Digger Indian may experience what to him may be a perfect temporary satisfaction. No one, however, would imagine that he experiences the same ethical enjoyment which a Shelley or a Tennyson might feel in the full realization of his desires and ideals.

For illustration, the man who eats dogs, reptiles and raw meat, creates wholly different vibratory conditions in the organs of taste than the epicure whose sense of taste is keyed to correspondence with Nature's finest and most delicate foods. The savage whose food affinities are confined to the coarse in Nature, is absolutely incapable of those pleasurable sensations which the epicure enjoys. In the same way, the man of cultivated taste sickens at the mere thought of dog meat and reptiles as food. If the savage cannot enjoy the finer effects of finer food through the physical organs of taste, how much less is his undeveloped intelligence prepared to understand or enjoy the higher harmonies of the soul life.

Or, again, the savage whose physical ear is attuned to the discordant vibrations of the Tom-Tom cannot be expected to grasp or enjoy the vibratory harmonics of a Beethoven sonata rendered in the most approved style of the modern musician.

Both the savage and the scholar enjoy spiritual intuitions. They are not, however, the same class of intuitions. Each receives his intuitions of spiritual things in the degree of his own refinement, and the acuteness of his spiritual senses. As a result, the æsthetic natures of the two vary as widely as the degrees of refinement in their physical and spiritual organisms. Nor does the undeveloped intelligence of the savage formulate moral conceptions which are possible to the scholar. Neither is he able to receive the same suggestions from other intelligences, either here or upon the spiritual plane, as the man of developed brain and cultivated moral principles.

Thus, Nature conditions man, and man so conditions himself, that The Struggle for Happiness is an infinitely varied and unequal struggle, and the capacity for happiness is an infinitely varied and unequal measure.

The Struggle for Happiness is forever subject to the mandates of intuition or those of reason. This means that happiness is a state or condition which depends upon both impulses that are involuntary and judgments that are voluntary and independent.

When primitive man undertakes to shape his course by reason, as well as by intuition, he becomes an independent experimenter in the midst of unknown and seemingly hostile conditions. Thus, he appears to bring confusion into the orderly scheme of Nature.

The dark early history of man covers that period when human intelligence abandoned purely intuitive methods before his rational knowledge was sufficient to keep him in harmony with the general laws. As a result, the general harmony which prevails in the animal world is lost to man until he becomes a rational and independent demonstrator of the harmonic principles in Nature.

Thus, throughout the ages, The Struggle for Happiness has shaped itself according to the degree and quality of individual development, and the degree of correspondence between reason and intuition.

The inherent elements of masculine and feminine nature condition man and woman to pursue the Struggle for Happiness along different lines.

It must be recalled, that man represents the positive and aggressive energies of all the life elements, while woman represents the receptive or absorbing energies of the same elements. The same inherent principle, therefore, which conditions one atom as positive and another as negative also conditions one intelligent soul as positive and aggressive and the other as receptive and pacific in its nature.

Therefore the same character of energy that impels an unconscious physical atom to seek vibratory equilibrium in another atom, impels the self-conscious soul to seek conditions and persons in harmony with itself.

In the Struggle for Happiness man represents all aggressive energies and elements, and seeks an individual and ethical content in terms of his essential strength, physical, spiritual and psychical. In the same way, woman, representing all receptive energies and elements, seeks an individual ethical adjustment in consonance with her essential qualities of receptivity, absorption and self-surrender.

The outcome of these two principles, running side by side through evolution, was inevitable. By reason of these innate energies and qualities, the Struggle for Happiness has been characterized by force on the masculine side and by self-surrender on that of the feminine.

Perhaps no better definition can be found for the aggressive spirit in which masculine intelligence seeks its own satisfaction than the spirit of Conquest by Force, which stands for war. Perhaps no better definition can be found for the non-resistant spirit in which feminine intelligence seeks its own content than the Spirit of Self-Surrender or Peace.

The one, therefore, the masculine, stands for acquirement and conquest by force, while the other, the feminine, represents accomplishment by self-surrender.

As already shown, the positive and receptive natures of masculine and feminine are accentuated in each higher kingdom.

Thus, positive and unconscious energies of mineral and vegetable substance merge into the positive conscious energies of animal intelligence. These, in turn, re-enforced by a higher element, give rise to the positive and aggressive psychical will of man. So on the receptive side of Nature, the powers of absorption increase from the mere material receptivity of mineral and vegetable atoms to the self-conscious psychical desire of the woman.

Thus we have, as the two dominating factors in the psychical Struggle for Happiness, the aggressive will of man on the one side and the absorbing desire of woman on the other. To put it in another way, we have on the one side a forceful or ambitious masculine intelligence, and on the other an intelligence that is pacific and unambitious in its nature.

Thus, it will be seen that the sex principle shapes the higher Struggle for Happiness as it does the lower struggles for nutrition and reproduction. It dominates the psychical as well as the physical plane, and governs ethical content as well as the physical functions.

Man pursues his ambitions and ideals in conformity to that strong psychical nature which dominates a strong physical body and strong physical appetites and passions. Moved by the spirit of conquest, he seeks to forcefully wrest his physical comfort and psychical content from Nature and his fellow man. The effect of this masculine inclination to conquer by force has been to largely concentrate his energies upon the physical plane of action. The concentration of will necessary to forceful acquirement keeps him close to the earth plane. As a result masculine activities are characterized by the physical conquest of physical and material nature, and by the attainment of temporal power in the affairs of men. This same spirit of conquest also charcterizes masculine sports and even the masculine pursuit of love. Man conquers, acquires, recreates and loves in conformity to his aggressive nature.

War is not a natural necessity in the struggle for nutrition. Nutrition merely furnishes masculinity a pretext for war. In fact, the struggle for nutrition is best conserved by pacific and cooperative methods, rather than by warring and competitive ones.

War is simply the physical, visible sign of the innate masculine constitution—the spirit of conquest by force.

As history shows, this conflict of masculine strength, this struggle for supremacy begins in personal, physical combat. This personal struggle rises to the dignity of rationally organized and directed warfare. This, in turn, merges into purely intellectual controversy, or into contests scientific and philosophic in their nature. In its last analysis, this primitive masculine ferocity is transformed into noble emulation along the higher lines of ethical principle, thus furnishing the world its ideals of perfect manhood.

This contest for individual supremacy, as between man and man, and between schools of men, appears to be the very life of masculine activities. To measure swords with his fellow man, first literally, and then intellectually, is an impulse inherent in man's nature. War, therefore, is not a natural necessity, only in so far as it represents the primitive impulses of barbarous masculine nature. Scientific, philosophic and religious controversy are unnecessary modes of education, except for the fact that undisciplined masculine intelligence conceives of no better way to impart its truths.

The history of primitive war is not the history of hostile environment moving upon empty stomachs. Instead, it is the history of fiercely ambitious intelligences moving upon each other for the love of conquest, supremacy and power.

If one doubts that the actual foundation of war lies in masculine temperament and not in environment, let him imagine what had been the results if men were of the psychical constitution of women. Nobody will believe that the struggle for nutrition would have engendered war among women. If man represented the same energies which constitute femininity, flags of true and arbitration treaties would have taken the place of invasion and conquest.

The earth is and has been so prolific in nutrition for all life that the theory of war, as a natural necessity in the struggle for nutrition, amounts to an absurdity.

While this is the history of masculine development, exactly

the reverse appears in the feminine Struggle for Happiness.

Woman, from the beginning, shapes her desire in conformity to her weaker body and her weaker physical appetites and passions. Conditioned by Nature in every element to non-resistance, her intelligence seeks to conciliate force and to achieve her desires through self-surrender.

The obscuration of woman in the earlier reaches of civilization is due as much to her own innate nature as to the tyranny of man. During this long dark period she permitted herself to be overwhelmed by the force and ferocity of the masculine will and masculine passion. She was enslaved mentally as well as physically. Her non-resistant spirit, as well as her physical body, was overshadowed by that of her savage mate. Not man alone, but woman also, is responsible for her long suppression and obscuration, for her physical servitude and mental bondage.

Dominated by desires, rather than by ambitions, the ideals of woman did not, in the beginning, necessitate aggressive methods. Lacking the masculine thirst for power and supremacy, she was not impelled to either the forceful acquirement of knowledge or the forceful control of environment. Her desires did not necessitate the same concentration of intelligence as to the general affairs of the world. Her rational powers were, therefore, permitted to lie dormant while her impulses largely shaped her destiny.

Bound to maternal duties and the individual relations of life, woman employed her intelligence along the lines of least resistance. As a result, her energies and emotions are more directly employed in the development of the love relations which are personal and individual to herself.

Out of this innate feminine love nature springs that silent, bloodless warfare of rivalry and jealousy as between woman and woman. Between her love of love and her love of beauty, woman has occupied herself too much in making herself attractive to men and an object of envy to other women.

If it can be said that masculine nature tends too much to pride of intelligence, it can be truthfully said that feminine nature tends too strongly to vanity of personal appearance and charm.

Where man goes to war in support of his convictions and opinions, woman enploys her intelligence to increase and preserve her personal beauty. The one is bent upon establishing his opinions through his strength, the other is consumed with the desire for personal admiration. The one is seeking a personal power, the other a personal love relation.

These are the paths along which the sexes travel. These are the general principles upon which man and woman conceived their ideals of happiness.

In both sexes, however, the low rivalries of undeveloped natures are slowly transformed. Masculine combats and contests lose their ferocity and virulence, while women, broadened by rational and moral development, finally rise above the petty rivalries and vanities and jealousies in the love relations of life.

As already explained, the more delicate physical organization of woman has its basis in the non-resistant spiritual principle. This means that woman, generally speaking, is in closer touch with and more susceptible to the purely spiritual side of Nature than is man. This greater susceptibility of woman to the vibrations of spiritual material and to the mental suggestions of spiritual intelligences, has two definite results. It gives to woman, first, her finer æsthetic nature, and next, conditions her to receive moral suggestion from the other side of life. This means that woman becomes the æsthetic factor in society and the natural monitor and inspiration along lines purely ethical.

While feminine intution is a source of certain kinds of knowledge, that knowledge has to do with spiritual relations, rather than with physical conditions. For this reason, woman is credited with using her intuitions rather than her rational powers, with being a creature of impulse rather than of reason. While this criticism is largely a just one, it is nevertheless this feminine power of spiritual intuition which serves to keep the masculine half of humanity in touch with the spiritual side of Nature, and to emphasize the power and importance of the love relations.

Conditioned by Nature to seek her happiness through physical, spiritual and psychical non-resistance, the physical servitude, mental subserviency, and petty vanities and deceptions of woman are explained. Dependent upon her attractions instead of her strength for her victories, moved by impulses rather than by judgments, the subtle and evasive tactics of a woman are easily understood. Feminine intelligence has served feminine weakness in its long Struggle for Happiness by methods which can only be defined as "feminine."

For example, we find illiterate women who are strangely cunning and resourceful. We find highly developed women whose diplomacy and tact are the wonder of the masculine mind. Though her receptive intelligence and weaker will leave woman the subject of masculine aggressiveness, at the same time they equip her to outwit brutality with cunning, and to disarm tyranny with tact. "Cunning" and "tact" are the especial subtle devices by which feminine intelligence guards feminine weakness and circumvents masculine will and logic. This ready weapon of "woman's wit" has, from the beginning, served her when opposed to masculine strength, masculine will and masculine reason.

How unlike are the methods of men and women, irrespective of time, race or development. Man, whether pursuing his natural tendency for war, whether engaged in commercial, political or scholastic pursuits, brings to bear all of the aggressive elements of his nature. Even in love, as well as in war, the masculine method is bold, self-assertive and imperious. It is, therefore, a natural principle and not a social custom nor a legal code that impels the individual man to seek the woman of his own choosing, and impels the individual woman to wait for his coming.

There are, of course, exceptions to the general law of sex. There are abnormal men and women. There are "degenerates" of both sexes. There are many curious perversions of natural law. There are honest fanatics and foolish experimenters. These, however, are quickly classified and removed from the ranks of those the world calls "normal men and women."

Briefly stating the psychical Struggle for Happiness as a sex principle, it appears to be the struggle for knowledge, wealth, fame and power on the masculine side, and a struggle for love, beauty, harmony and pleasure on the feminine side. Who shall say that Nature has not wisely apportioned each to its task? Who shall deny that the sum of masculine ambitions and feminine desires constitutes completion and happiness when intelligently joined?

The evolution of happiness is bound up in this reciprocal scheme of sex. This principle, and none other, conditions the terms upon which an individual equalization and balance of masculine and feminine nature shall obtain. Setting out with fierce aggressiveness on one side and stupid submission on the other, these differing intelligences have run a long and painful course of tyranny and slavery in the world. In their final analysis, however as manly strength and womanly grace, they justify the painful processes of evolution.

At the beginning the masculine and feminine ideals of happiness were as different as their methods were unlike. The savage looked to war, conquest and control. The slave looked only to the amelioration of her condition through conciliation of and submission to her master. Her master and her children constituted the only world she could hope to influence to her advantage.

In its later and higher aspects, however, the Struggle for Happiness appears as the rational effort of man and woman to effect a harmonious relation as between themselves. It is a mutual effort to bring their individual ideals into correspondence and harmony.

In such a relation the spirit of conquest and the spirit of peace constitute supplementary powers of intelligence. The masculine ambition for supremacy and control finds its satisfaction in the feminine desire for self-surrender. Reason and intuition find their natural relation and employment, while both thought and feeling unite to promote the usefulness and the beauty of life.

In such a relation, and in such alone, masculine will and feminine desire will accord as to all material interests, intellectual occupations and ethical principles, thereby effecting that harmony of activity which the soul recognizes as a state of happiness.

A few years ago Mr. Drummond stated the object of life by a question which he answers himself. "Why do you want to live to-morrow?" is his question. He answers: "It is because there is some one who loves you and whom you want to see to-morrow and to be with and to love back. There is no reason why we should want to live on than that we love and are beloved."

Here is the philosophy of life in a paragraph. It will be observed that Mr. Drummond does not say here that man desires to live in order to fulfill his destiny in the struggle for nutrition. Nor does he hint that woman desires to live on that she may "pay the eternal debt of motherhood." On the contrary, he here distinctly recognizes that the only motive, inspiration and purpose of individual life are the attainment of happiness through the love relations.

When Mr. Drummond declared that "Love is the greatest thing in the world," he stated the case of Nature fairly. He only failed when he attempted to trace the origin of love.

How shall the Struggle for Happiness be accelerated? How shall the goal of masculine ambition and feminine desire be earliest reached? These are questions the individual is entitled to ask of science and philosophy.

Since the wisest of earth have agreed that human happiness rests, primarily, upon the love relations of life, this becomes at once an individual and personal question. It resolves itself into a question of personal Intelligence, Courage and Perseverance.

The state cannot legislate upon the question of individual happiness. Law cannot compel it. Governments can, at best, merely protect the individual in his right to its pursuit and enjoyment.

Each individual is the architect of his own destiny. He is the builder or destroyer of his own happiness. There is no royal road to happiness, any more than there is to knowledge, power or fame. Rational happiness necessitates rational knowledge of the laws of life. It necessitates rational conformity to spiritual principles. It necessitates legal sanctions for natural relations.

This means that the Struggle for Happiness is a matter of evolution and not revolution. It means that Nature is developing the powers of the individual soul just as it does the functions of the individual body, through and by an infinite series of experiments and adaptations which finally mold and condition it for its noblest destiny—Happiness here and hereafter.

CHAPTER XIV.

MASCULINE REASON AND FEMININE INTUITION.

Neither physical science nor speculative philosophy accounts for the differing intellectual processes of men and women.

Both, however, agree that a difference exists. To man is universally credited the stronger rational powers, to woman the keener intuitions.

Just what Reason is, and just what Intuition is, are problems not, as yet, solved by materialistic science nor by speculative philosophies. Physical science does fairly well when it declares that physical effects have physical causes. When it enters the domain of pure intelligence it becomes speculative.

"All the laws of physical evolution can never explain the first genesis of mind." So declares the Encyclopedia Britannica in its summary of the Darwinian theory.

Physical science, after accepting the digestive organs as the cause of mental phenomena, proceeds, tolerably well in its analysis of Reason. It traces fairly well the relation between certain physical causes and certain rational operations of the primitive mind, as relates to things tangible and visible to the physical senses,.

For example: It traces a relation between the rigors of climate and the rational act of the savage who constructs a hut for shelter. It traces cause and rational effect when he converts the skins of animals into clothing. It traces cause and effect in the act of trapping an animal for food purposes. Here the relation between the physical necessity and the rational act is clear and unmistakable.

There are, however, other phenomena connected with physical acts which are not explainable on the same grounds. For illustration, By what operation of intelligence, or for what rational cause does the same savage desire and seek a certain class of foods? In this case he acts by Intuition, since neither the savage himself nor physical science can rationally explain the law governing the selection of foods.

Those intelligent methods which are purely rational are intelligible to Reason. As yet, however, Reason has not satisfied itself as to the nature and meaning of those intelligent processes which we define as intuitive.

For example: Our rational judgment of an individual may directly contradict our intuitions concerning him. Though Reason may find no fault in him, Intuition may condemn him and we may feel the man to be false or vicious whose public life seems fair and honest. Or again, a jury pronounces a man guilty upon the circumstantial, or what we would term the rational evidence. At the same time, every member of that jury may have an "impression," or Intuition, of the prisoner's innocence. On the other hand, the accused may be acquitted upon the rational evidence, and yet, somehow, impress every juryman with a feeling of his guilt.

Here we have the double process of intelligence, viz., Reason and Intuition. Here are rational and intuitive processes simultaneously moving the intelligent Ego to two diametrically opposite decisions. And this is one of the common experiences of daily life. At every turn, at every stage of development, human intelligence deals with the opposite phenomena of Reason and Intuition.

Each one of us recognizes in himself the play of these two principles. We continually alternate between acts governed by Reason and those governed by Intuition. The one act is rational, the other is what we define as an impulsive or emotional or intuitive act. We trace the effects of these dual processes in all acts that have been recorded as human history. We take note of them in all current affairs, in the lives of our neighbors and acquaintances, and in our own families.

The history of intellectual development is the history of Reason in apparent conflict with Intuition. The achievements of man are, however, the combined results of both activities. Human achievements, intellectually speaking, are measured by the proper exercise and balance as between the two. The prophecies (or impulses) of Intuition, practically carried out by Reason, constitute intellectual development. Intuition furnishes man his ideals of achievement and happiness. It is individual Reason, however, which equips him to realize those ideals.

Each one of us feels that he understands his own rational processes. We know why we reach such or such a rational opinion. Most of us, however, are wholly at sea when we attempt to analyze our own intuitions. Most of us are confounded when we would explain that subtle monitor which so often contradicts and overrides Reason and deflects us from our rational conclusions.

It is this "inexplicable" activity of the intelligent Ego which refutes every theory of a mechanical and physical basis of intelligence. It is this subtle monitor which keeps the individual man in touch with the spiritual plane, and fortifies even low grade intelligence against skepticism of spiritual things.

The intelligence of this age demands of science an explanation of these apparently conflicting phenomena. The intelligence of man demands explanation of its own activities which shall coincide with the universal experience of intelligence itself. It rejects those theories which separate human from animal intelligence by nothing except a matter of additional feeding, breeding and battle.

Physical science fails to demonstrate the missing link between animal and human minds, just as it does between animal and human organisms. Nature does not furnish either hybrid physical or mental types.

If the kingdoms of Nature were not separated by unlike life elements there would obtain infinite variations and experiments in the physical organism. As with the body so with intelligence. If the phenomena of intelligence were separated only by time and experience, there would be mental types of infinite variation between animal and man. There would be no distinct line of demarcation between them.

This, however, is not the fact. An entity is distinctly an animal or a human. The intelligence of a living entity is conditioned to a rational and moral development, or it is not. No animal infant has human capacities of mind or of conscience. No normal human infant lacks those capacities.

Science, therefore, is called upon to analyze human intelligence in terms of human nature rather than in those of animal nature. It is asked to give an intelligible explanation of what we term "Reason" and what we know as "Intuition."

This it does in accordance with its knowledge of two material organisms and two material planes of vibratory action.

As already explained, every individual earthly Ego inhabits and operates two material bodies, the one physical and the other spiritual. As already explained, each body is provided with sensory organs adapted to the vibrations of matter upon its own plane. It is through these two material instruments that the third and highest entity, the intelligent soul, operates for the acquirement of knowledge and the attainment of happiness.

When the Ego operates upon the physical plane it must be guided by the physical organs of sensation. It must depend upon the reports which are conveyed to it through the medium of the physical nerves and the physical brain. When operating upon the spiritual plane the intelligent Ego employs the spiritual organs of sense and depends upon the reports conveyed through the spiritual nerves and spiritual brain.

When the physically embodied man uses the knowledge he has independently acquired by physical means as a basis of action, he is said to exercise Reason. When he uses the knowledge acquired (unconsciously) by spiritual means he is said to employ Intuition.

Thus, Reason has to do with the conscious acquirement and exercise of knowledge upon the physical plane, while Intuition has to do with the unconscious acquirement of knowledge upon the spiritual plane.

Intuition is the universal spiritual faculty of all physically embodied intelligent entities. Reason, however, is an individual power which distinguishes the intelligent activities of man alone. Rational Intelligence is the peculiar capacity of the soul endowed human.

The activities of man are, therefore, based upon both Reason and Intuition. It is man alone who operates intelligently through both the physical and spiritual organs of sensation. Only the human physical brain registers the vibrations of physical matter with sufficient exactness and nicety to enable the Ego to form independent (or rational) judgments concerning those vibrations, and to shape his course accordingly.

All men and all women have these double capacities of Reason and Intuition. That is to say, both men and women may enjoy spiritual intuitions and may form independent rational concepts. This returns us to that fundamental principle of sex which conditions man to develop the rational powers of intelligence more forcefully, and conditions woman to greater reliance upon her spiritual faculties.

The coarser the physical body, the coarser the spiritual body. The finer the physical body the finer the spiritual body. The coarser the physical body the less freely do the spiritual sensory organs receive and register the vibrations of spiritual material. This means that coarse physical conditions interfere with the operations of the spiritual faculties.

A physically refined man experiences keener intuitions than the man who is physically gross. This general law aids the ascetic to achieve spiritual powers, such as clairvoyance and clairaudience. By this same general law women, as a class, are naturally more "spiritual" or intuitive than men.

Certain of our modern writers have erroneously explained this operation of the Ego through its spiritual sensory organs, as the operations of a "sub-conscious self." This gives an erroneous impression. There is but one intelligent Ego, one "I," one "Self," operating through two sets of sensory organs. This spiritual process of intelligence is, to the untrained, a super-conscious act; to the trained student, however, the use of the spiritual senses constitutes a perfectly conscious and rational act.

It follows, therefore, that the finer the physical organism of a man or woman, the more nearly it approaches the spiritual plane of vibrations, or, as we say, the more intuitional the individual becomes. We have a common (and correct) habit of referring to persons long ill and to the ascetic and to very delicate persons, as "spiritual." Even the skeptic knows what is meant when we say that such or such a person is "spiritual" in appearance, or in manner, or in character. He could not, however, satisfactorily explain wherein this "spirituality" consists.

Men and women may, at will, cultivate either the rational or the intuitional faculties or both. They may, at will, neglect both. Cultivation of the rational powers to the neglect of the intuitive, sharpens the intelligence in its activities upon the physical plane. It exercises the physical brain and strengthens it in the acquirement of knowledge. Cultivation of the rational powers to the exclusion of the intuitive inclines the individual intelligence to materialism. Dependence upon the intuitive faculties to the neglect of the rational powers sharpens the spiritual faculties, but leaves the physical brain power undeveloped. Dependence upon Intuition alone, promotes superstition. The neglect of both processes of intelligence means savagery and stupidity.

Darwin, himself, furnishes the best possible illustration of rational intelligence, or of intelligence focused upon the physical plane, in pursuit of knowledge of physical things. Here is intelligence operating through the physical senses and the physical only, and refusing all aid and testimony of the intuitions. Here is rational intelligence alone. As a result we have only bald facts of physical Nature, and materialistic dogmas to account for those facts.

Joan D'Arc, on the contrary, represents (during her prophetic life) Intuition. She represents intelligence receiving its impressions and drawing its conclusions from the spiritual side of life only. Here is intelligence wholly unaided by the rational and independent processes of the physical brain. The Maid of Orleans perfectly illustrates the principle of spiritual mediumship.

We have also another and a higher expression of intelligence. Plato, Shakespeare, Emerson and, in fact, all great intelligences who have won the greatest earthly fame as poets, philosophers and divines, have been those who exercised both Reason and Intuition. Such men illustrate the conscious exercise of the rational powers and a super-conscious exercise of the intuitive powers. Intelligence so balanced in its activities, inclines to the higher poetic and philosophic thought.

There is still another and higher order of intelligence possible to the earth plane. This, however, is a distinction very difficult to explain. It is needless to say more than that by a purely scientific course of training (already referred to) a man may exercise his intelligence rationally upon both planes of existence. That is to say, the intelligence of a physically embodied man may self-develop to the point where its processes upon both planes are self-conscious and rational. This illustrates the highest possible development of a physically embodied intelligence. A Buddha or a Christ illustrates such development.

Reason and Intuition represent the struggle of an individualized intelligence which is seeking knowledge of itself and individual happiness upon the physical plane of existence.

Masculine intelligence tends to rational development, while feminine intelligence more generally employs intuitive methods.

These are facts which are noted, but not explained, by modern physical science, nor by any of the modern philosophies. The higher science solves this problem by the same law of polarity which constitutes the text of this work.

As far as finite science penetrates, it finds matter undergoing transformations along intelligent lines. It finds that intelligence is universally positive to matter. That is, matter is universally acted upon by a general intelligence. What is true of universal intelligence is also true of individual intelligence. While general intelligence governs the unconscious operations of man's physical body, it is the individual or psychical intelligence of man which coarsens or refines or strengthens or weakens its material instruments accordingly as its individual methods conform to or contravene general laws.

This brings us to a consideration of that principle which impels masculine intelligence to seek rational development, and conditions feminine intelligence to follow intuitive methods.

It is the intelligent soul of man, rather than the physical body, which is seeking to know itself and to realize its ideals upon the physical plane. It is the positive masculine soul and the receptive feminine soul which are seeking individual and ethical content in accordance with their differing elements and energies.

The Ego in seeking this happiness is continually governed by that law of polarity which conditions it either to aggressiveness or to non-resistance.

All activities are, primarily, activities of the soul. It is, therefore, the soul itself which, acting through its physical and spiritual instruments, gives rise to Reason and Intuition.

Man is the positive and aggressive will force of the human family. He directs his intelligence toward the gratification of that imperious will. He concentrates his powers of intelligence along lines which are to gratify the spirit of conquest. By reason of this he becomes the positive intellectual as well as the positive physical factor in the evolution of man. Everything on the masculine side of Nature exhibits this positive and aggressive energy. Positive masculine force governs generation on the physical plane and also what we term organization and creation upon the intellectual plane. Everything on the masculine side is arrayed for that struggle for supremacy which results in intellectual as well as physical contests.

Moved and inspired by this innate force of will, the intelligent soul of man self-develops naturally along the path of acquirement by force. In fact, the masculine intelligence appears to seek the lines of the greatest resistance. From the beginning masculine intelligence has sought to gratify itself through obtaining supremacy, power and control as among other intelligences.

Acquisition, whether of material things or of power and control, necessitates definite knowledge of that which is to be conquered, acquired or governed.

Necessity for knowledge stimulates the intelligence to an acquirement of knowledge. This necessity for knowledge and the effort to gain it compel concentration of the intelligence. Knowledge so acquired, for immediate material uses, is always practically applied. It is this concentration of intelligence, this acquirement of knowledge and this practical application of knowledge which together constitute the rational process.

Masculine intelligence accepts this physical world and this

physical life as his field of conquest. To this end man concentrates all energies of will and intelligence for the achievement of success among his fellow men. Such conquest and acquirement seem possible only to the masculine half of humanity.

By reason of this forceful pursuit and practical application of knowledge, man has been well named "The Hunter for Truth."

On the other side of Nature, woman represents the principle of intelligent desire. From the beginning, she has directed her intelligence toward the gratification of her pacific nature. She exercises her intelligence along the lines of least resistance. That is, she exercises it through the spiritual faculties and super-conscious processes, rather than by those rational methods which necessitate greater concentration. As a result, the intelligence of woman is more affected by spiritual than by physical vibrations, and by spiritual rather than physical influences, conditions and relations. She centers her intelligence where it enjoys the greater harmonies. This means that she centers her intelligence upon the love relations which are personal to herself.

Thus, she becomes the pacific intellectual, as well as the non-resistant physical factor in the evolution of man. Everything on the feminine side of Nature exhibits this pacific quality in relation to its masculine complement. Everything on the feminine side is conditioned to self-surrender. The feminine half of humanity is not merely dominated physically by masculine physical force, but by the masculine will and force of intelligence. On the feminine side there is no such thing as a struggle for supremacy through force of will. As between women themselves there is no exhibition of that spirit of conquest which men display among themselves and towards women. Feminine nature does not naturally nor voluntarily incline to physical contest. Women are roused to forceful combat only where their personal or love relations are threatened.

By reason of these facts the feminine mind is stimulated to forceful activity mainly in defense of the spiritual or the love relations of life. Her desire for knowledge is not to gain a supremacy among other women. It is rather that she may enjoy the love relations and find self-content in spiritually harmonious conditions. If man is named "The Hunter for Truth," woman may well be characterized as "The Searcher for Love."

From the beginning, feminine intelligence has exercised itself toward the gratification of an absorbing love nature. This object does not compel the same exact knowledge of physical things and physical conditions that man must acquire to accomplish masculine ends and ambitions.

Feminine ambition does not look (primarily) for its satisfactions to material possession nor to temporal power or fame. Seeking love as the first object of life, the individual woman is not induced to acquire practical knowledge as a means to this end. She lacks that stimulating ambition which first calls for, and then maintains, a concentration of intelligence upon the physical side of life.

For this reason woman has not kept pace with man intellectually, but has depended more largely upon the easier (and super-conscious) exercise of the spiritual faculties. She therefore maintains a closer touch with spiritual things and acts from her intuitions from that plane, rather than from rational judgments as to conditions upon the physical plane.

She relies, therefore, upon knowledge gained by the least effort of will. This, in effect, retards rational development and gives to masculine intelligence the greater control of the physical and practical side of life, through his greater knowledge and stronger will.

Spiritual intuitions translate themselves to intelligence as unaccountable impulses and emotions, so that one who acts from Intuition instead of Reason is termed impulsive and emotional. Such an individual is quick to love, to hate, to hope or to fear, but is slow in framing reasons for such impulses and such emotions.

Because of this relative difference in the intelligence, woman comes to be regarded as the emotional factor in the home and in society, and man acquires the distinction of being the more rational element in social development.

Thus, women are looked upon as "sentimental," while men pride themselves as being thoroughly "practical."

Thus, it becomes apparent that this problem of masculine and feminine intelligence is not one of inferiority or superiority. It is merely a question as to difference in kind of intelligence. The distinction between masculine and feminine intelligence is no more an invidious distinction as to woman, than is the difference in their physical organisms. In fact, the difference in the intellectual activities of the sexes rests upon the same principle which differentiates their physical bodies, viz., force of will on one side and desire without will on the other.

This question of woman's mental inferiority is fairly tested in our advanced coeducational system. Women may well turn with pride to their record in such colleges and universities as have been opened to them by the improved masculine intelligence. These institutions do not report woman as the mental inferior of her masculine class-mate. They do not show her as inferior in the acquirement of knowledge. On the contrary, she challenges the young men and frequently wins in the race for honors.

This, however, has to do with woman as an absorber of knowledge, and not as the pioneer in an independent acquisition of hitherto unknown facts. Women as learners and teachers have fully demonstrated their intellectual equality. It remains for them, however, to prove their strength in original acquisition. It remains for woman to show that positive force of intelligence which governs the organization and classification of new facts into definite systems of law, science, art, philosophy and mechanics.

Thus far, it must be admitted, the masculine intelligence almost entirely represents what we define as the creative power in the intellectual world.

It is, therefore, not in degree, but in kind that masculine and feminine intelligence differ. It is not in value, but in force that they are unequal. Though different in kind they must be reckoned as equal in value in the intellectual progress of the world; for when we take into account the ethical value of woman's intuitive intelligence, it appears that the scheme of Nature is to improve the individual man and woman through a mutual exchange of intellectual powers. This continual exchange must be taken as the prophecy of a final perfect balance or completion.

The especial errors of masculine and feminine nature are also referable to the inherent principles of aggression and non-resistance. The sins of masculine intelligence are mainly the sins of *commission,* while those of woman are mainly the sins of *omission.* Masculine will inclines to the tyrannic use of force. Feminine desire inclines to inertia and self-surrender. One is the error of strength, the other that of weakness.

The misuse and misdirection of masculine force are nowhere better illustrated than in his treatment of woman. He seeks to control her not only physically by physical force, but he seeks to control her intelligence by his own force of will. This tyrannic masculine nature exhibits itself also wherever men are able to control other men by force of body and will.

On the other hand, the greatest error of woman is her weakness, her submission and self-surrender to the physical and intellectual tyranny of man.

Prostitution, the venal sin of woman, is the sin of weakness and self-surrender to the aggressive demands of man. Legal oppression and marital slavery were never made possible except by the weak, stupid submission of women themselves.

The non-resistant nature of woman is responsible for another obstacle in the path of her development.

The closer touch of woman to the spiritual plane exposes her to certain dangers which men largely escape by their greater aggressiveness. This danger is an indulgence in spiritual mediumship, for woman's tendency to submission is as much of a temptation on the other side of life as it is on this.

Saint Paul is accused of discourtesy to woman in that he admonished her to silence in the church. The probabilities are that he simply acted upon his knowledge of the then stage of rational development, his knowledge of the emotional nature of woman and his acquaintance with the laws of spiritual control. He doubtless realized that woman was not then prepared to withstand all of the influences she would invite from the spiritual plane. His purpose was probably to protect her from hysteria, mediumistic control and possibly obsession and insanity.

Even after nineteen hundred years women are still so suceptible to spiritual forces that they can scarcely bear the strain of public life. Nervous prostration is the bane of our ambitious women who are publicly contending with the spiritual conditions of our present stage of development.

Even women will admit that it is mainly women who indulge in nervous and emotional conditions. The agonies of hysteria and the ecstasies of religious exaltation are largely given over to women. Even women will concede that their natural tendency is to view life through the emotions rather than through cold reason.

In this connection it must be explained that each sex in itself illustrates infinite gradations of will force. It is true that masculine nature is positive to feminine nature as a whole. Each sex, however, embraces individuals of differing force which would make them positive and negative to one another. There are certain women who, in some respects, are positive to certain men. The union of two such individuals is the most flagrant breach of nature. This means discord without hope of betterment. It means lifelong contempt on the woman's side and unending humiliation to the man. It is the relatively negative of both sexes who become spiritual mediums; of these, however, the proportion of women is much the larger.

The susceptibility of women to a stronger will power obtains in their relations to both planes of life. This is illustrated universally in all normal conditions of human life. In savagery and civilization women generally yield to men in matters of both public and private control of the state and the home. It is the masculine intelligence which furnishes laws, organizes society and enforces obedience. Feminine intelligence universally submits to those laws and sustains social organization by a passive obedience. When the laws of men oppress women, they have heretofore had no recourse except to gradually improve masculine nature through the power of pacific influence. What the woman of the future will do remains to be seen.

Even the high type women of the Anglo-Saxon race yield to the will of men, publicly and privately. These women do not fear physical violence. They yield simply because men have the stronger wills. They yield to a purely intellectual coercion.

On another point germane to intellectual development, men and women appear to be confused. This relates to what we know as "æstheticism" and the "artistic temperament." What constitutes the artistic temperament, what gives rise to æsthetic tastes, and what the explanation of genius, have long been puzzling questions. These are phenomena, however, which are explainable under spiritual laws and principles.

The individual whose spiritual organs of sensation are susceptible to spiritual vibrations, enjoys a super-conscious perception of sound, color, form and arrangement upon the spiritual plane, as well as the conscious perception of such vibrations upon the physical plane. This susceptibility to the higher harmonies of spiritual nature, gives rise to what we term our ideals, and induces the artistic temperament.

Æsthetic tastes are the result of this artistic temperament. Æstheticism is, of course, marked by a greater or less degree of refinement according to the general spiritual development of the individual. When susceptibility to spiritual harmonies includes but one class of vibrations we then have the specialist; that is, a music lover, the lover of color, or one whose tastes tend to form, arrangement and design.

The artistic temperament is very common, and æstheticism is a very general pleasure of the intelligence. There are, however, few geniuses or artists. Art demands a higher order of intelligence than is involved in the mere perception and enjoyment of the beautiful and delightsome in Nature.

A genius is one who has the ability to rationally translate and physically reproduce his spiritual perceptions and intuitions of the beautiful in Nature.

An artist is a genius plus the industry to execute.

Neither æstheticism nor art is the basis of moral rectitude. Neither susceptibility to beauty nor ability to physically interpret such beauty creates moral sentiments. Morality is wholly a question of relationships between intelligent beings. The keenly intuitive are not necessarily of fine moral sense. Indeed, the biographical history of art suggests the contrary.

When an individual of keen spiritual intuitions lacks the ability to rationally translate his "impressions" and "visions" and "inspirations," we have only the person of æsthetic tastes, the appreciator of art and the patron of artists. We have not the composer, the sculptor, the poet, painter nor designer.

It may be safely said that æstheticism is the ruling vice, as well as the ruling gift, of woman. Women, as the more highly spiritualized and intuitional half of humanity, are therefore the more æsthetic portion of society. After woman's love of love, her love of the beautiful becomes her great temptation. Mere æstheticism, unaccompanied by either a high order of intelligence or a keen moral sense, is a misfortune rather than an accomplishment.

If love of love is one factor in prostitution, love of the beautiful is the other. The love of dress, adornment and decoration is responsible for a large share of the immoralities and sins of women. Fashion is essentially the feminine folly that offsets the coarser physical indulgences and sports of men. Women spend as much time, money and energy in satisfying their æsthetic tastes, and in the effort to be beautiful, as men do in the gratification of their coarse physical appetites and passions. Where the one finds pleasure in the adornment of her body and the decoration of her house, the other seeks his satisfaction in the vices of gluttony, drunkenness, gaming and in sporting generally.

It should give women cause for thought, that a large part of the world's industry is a contribution to that character of æstheticism which neither reason nor conscience can commend.

The pursuit of the Good, the True and the Beautiful, is rightly said to be the highest occupation of the soul. These pursuits may well be said to cover the whole field of proper intellectual activity and development. That which is Good has to do with *ethical principles*. That which is True has to do with *scientfic fact*. That which is Beautiful has to do with the *harmonies of material Nature*.

True to the sex principle, the masculine mind more naturally seeks that which is true. That is to say, it seeks that which is scientifically true. Feminine intelligence, on the contrary, just as naturally seeks that which is beautiful. Man, therefore, by a natural law, becomes the utilitarian, while woman as naturally becomes the æsthetic factor in society. These tendencies are curiously illustrated by the criticisms the sexes so freely pass upon each other. Men condemn women for their vanity and emotionalism, for want of Reason, and for lack of practicality and exactness. Æsthetic woman, on the other hand, is daily and hourly offended by the coarse physical tastes and coarser physical habits and the brutal physical sports of men.

It is interesting to note that the average woman is more offended by these things than she is by mere masculine immorality.

Women are in danger of overestimating their own more delicate tastes and habits, attributing them to moral superiority, when the real cause is simply a finer spiritual organization.

If any one doubts this, he has only to study woman of the under world whose love of the beautiful has contributed so largely to her downfall. Woman does not morally improve the world by her æstheticism. Those tastes merely serve to make life picturesque and pleasant and beautiful. The moral ideals of the world do not depend upon color, sound, form or arrangement. They are, instead, the result of right relations between intelligent beings, and a right understanding and practice of the principles of equity, justice, mercy, sympathy and love.

Æstheticism is one thing and ethics quite another. One has to do with the senses, the other with the growth and development of the intelligent soul. The one has to do with harmonious vibrations of matter, the other with principles which govern the relations between intelligent beings.

Woman is a moral factor, but not by reason of her æstheticism. The more sensitive spiritual organization conditions her to receive suggestions made by intelligences upon the spiritual plane more readily than man. To these experiences and not to her æsthetic tastes woman owes her quick intuitions of moral right and wrong. It is her ready, rational and practical application of those intelligent suggestions which constitutes woman the moral strength of earthly society.

Her ready intuition upon moral questions has long been a puzzling characteristic of woman. How frequently women are heard to give an instant decision on a question involving moral principle, which decision she confesses came not by Reason but by Intuition. Trusting to Intuition rather than to Reason, she too often neglects the slower, more difficult, but more independent process of a rational judgment.

A woman's "Because" covers a multitude of spiritual impressions. A woman will act rationally upon a suggestion which she cannot rationally understand nor explain. A woman's "Because" frequently leads to a better decision than a man's independent Reason.

This closer touch with the spiritual side of Nature, therefore, safeguards woman in the ordinary affairs of life, where she is unable to make independent decision. These conditions and experiences equip her with a faith in spiritual things which the cold Reason of man cannot override. She is, therefore, religious by nature, and constitutes the link that binds man to the consideration of spiritual things.

This quicker apprehension of spiritual relations conditions woman to search for beauty, for harmony and love, rather than for material possession and scientific fact.

Physical science notes these fundamental differences in the mental constitution of the sexes. It does not, however, pretend to explain them, except by a general surmise that this force on one side and passivity on the other have been evolved by the requirements of nutrition and reproduction.

Darwin says: "The chief distinction in the intellectual powers of the two sexes, is shown by man's attaining a higher eminence in whatever he takes up, than can woman, whether requiring deep thought, reason or imagination."

This really means that man concentrates and projects his intelligence forcefully in pursuit of knowledge, rather than that he necessarily possesses greater intelligence than woman. Nor does it mean that he necessarily possesses greater "imagination" (spiritual Intuition) than woman. Does it not merely imply that he has acquired a more masterful and forceful style of publicly expressing those imaginations or Intuitions?

He says further: "It is generally admitted that with woman the powers of intuition or rapid perception, or perhaps imitation, are more strongly marked than in man."

He adds: "Some at least of these faculties are characteristic of the lower races and therefore of a lower civilization."

Had Darwin understood the spiritual principles and spiritual facts of intelligent life, he would have known that Intuition is a faculty possessed by animals as well as the lower races. He would have known that what he terms instinct in the animal, is in reality the intuitive process of intelligence. Nor did he observe that the powers of Intuition are as varied as those of Reason. He did not observe that human Intuition covers an immeasurably broader field than animal Intuition. Nor does it appear that he discovered that the intuitions of a high type woman are infinitely keener than those of a woman of low development. His statement, however, as to man's superior rational powers and woman's keener perceptive faculties, is, in a general sense, correct.

History and universal experience support this statement.

History does not furnish a woman who ranks, intellectually with Buddha, Confucius or Zoroaster. There is no feminine Abraham, Moses, Solomon, Daniel or Isaiah. There is no woman who compares with a Christ, a St. John or a St. Paul. No woman's name stands with that of Solon, Pythagoras, Socrates, Plato or Seneca. There is no feminine Homer, Ossian, Dante or Milton. No woman has risen to divide the honors with a Beethoven or a Michael Angelo. The sculptured beauty of ancient Greece was masculine art. No woman has entered the lists with Euclid, Copernicus, Gallileo, Newton, Franklin, Darwin and Edison. There is no woman Shakespeare, Bacon or Blackstone.

In the realm of religious reformation and enlargement, masculine intelligence has led the way. It was the intellectual force of a Luther, Calvin, Knox and Wesley which led the organized bodies of religious faith to higher levels of spiritual liberty.

Speculative philosophy has gained nothing from feminine intelligence. Schopenhauer, Kant, Hamilton and Spencer have no feminine counterparts. By reason of her innate nature and keen spiritual perceptions, it could not be expected that there will ever arise a feminine skeptic to rank with Paine, Voltaire and Renan.

Even in the fields of poetry and fiction, the peculiar domain of the spiritual perceptions, man continues to lead. Hugo, Scott, Bulwer, Balzac, Thackeray and Dickens, have no feminine rivals in the fields of imagination. George Sands, George Eliot and Mrs. Browning are, perhaps, the most forceful women writers who have thus far appealed to public favor.

Thus, in every department of intellectual development, poetry, religion, science, art, philosophy, law and literature, masculine intelligence dominates.

The higher science traces these mental differences and inequalities of achievement to the universal principles which govern sex itself, viz., the principles of force and non-resistance.

All that has been said is not meant to brand woman as a mental inferior. There can be no question of superiority between two indispensable principles of intelligence. These tremendous facts of intellectual development mean something, and that meaning lies in the essential nature of man and woman themselves. What these illustrations imply and all that they imply, is the fact that there is an inequality of will force between man and woman. By this inequality one is conditioned to more aggressive exercise of body, spirit and intelligence than the other. This inequality of will displays itself in the intellectual world just as inequality of strength displays itself in the physical world.

The intellectual evolution of man and woman is an expression of the positive and receptive energies and activities of all the life elements. The individual man and woman represent the accumulated gains of all evolution. The individual man and the individual woman, as the independent exponents of Reason and Intuition, represent the ascent of intelligence through all of the life elements.

Man as individual will, and woman as individual desire, constitute the highest creative powers of intelligence upon the physical plane.

Man as generator and woman as nourisher of life represent the operation of the lower life elements, but man as Reason and woman as Intuition stand at the apex of intellectual evolution. Man as "The Hunter for Truth" and woman as "The Searcher for Love" together exemplify the highest uses of intelligence upon the physical plane.

This characterization of man as Reason and woman as Intuition, is not meant to imply that men are without Intuition nor that women are without Reason. Nor is it meant that man alone develops the rational powers, nor that woman alone develops the love relations.

What is meant, and all that is meant in this connection, is to clearly state those inherent principles of masculine force and feminine receptivity, which, in the higher development, assign man to the more forceful exercise of intelligence and conditions woman to an absorbing rather than a creative character of intelligence.

While it is true that the masculine mind inclines to rational methods, yet religion, art, poetry and romance embody the spiritual perceptions of men. It is also the spiritual nature of man which rises above lust to love, and co-operates with woman in the home relations and in the social philanthropies.

It is true that feminine nature relies chiefly upon spiritual intuitions. It is equally true that woman's ability to reason enables her to practically apply those intuitions. Her rational powers enable her, when she will, to receive and absorb knowledge, to co-operate intellectually with man, and to reconstruct and reform society. It is, indeed, woman's rational application of spiritual perceptions that yield her an indisputable influence in the world.

It was inevitable that masculine intelligence should seek conquest and achievement by force. It was inevitable that the feminine intelligence should seek to accomplish by surrender. It was inevitable that one should look to material acquirement, to temporal power, to knowledge and fame for happiness. It was just as inevitable that woman should look to æsthetic pleasures, social harmonies and the love relations of life for her happiness. It is, therefore, clear that the masculine mind should lead in rational development and that the feminine mind should more readily determine spiritual relations and ethical principles.

It is natural that man should become the master of this material world, thus contributing to the comforts, ease and pleasures of life. It was just as natural that woman should maintain those spiritual principles upon the physical plane which give sweetness and beauty and value to living itself.

Thus, it appears that spiritual principles and not physical circumstance, accident and hostility, account for this divergence of masculine and feminine intelligence. It is not the struggle for nutrition that evolves intelligence. It is not the struggle for reproduction that evolves love. It is not enforced competitions that create masculine nature. It is, instead, the forceful life elements which give rise to competition and constitute man the conqueror of material things, the controller of temporal affairs, and the developer of rational intelligence. It is not enforced sacrifices that create feminine nature. It is, instead, the inherent elements of receptivity which constitute woman the sacrificial physical factor, the pacific intellectual power, and the developer of social ethics.

These, then, are the differing activities of sex intelligence, which are seeking equalization and completion in each other.

The present is an epoch in the history of intellectual evolution. There are to-day indisputable evidences of that principle which is forever seeking to equalize and harmonize all positive and receptive elements and energies.

At no period of the world's history has the intellectual relation of the sexes been of so much importance. At no period has there obtained such equality of intelligence between them. At no previous period has the best masculine intelligence given such rational consideration to spiritual things. Never before in history has feminine intelligence attained such power by purely practical and rational methods.

The most highly developed men of the superior races mark that point of development where the masculine nature is modified and refined through accelerated spiritual perception. On the other hand, the most highly developed woman of to-day marks that evolutionary stage where feminine nature is strengthened to independent and rational methods of thought and action.

The so-called "new woman" is only the universal woman with a stronger will, better controlled emotions, better reasoning powers and a larger knowledge of herself and the world.

Thus, the best manhood of to-day, without losing its essential character as force, is softened to an appreciation of the spiritual principles in Nature. The best womanhood, without losing its essentially pacific nature, is raised to the possibility of more forceful activity.

Man, without losing his will to conquer, acquire and achieve, is able to perceive that the highest achievement lies primarily in perfect relations of individual life. Woman on the other hand, without losing her desire for love and personal happiness, is risen to a rational consideration and an altruistic interest in the public progress and in the love relations of all human life.

The best womanhood of to-day is arrayed for a peaceful crusade. That crusade is conducted in the name of education, industry, art and of equality, altruism and love. She thus moves into higher activities while maintaining the feminine principle of accomplishment by pacific methods. She stands for arbitration, not for war; for principles and not for policies. She puts questions of morality before questions of expediency. She seeks domestic equality rather than political power. She stands for mercy as well as for justice. She relies upon the power of love rather than the power of legislation. She advocates advancement by reciprocity and not by subjugation.

In brief, the best womanhood of to-day represents the search for love conducted by the light of reason as well as by the impulses and emotions.

Nor must it be imagined that feminine intelligence is not power. The history of intellectual development is the history of pacific influence modifying and overcoming the brutalities of intellectual force. Intellectual development necessitates this receptive element even as it does the aggressive. Nature has decreed this complementary and co-operating struggle as the price of a final perfect balance and completion.

Masculine intelligence organizes, while feminine intelligence maintains those organizations. While it is the masculine mind which evolves law, it is the feminine intelligence which preserves law by non-resistance. Receptively is not negation. A pacific intelligence is not lack of intelligence. The power to absorb is as distinctly a power as that aggressive energy which creates and organizes. Man does not merely act upon feminine nature. He receives as well. He does not merely exert force, but receives that which modifies force and gives rise to new ideas. The relation of man and woman is not merely that of aggressive will acting upon passive desire. It is rather a relation in which will is softened by desire and desire is strengthened by will.

It is true that the positive never becomes the receptive, even in intellectual life, nor the contrary. Each, however, receives from the other that which adds to or strengthens the weaker part. The savage will of man is slowly tempered and raised to manly courage. The stupid desire of the slave is slowly strengthened and raised to womanly tenderness and grace. Man loses his ferocity, not his force. Woman loses her stupidity but not her pacific nature. It is, in fact, these ceaseless co-operations which spiritualize the masculine mind and rationalize the feminine.

In this stupendous scheme of physical, intellectual and moral evolution, these dual principles and powers are absolutely interdependent. Men, cut off from natural association with women and grouped by themselves, degenerate and revert to the primitive state of lawlessness and ferocity. Women, denied the natural association with men, degenerate with equal rapidity into physical and intellectual passivity and inertia.

Nature proves:

(1) That the secret of material conquest and rational development is force of intelligence.

(2) That the secret of spiritual development is receptivity of intelligence.

It is not aggression alone nor receptivity alone that evolves.

It is not will alone nor desire alone that generates and reproduces. It is not Reason alone nor Intuition alone that educates and develops.

Aggression without receptivity is mere destructiveness. Will without desire is waste. Reason without Intuition is intellect without inspiration or ideal. On the other hand, receptivity without aggression is stagnation. Desire without will is impotency. Intuition without Reason is impulse without direction or purpose.

Evolution involves, not merely force, but absorption of force. Generation involves, not merely the will to generate, but the desire to nourish. Intellectual development involves not merely the activity of Reason, but Reason illuminated by the activities of Intuition. Reason without inspiration is hardening. Intuition without judgment is disintegrating. Rational conceptions without spiritual perceptions engender skepticism and dogmatism. Spiritual perceptions without rational conceptions mean superstition and dogmatism.

It is only an equal development and harmony of these intelligent powers which constitute the properly balanced individual.

A well-balanced intelligence must know the uses of intelligence upon both planes of existence. The attainment of such a state is more rapidly effected through the perfect individual relation of man and woman.

The true intellectual relation of man and woman is that of master and pupil, and this is the universal ideal which intelligent men and women have always cherished.

This statement is made without fear of challenge from the highest type of either sex. Man as master and woman as pupil of the master, is the relation that every man seeks and every woman craves. This is the relation which gratifies masculine pride of intelligence and furnishes woman the intellectual strength upon which she loves to lean.

Any other relation means disappointment, humiliation and discord.

The man who find himself mated to either an intellectual superior or one who disregards his opinions, is humiliated and disappointed. On the other hand, the woman who binds herself to a mental inferior is equally disappointed and disgusted.

In the individual relation, as well as in the general work of the world, the masculine mind should be the pioneer. This demands that force which goes into unexplored regions, which conquers obstacles, collects new material and organizes and arranges its facts systematically. In the individual relation the feminine mind should be to the masculine just what the universal intelligence of woman is to the universal intelligence of man.

This means that it follows the pioneer, and absorbs, digests and utilizes that which has been collected, classified and systematized.

This is the one and only relation between man and woman which gives dignity and value to earthly life and absolutely satisfies both Reason and Intuition. This is the one relation which outlasts physical life and perpetuates its conjoined activities in a higher world than this.

The radical "new woman" illustrates the sex principle intellectually as clearly as does the conservative "old woman." Her vehement demands for equality are misunderstood. She is not asking that law shall abrogate the decrees of Nature by adopting the same measures and the same standards for both men and women. This demand for equality is, at its foundation, merely the demand for equality of opportunity and for recognition in those enterprises and activities which she feels competent to enter. No true woman would repudiate the natural leadership of strong and masterful men. The protest is not against such men. It is rather against the dictation and domination of men whom she knows to be her intellectual, as well as moral, inferiors.

The protest of the "new woman" is, in fact, a protest against the lamentable scarcity of masters.

This highest relation in which man and woman are to realize the ideal is being slowly but surely wrought out by time.

The intellectual dependence which man seeks from woman, and the intellectual equality of which woman dreams, are one and the same thing, and are already foreshadowed in the higher races. The processes of the higher evolution are slowly but surely adjusting these two powers into perfect balance.

This intellectual and moral interdependence, reciprocity and companionship of two intelligent beings, is the highest ideal of which the human mind is capable, and a philosophy which conserves Nature declares that this ideal is already working itself out as a very practical, tangible, earthly reality.

This ideal is realized wherever the principles of aggressive and receptive intelligence strike the true balance in any individual man and woman.

Every woman acknowledges that principle when she finds the Master.

CHAPTER XV.

THE SPIRITUAL BASIS OF LOVE

"All the laws of Physical Evolution cannot explain the first genesis of mind."

Thus declares the reviewer of "Evolution" in the Encyclopedia Britannica.

If this applies to mind, how much more forcefully it applies to the phenomenon of love.

By love is meant that highest form of intelligent attraction and intelligent sympathy which exists between man and woman, between children and parents, and between relatives and friends.

This term also includes those peculiar individual sympathies and affections which exist between animals in their own relations. It also includes those sympathies which bind them to human beings.

Those preceding chapters on the genesis of physical life, the basis of evolution, and the natural law of selection, go to show that individual life and individual intelligence come into the world by the operation of a universal law of affinity, or the law of correspondence between positive and receptive energies.

This chapter is intended to show that love comes into the world by the same intelligent spiritual principle. The orderly and sequential development of love under this spiritual law of affinity, constitutes the Harmonics of Evolution.

The great common sense of the world is beginning to demand of science that it shall give an account of universal phenomena which shall accord with universal experience and with the common impulses, aspirations and ideals of mankind. Every intelligent student and observer of life knows that love is a universal phenomenon attaching to intelligent, sentient life under normal conditions. Animals as well as humans love and seek love in others of their kind. Animals as well as humans may entertain sex love that is not physical passion. Animals like humans may love their offspring, may experience friendship for others of their own species, or of other species. They may love human beings.

Not man alone, but the animal also, is engaged in that universal struggle for individual adjustment and an ethical content which depend primarily upon the love relations.

Physical materialism no more accounts for love than it does for the genesis of life or the individualizing of intelligence. Feeding, breeding and battle no more account for love than they do for matter, motion and space. It undertakes, however, to account for it. That is to say, physical science claims to have at least discovered the basis of love between man and woman. Familial love, it casually explains as a matter of physical inheritance, or a habit. Friendship is translated as self-love or love of approbation. As to altruism, physical science frankly confesses that it has no theory. So convinced was Darwin of its abnormality that he deplored that character of philanthropy which cares for the unfit children of men. He saw in this a direct defiance of Nature's primal purpose, viz., preservation of species through a battle of the strong against the weak.

Physical science, however, does claim to have found a natural physical basis for that love which exists between man and woman, and whose power and influence give life its permanent values. Scientific skepticism finds the basis of this character of love in an "overshadowing instinct for reproduction."

Darwin's contribution to science is unquestioned. His facts are of tremendous value. In theory alone he is disappointing. It is not physical fact, but intellectual dogma, that has degraded the sex relation to a "biological need."

It is, therefore, not fact, but mere opinion, that the higher science encounters in a discussion of this question.

The position of physical science as to the origin and nature of love, will be better understood by direct quotations from recognized authority. Such authority declares:

(1) "Man is a mammal like any other, and only distinguished from the animals of his class by a greater cerebral development."

(2) "Generation is the outcome of nutrition."

(3) "The tyrannic need of procreation is the overwhelming and overshadowing principle of individual human development."

(4) "The institution of marriage has had no other object than the regulation of sexual unions. These have for their aim the satisfaction of one of the most imperious biological needs—the sexual appetite."

(5) "The prime cause of marriage and the family is purely biological; it is the powerful instinct of reproduction."

(6) "If we are willing to descend to the foundation of things, we find that human love is essentially rut in an intelligent being."

Here, in brief, is the ultimatum of physical science as to the love principle in Nature. Thus is man reduced to the level of the animal. Thus is he accepted as a result of physical functions.

Not physical facts of Nature, but Darwin's interpretation of those facts, misled even so great an intelligence as Häekle. Here was a man of keen spiritual intuitions as well as of rare rational powers. It was always under protest that he forced his reason to accept what intuition denied. The result was, a theory as to the basis of love which satisfied neither Häekle himself nor convinced the world.

When contemplating the sex attraction of two microscopic cells and the transcendent powers of love in the human world, the great naturalist is led to say:

"We glorify love as the source of the most splendid creations of art; of the noblest productions of poetry, of plastic art, and of music; we reverence in it the most powerful factor in human civilization, the basis of family life, and, consequently, of the development of the state; * * * so wonderful is love, and so immeasurably important is its influence on mental life, that in this point, more than in any other 'supernatural' causation seems to mock every natural explanation." Häekle also says: "On the other hand, we fear love as a destructive flame; it is love that drives so many to ruin; it is love that has caused more misery, vice, and crime than all other calamities. * * * Notwithstanding all this, the comparative history of evolution leads us back very clearly and indubitably to the oldest and simplest source of love, to the elective affinity of two differing cells."

In a certain sense Häekle was right. The principle of elective affinity is, in fact, the true love principle. The naturalist, however, drew his conclusions without any knowledge of the spiritual side of Nature. To him this conjugation of cells was therefore a physical act only. Love was therefore to him the efflorescence of that physical affinity. Moreover, had he possessed actual knowledge of the true love principle, he could not have confused the constructive nature of love with the destructive nature of uncontrolled physical passion.

On the contrary, he would have realized that it is lust and not love that drives to ruin. He would have known that a principle of harmony never yet engendered either misery, vice or crime.

With what satisfaction such an intelligence as that of Häekle had pursued the higher science. What pleasure such a soul would have in the analysis of a principle which accounts for love rationally and in accordance with his highest intuitions. He would have then understood that there is nothing supernatural in love, nor yet anything that seems to degrade our highest ideals of it. On the contrary, an actual knowledge of the principle of elective affinity would have revealed a natural and purely scientific pathway of love, from its lowliest point in atomic activity to its summit in the life of the soul.

The Darwinian theory has brought confusion and dismay to many other honest and intelligent minds during the past half century. Unable to dispute his facts, they have felt compelled to accept the theories against which every intuition rebelled.

As long as the mind confines itself exclusively to the operations of the physical functions, those theories appear logical. The moment, however, that the mind is diverted to other equally palpable facts, the commonest facts of intelligence, morality and love, that moment physical materialism fails and we must construct other theories to account for things.

The conflict between reason and intuition is never so sharp

nor so clearly apparent as in the case of those scholars who accept Darwinism with such unmistakable loathing and humiliation. The man who stops for a moment to consider the motives and ideals of his own life, can scarcely bring himself to believe that those motives, ideals and aspirations had their root and nourishment in the absolutely unconscious and purely selfish demands of nutrition and reproduction. He cannot bring himself to feel that what he cherishes as his individual power and capacity, and what he experiences as his individual ambition or patriotism or love or altruism, is the mere efflorescence of a blind digestive apparatus, or a blind procreative passion. He cannot believe that all he is—as an individual—is but an ephemeral combination of matter, an infinitesimal physical contribution to species.

The higher science formulates two grave charges against scientific skepticism, viz.:

(1) That it obscures the Individual in Nature.

(2) That it levels intelligent love to "the instinct of reproduction."

Though physical science asserts it as a fact, it is not able to show that man, even physically, is an animal only removed from the ape in point of time and additional feeding and breeding. Physically, man is an animal. Structurally, he is related to the ape. However, he is not an ape either physically, structurally, mentally or morally.

Though the ape appears to be a rudimentary man, he is not a man. There are differences physiologically, as well as mentally and morally, between the highest ape and the lowest human thus far discovered.

It must be remembered that actual proof of Darwin's theory rests upon this still "missing link." Until this is found, the whole elaborate theory means nothing in science. Even in a physical sense Darwinism fails to bridge the gulf between animal and man.

Since the writings of Darwin physical science has discovered a most interesting and important fact which bears directly upon this point. It is now discovered that the brain measurements of men and monkeys disclose a radical difference in actual quantity. Between the highest type monkey and the lowest type man the balance of actual brain matter is largely on the side of man. The ratio of difference is about 60 to 100.

More than this, man, one of the weaker mammals, is born unclothed, and practically defenseless from the elements and the stronger animals. From birth to death man is the one being who must live, attain, enjoy or suffer in the exact ratio of his own independent and rational self-development.

The resemblances, however, between the higher apes and low type men are distinct enough to logically show the physical relationship. If we were to compare an ape and a man, merely as a physical organism, the Darwinian theory would have much weight. The movement, however, that we compare an ape and a man as individual entities that moment the theory fails.

The distinction between man and animal is the absolutely impassable gulf of rational and moral capacity.

Thus far finite science has discovered no natural link between an intelligent being endowed with the higher capacities of the soul element and one that is not. In the very lowest reaches of human society the child is born with capacities which no animal possesses, viz., capacities for rational and moral development. In this particular the lowest human transcends the highest animal just as the nucleated life cell transcends vegetable substance in its energies and capacities.

The demarcation between the lowest human and the highest animal is even more distinct than that between the lichen and the rock, or between low animal forms and certain vegetable growths.

It is true that an undeveloped human resembles an animal. It is also true that he may live on indefinitely, looking mainly to animal appetites and passions for his satisfactions. Indeed, the neglect of the higher nature reduces man below the animal, so far as external conduct and habit are concerned. The very fact that he does possess the power of individual reason, means that he has the capacity to disregard general principles, choose his own course, and thus apparently fall below the level of the brute.

Nevertheless, the low type human—unlike the ape—may at

any stage of beastliness and degradation, rise from that condition to one of rational and moral life. By force of the inherent powers of the soul he may at any time abandon animalism and assume the higher rôle of the human. Proofs of this are those facts which show the rapid development of low born children under civilized systems.

If man were, indeed, merely an improved animal, two conditions would obtain:

(1) The earth would teem with hybrids, physical, mental and moral. There would exist an infinite series of experiments between apes and men, entities which could be classified as neither animals nor men.

(2) Low type men could no more be suddenly raised to rational and moral standards by highly developed systems, than could the ape.

Neither of these conditions obtains, but the reverse is the fact.

Physical science has no difficulty in distinguishing men from apes. One is distinctly human and the other is distinctly animal. No amount of culture will raise the chimpanzee to the rational and moral plane of man. Every normal human infant, however, is susceptible to both rational and moral development.

For illustration, negro children, offspring of the lowest full blood Africans, show remarkable development under a system of education. That advance is as marked physically as it is intellectually and morally. The first generation shows an improved head, enlarged brain and modified features. The physical animal resemblances are wonderfully modified, while the mental and moral superiority to low type parents appears magical.

This change, however, physical, mental and moral, is not due to education alone. It is due to the latent capacities of the intelligent soul. Education merely develops those latent possibilities of the soul. It does not create the soul itself.

By reason of this higher element the energies and activities of man are fundamentally superior to the energies and activities of the animal. The Darwinian theory lacks the one important link which alone could verify its elaborate speculation. That link is still missing. Neither physical science nor any other science has discovered or will discover such link between man and the ape. It does not exist in this world nor in the next.

Does it not seem singular that a science which recognizes the fundamental differences between a rock, a tree and an animal, fails to recognize the same fundamental differences between a Plato, a Shakespeare, a Darwin, and the ape?

A science that does not take into account the psychical element in man, necessarily analyzes him as it would the pure animal. It is therefore bound to closely relate the highest activities and attributes to man to the activities and attributes of the animal. Interpreting man through his physical functions, means the interpretation of love, as well as intelligence, through the same causes. Physical science, from its present point of view, can do no better than to define love as "essentially rut in an intelligent being."

Nature, however, contravenes this position. Science that deals with two worlds of intelligence, love and ethical activity instead of one, arrives at a far different conclusion. The higher science which takes into account the fact of love in two worlds, holds this phenomenon to be the expression of a spiritual principle and not that of a physical passion.

The differences and distinctions of the two schools will be better understood by sharp contrasts of their basic propositions. Physical science, as already stated, lays down as fundamental:

(1) That man is a mammal like any other, only distinguished from the animal by greater cerebral development.

Nature, on the contrary, demonstrates that man is a mammal, but fundamentally different from all others. He is distinguished from the animal by one additional element which engenders capacities, superior to those of any and all animals.

(2) Physical science claims that generation is the outcome of nutrition.

Nature, however, demonstrates that generation is the outcome and incident of the universal struggle for vibratory correspondence, which is the struggle for self-completion.

(3) Physical science declares that the tyrannic need of pro-

creation is the overwhelming and overshadowing principle of individual human development.

Nature, on the contrary, demonstrates that the natural necessity for self-completion is the overmastering and overwhelming principle of individual human development.

(4) Physical science declares that the institution of marriage has no other object than the regulation of sexual unions.

Natural Science, however, shows that the institution of marriage has for its object the rational regulation of the sex relation as a necessary part of the struggle for completion.

(5) Physical science declares that the prime cause of marriage and the family is purely biological. It is the powerful instinct of reproduction.

Nature, however, shows that the prime cause of marriage is spiritual rather than purely biological; it is the powerful impulse for vibratory correspondence or for sympathy.

(6) Physical science declares that if we are willing to descend to the foundation of things we find that human love is essentially rut in an intelligent being.

Natural Science, however, finds that if we are willing to investigate the spiritual principle of affinity, we find that human love is essentially an harmonic relation established between two intelligent beings.

Thus, while agreeing as to physical fact, it is yet possible for scholars to differ so widely as to formulate philosophies diametrically opposed. It must be remembered, however, that the one school bases its philosophy upon physical facts alone, while the other takes into account both physical and spiritual facts of Nature.

The higher science, taking into account as it does both spiritual and psychical principles, as well as physical functions, finds that love is an expression of the same principle which refines matter, increases vibratory action, generates life and individualizes intelligence.

The phenomenon of love which obtains among intelligent entities, is therefore as universal in principle and as normal in expression as are matter, motion, life and intelligence. It does not matter whether that love be manifested by animal or human, or whether expressed as sex love, or familial love, or as friendship.

Love is the sensation and emotion which accrue to conscious intelligence when vibratory correspondence obtains. It does not matter whether this affinity occurs in the higher or lower kingdom of intelligence, whether as between man and woman, between parents and children or between friends.

When an individual intelligence observes this principle as molecular action, he talks of the "law of vibration." When, however, he himself feels and enjoys the operation of that principle between himself and another individual, he calls that experience love.

Chemical union represents merely the vibratory correspondence of two material particles or atoms, while love and marriage represent the vibratory correspondence of material organisms, but they also represent the mutual sensations and emotions of two intelligent beings. The difference in effects between mineral marriages and human marriages can only be measured by the difference between the energies and capacities of unconscious atoms and those of self-conscious intelligent souls. Sex Love, humanly speaking, is a complex activity in Nature. It embraces first, all of the unconscious or involuntary affinities of the lower elements; and second, it includes the direct self-conscious impulses, sensations and emotions of individual intelligence. These impulses, sensations, emotions and satisfactions of love, are simply the values which accrue to an individual intelligence as it cooperates with the universal law of affinity.

Nothing but a rational contact with the spiritual plane and a rational study of life upon that plane, enables the investigator to comprehend the nature of that activity defined as love.

Love is an activity of individualized intelligence. It is therefore an activity common to both animals and humans. Love, however, in animal and human life, represents widely different values. Animal love rises no higher than the potencies and capacities of animal nature. Human love, though an expression of the same principle, is an activity augmented, ennobled and illuminated by the potencies and capacities of the higher soul element.

The fact that animals do love their mates, their offspring, and those of their own and other species, must be accepted as proof that love is a possibility and normal accompaniment of all conscious intelligent life. When the psychical energies have been added to animal passions, the love nature rises correspondingly in power and scope. The capacity for loving is increased just as the measure of ethical enjoyment is enlarged.

More than this, the individual love relation of man and woman embraces greater ethical possibilities than any other love relations in the human family. This is true by reason of the fact that only between individuals of opposite polarity can there exist an affinity and union which may be at once physical, spiritual and psychical. Nor are such close harmonies possible even spiritually and psychically except between representatives of positive and receptive elements. Men and women who have vainly sought for ethical content in children or in friendship, will sooner or later confess this law.

When the scientific investigator proves the fact of life after physical death, he proves that man is a spiritual being with intelligent occupations, as well as a mammal with physical functions. When he finds that rational and moral development continue in that other life, he proves that there are other generative agencies in Nature than the organs of digestion and reproduction. When he discovers that the spiritual world is inhabited by men and women, he discovers that sex is a universal spiritual principle, instead of a biological need. When he finds that men and women seek each other in that spiritual world with the same definite, exclusive desire they do here, he then realizes that sex love represents a higher necessity than that of reproduction.

On the spiritual side of life men and women seek each other with even greater desire than they do here. They seek each other, however, in response to the passion for intellectual companionship, and not from the passion of lust which too often takes the place of intelligent love in human marriages.

In that world, therefore, as in this, the same exclusive character dominates the love relations of the sexes. It is not *any* man or *any* woman, but the *ideal* man or *ideal* woman that the spiritual lover seeks.

It may be a comfort to those who are unhappily mated here, to know that spiritual life not only gives freedom, but equips the individual to much more easily form happy relations. It must be remembered, however, that a proper discharge of earthly obligation is in itself the development which best fits the spiritual man or woman to form higher and happier relations in spiritual life.

Whoever is a frequenter of the seance room is apt to gain information (and frequently misleading suggestions) concerning this spiritual law of affinity. No matter how depraved the spiritual life may be, the physically disembodied soon learn that the sex relation depends upon that principle of affinity.

The efforts of the low grade intelligences upon the spiritual plane to interpret this law through the medium of the seance room, have given rise to many serious misconceptions and are responsible for much immorality.

These are some of the facts which the higher science interprets to mean that sex is an eternal principle, that love is a spiritual activity of intelligence, and that marriage is a spiritual and psychical necessity as well as a biological need.

Thus, Nature brings a message to man that satisfies his intelligence and inspires the soul to still higher effort. In such a reading of nature, the individual man finds compensation for the struggle for nutrition, and the individual woman finds reconcilement to the sacrifices of reproduction. This reading of Nature and this alone dignifies the sex relation, gives value to individual existence, and explains human love as an intelligent need of the soul, and not a blind lust of the body.

Mr. Drummond differs radically from Darwinism as to the origin of love. The moralist believes he has found the physical basis of love in the enforced pains, penalties and sacrifices of maternity. The moralist, however, entirely agrees with the materialist, that sex is merely a physical device for reproduction. He agrees that the love relation of man and woman is, at the root, "a physical passion miscalled love."

Mr. Drummond, however, does believe that there is love in the world which is not physical passion, nor related to it, nor anything like it. He defines love as a phenomenon essentially made up of the virtues of pity, compassion, patience and self-sacrifice.

With this as a starting point, he next declares that love originated in the physical disabilities of the female half of life. And he finally claims that by reason of this fact, "love comes into the world at the point of the sword."

Briefly summarizing this new moral philosophy by brief quotations, we are told:

(1) "Everything in the moral world has a physical basis."

(2) "Man progresses, not by any innate tendency to progress, nor by any energies inherent in protoplasmic cells from which he first set out, but by continuous feeding and re-enforcement from without."

(3) "Life is controlled by its functions. Living out these is life. The activities of the higher life are determined by these same lines."

(4) Sex affinity "may be the physical basis of a passion which is frequently miscalled love * * * but love has come down a wholly different line."

(5) "Sympathy, tenderness, unselfishness, and the long list of virtues which go to make up altruism (or love) are the direct outcome and essential accompaniment of the reproductive function."

(6) "The goal of the whole plant and animal kingdom seems to be the creation of a family."

(7) "Before altruism was strong enough to take its own initiative necessity had to be laid upon all mothers to act as they did."

(8) "The object of Nature is to turn out mothers."

(9) "Man fulfills his destiny in the struggle for nutrition, woman fulfills hers in the industries and activities of home and in paying the eternal debt of motherhood."

(10) "Love came into the world at the point of the sword."

Thus, a new moral philosophy, attempting to build upon physical materialism, brings no loftier message to the world than the system it partly rejects.

It is with a feeling almost akin to resentment that intelligent minds, especially among women, follow this theory to the end.

Sex, "a physical device for reproduction;" sex love, a "physical passion;" the object of evolution, the "manufacture of mothers;" man's individual destiny, "fulfilled in the struggle for nutrition;" woman's individual destiny, "fulfilled in paying the eternal debt of motherhood;" the goal of individual life and endeavor, "the creation of an improved family;" competition and sacrifice the universal principles of progress!

This, in brief, is the message brought to the aspiring soul of man by the speculations of a scientific theologian.

The differences and distinctions between Mr. Drummond's philosophy and the higher science are more easily appreciated by sharp contrast of their basic principles.

Moral philosophy, based upon physical materialism, assumes:

(1) That everything in the moral world has a physical basis.

Nature, on the contrary, demonstrates that while everything in the physical world has a spiritual basis, everything in the moral human world has a psychical basis.

(2) The moralist assumes that man progresses, not by any innate tendency to progress, nor by any energies inherent in protoplasmic cells from which he first sets out, but by continuous feeding and re-enforcement from without.

Nature, however, demonstrates that man progresses by reason of inherent spiritual energies in protoplasmic cells from which he first sets out. His progress upon the physical plane is sustained by nutrition, influenced by environment, and perpetuated by reproduction.

(3) The moralist assumes that life is controlled by its functions. Living out these is life. The activities of the higher life are determined by these same lines.

Nature, on the contrary, demonstrates that life is controlled by the affinities which inhere in the vital elements. Responding to these is living. The activities of the higher life are determined by the higher affinities of the intelligent soul.

(4) The moralist assumes that sex affinity may be the physical basis of a passion miscalled love, but that love comes down a wholly different line.

Nature demonstrates that sex affinity is the spiritual principle of polarity which operates as positive and receptive energy, and as centrifugal and centripetal force, in the processes of evolution.

(5) Moral philosophy assumes that sympathy, tenderness, unselfishness and the long list of virtues which go to make up altruism, are the direct outcome of the reproductive function.

Nature demonstrates that sympathy, tenderness, unselfishness and all other virtues which go to make up love and altruism are the direct outcome and essential accompaniment of a spiritual principle of affinity.

(6) The moralist believes that the goal of the whole plant and animal kingdom is the creation of a family.

Nature, however, demonstrates that the goal of all energies and of all activities is the completion of an individual.

(7) The moralist finds that before altruism was strong enough to take its own initiative necessity had to be laid upon all mothers to act as they did.

Natural Science holds that before individual intelligence was strong enough to take the initiative, maternal solicitude appears as an involuntary response to the law of vibratory correspondence or affinity.

(8) The moralist declares that the object of Nature is to turn out mothers.

Nature, however, would strongly suggest that its chief object is to turn out men and women.

(9) The moralist holds that man fulfills his destiny in the struggle for nutrition, and woman fulfills hers in the industries and activities of home and in paying the eternal debt of motherhood.

The processes of evolution, as well as human experience, teach us that man and woman fulfill their destiny, primarily, in that individual relation which governs the highest possible development of each, physically, spiritually and psychically.

(10) The moralist has said that love comes into the world at the point of the sword.

This most serious charge against Nature, Nature itself contravenes. The study of love and the love relations among intelligent creatures, animal and human, clearly demonstrates the fact that love comes into the world by a natural law of affinity, which law becomes in intelligent life the natural law of self-consent.

Thus, Nature, when rightly seen and comprehended, corrects, one by one, the errors of specialists and speculators.

When science discovers that the individual love relations of this life are perpetuated in the higher life, it is very naturally inferred that those relations have their basis in spiritual principles and in psychical or intellectual attractions. When it is observed that the individual love relation of man and woman survives the ties of family and the bonds of friendship, it is taken to mean that this individual, spiritual relation, is of primary importance. It is also accepted to mean that man and woman are fundamentally incomplete in both worlds, except through that individual relation.

As its name would imply, Natural Science investigates and analyzes this love principle exactly as it does the natural law of motion and number, of life, and of intelligence, viz., by study and comparison of its modes, objects and effects in two correlated worlds of intelligent activity.

Such investigation and interpretation incidentally explain maternal love as an expression of the same principle of affinity which governs sex love. From its lowest involuntary activities to its highest voluntary expressions, maternal solicitude and maternal love represent that same natural law of consent which governs sex love and all other love relations.

In human maternity, as in human sex love, we have a relation which may express physical, spiritual and psychical sympathies. In lower life maternal solicitude is an intuitive (involuntary and irrational) response to this universal law of correspondence. In the higher life of intelligence, however, maternal love expresses a self-conscious, voluntary and rational activity of individual intelligence, as well as involuntary affinities and impulses. Maternal love in the human kingdom is, therefore, a spiritual and psychical relation, as well as a physical bond, and represents the higher affinities of spirit and soul, as well as the bonds of the flesh.

Such an interpretation further proves that love comes into the world, not by the physical pains and sacrifices of maternity, nor by force nor compulsion nor sacrifice of any kind or character. On the contrary, this reading of Nature merely discloses maternal love as but one of the expressions of a universal law of affinity. It explains maternal solicitude as merely the natural expression of the receptive feminine nature, and is no more the outcome of maternal pain and sacrifice than masculine aggressiveness is the outcome of the struggle for nutrition.

Though maternal love obtains in the lower kingdoms of intelligence, it is the human mother who enjoys those increased capacities for loving which dawn with the induction of the highest life element, the soul.

Human mother love appears more and is more, in both volume and effect, than the maternal love of an animal. That is to say, the human mother has the capacities for those higher affinities which the animal lacks. A tigress just as truly loves her infant as does the human mother. There exists between them close affinities of both flesh and spirit. But who is there in this age of enlightenment that can fail to note the difference between tiger love and human love.

Physical science has a peculiar way of interpreting "selection" that would seem to imply that somehow both animals and humans, in their individual love relations and maternal activities, are really engaged in the effort to preserve species only.

For instance, we are told that "the preservation of the species was before everything else the object of selection." Also that "the first necessity of societies is that they endure." Again we are told that the price of the endurance and perpetuation of species depends upon the care which mothers bestow upon their offspring. It is therefore concluded that the manufacture of mothers is the first object of Nature.

While this assumes to speak for Nature, how shall the individual impulse and intent of the mother be understood?

In truth, here is where physical science wholly fails. It assumes a purpose for Nature, and in that assumption it loses all sight of the individual.

The maternal love of the tiger is a purely individual emotion. The tigress has nothing but this individual impulse and individual intent to bind her to her offspring. She has no remote intuitions as to the preservation of species, nor of the natural necessity for good mothers. She has not the faintest perception of a moral responsibility to her child. On the contrary, she is bound to it solely by a fierce, irrational affinity or passion. She is bound by tiger love. She is simply impelled, not compelled by that love, to feed, to caress and to guard it.

Unless we hold that love is a mere habit of enforced sacrifice, there is absolutely nothing in physical maternity to create or occasion love. There is, in fact, everything to engender dread, opposition and resentment on the part of an intelligent human being, forced to thus suffer for the race. Love is not a habit. It is not an inherited result of physical discomfort and sacrifice. To associate love with any idea of compulsion, is to entertain an absurdity. The very nature of love forbids the thought. Fear and endurance and patience and self-repression may come by compulsions and sacrifices, but love, never. The attempt to so relate it is an offense against Nature. It contradicts intuition. It confuses reason. It belies experience.

Love is as involuntary as breathing. It is the instant and involuntary response of the individual to the universal law of harmonics.

There is no known principle or process in Nature that can *compel* love between the very meanest of Nature's children. Love is the exact reverse of compulsion. It is the one phenomenon in Nature which, from its lowest to its highest expression, defies every phase of force, whether that force be physical, spiritual or psychical.

Maternal love survives physical death; which proves, as in the case of sex love, that there is involved a spiritual principle and a spiritual relationship as well as a physical relationship.

Science, therefore, assumes that maternity is a spiritual activity of feminine nature which conserves the perpetuation of the race. It is found further, that mother love in the spiritual world, even as it is in the physical world, is an incidental relation of life, rather than the purpose of living.

In that life, as in this, intellectual activities and ethical enjoyments constitute the occupation of intelligent beings. Intellectual ambitions and moral purposes and ethical satisfactions are, therefore, held to be the governing causes in the evolution of man upon both planes of existence.

From the vantage ground of its broader investigation, the higher science is justified in assuming that neither preservation of species, the creation of a family, nor the manufacture of mothers, has been the inspiration of that Great Intelligence which guides the infinite scheme of evolution. On the contrary, all of the laws of Nature on both sides of life, combine to show that the primary object of this Great Intelligence has been the perfecting and completion of the individual man and woman.

Science claims that this primary object of Nature is distinctly foreshadowed in that spiritual principle in Nature which impels the individual entity to seek vibratory correspondence in another like entity of opposite polarity.

Under such interpretation, completion can be effected neither in a physical struggle for nutrition nor in a physical struggle for reproduction. Instead, completion, as designed by Nature and sought by man, involves, primarily, the establishment of a perfect relation between the individual man and woman. In this greatest struggle, therefore, the struggle for self-completion, man and woman pursue their destiny here and hereafter.

Thus, the universal individual is spiritually impelled, and not physically compelled, to love. Correspondences, co-operations and harmonics, and not compulsions, competitions and sacrifices, are the natural pathway of love. Thus, destiny is fulfilled, not in contributions to the body nor to progeny nor to species, but in the highest possible individual relations and attainments. Destiny is fulfilled in intellectual and moral activities, and not merely in following out the lines laid by the physical functions. Individual completion through individual love relations, is the primary object of living.

Thus, the higher science finds a spiritual basis for love, instead of a physical one. It lays down a law of fulfillment in the place of the law of sacrifice. It formulates a philosophy of individual development instead of one of individual repression. This philosophy looks, not to sacrifice and resignation, but to individual happiness, as the goal of intelligent life and endeavor.

Thus, one by one the higher science corrects the errors of physical science, and of a moral philosophy largely based upon those errors.

Thus, by a long series of carefully proved facts, it clearly demonstrates that everything in this physical world has a spiritual basis, and that love comes into the world through this spiritual law of consent, and not through physical compulsions, "at the point of the sword."

CHAPTER XVI.

PHYSICAL SCIENCE CORROBORATES NATURAL SCIENCE.

The purpose of this chapter is to show that physical science unintentionally corroborates those fundamental princpiles laid down in this work as follows:

(1) The evolution of man is based in spiritual principles and forces.

(2) Sex represents the spiritual principle of polarity, or the principle of centrifugal and centripetal force.

(3) The male entity represents the aggressive or centrifugal force, while the feminine entity represents receptive or centripetal force.

(4) The greatest struggle in Nature is the struggle for completion through vibratory correspondences.

(5) Sex selection illustrates this struggle for completion in the higher kingdoms.

(6) The expression of this principle in sentient, intelligent life appears as an intuitional affinity, or an individual preference or choice.

(7) Love is the expression of this principle operating through and upon individual intelligences.

(8) This struggle for completion and ethical content governs animal marriage as well as human marriage.

(9) Two animals, as well as two humans, may fulfill this law of correspondence.

(10) Such marriage constitutes a free, natural, monogamous, and indissoluble love union. Mathematically speaking, such a marriage is a vibratory harmonic. On the physical side it is

passion, on the spiritual love, and in an ethical sense it is content or happiness to individual intelligence.

Most of the quotations selected are from "The Evolution of Marriage." The author of that work, M. Letorneau, is a recognized authority of the Darwinian school. His position as to the basis of evolution is clearly stated. He says: "The great forces called natural are unconscious; their blind action results, however, in a world of life, choice, selection and a progressive evolution."

Notwithstanding this very definite agreement with physical materialism, the author almost immediately introduces a word which contravenes this interpretation of Nature. The work opens with a careful analysis of love and marriage in animal life. In referring to procreation among superior animals, the physicist says: "In their case the act of procreation is a real efflorescence, not only physical but psychical." Again, still discussing the same subject, he adds: "It is important to bear in mind that all this expenditure of physical and psychical forces has for its motive and result the conjugation of two differing cells."

That which is important in this connection, is, not to show that this "psychical" force has another origin and motive than the conjugation of two differing cells, but that this conjugation of two physical cells requires an expenditure of two kinds of force, even in animal life, viz., physical force and "psychical" force.

Now, "psychical" force is distinctly not "physical" force, and yet physical science claims that Nature has provided only blind physical forces.

The anthropologist says elsewhere that, "In pairing season the psychic faculties of the animal are over-excited." So throughout the entire work are the words "psychic" and "psychical" repeatedly used. Reference to "psychical" causes and "psychical" phenomena is indulged as freely as if the author believed that half the facts of Nature are due to "psychical" causes.

This is one of the common contradictions and inconsistencies of our authorities of the modern school of physical science. It is a contradiction and inconsistency which none of them explains. There is, in fact, but one explanation, and that a very simple one. No person of average intelligence can study living, sentient, intelligent Nature and escape the conviction that there are two classes of phenomena in existence, viz., physical phenomena and "psychical" phenomena.

The student finds that one part of the phenomena of life appears to have a logical relation to, or basis in, the visible and tangible physical functions of Nature. He finds, on the other hand, that the other part of life's phenomena, though equally self-evident, is wholly intangible to physical sense and wholly elusive under all physical instruments and physical tests.

The physicist refers to the "psychical" phenomena of sex and the "psychical forces" in generation and reproduction, because he finds no other words to define those self-evident conditions. He is driven to the use of the term "psychical" merely because he recognizes forces which are not physical. It will be recalled that the word "psychical" is derived from the Greek "psyche," meaning the soul. The physical scientist probably did not intend by the use of that word to recognize spiritual forces, nor to acknowledge a soul element in Nature.

What he does, however, is to confess that intelligent creatures employ forces which must be recognized as super-physical. By such admission he corroborates the higher science. The pysicist uses the word "psychical" to define super-physical forces, while the Natural Scientist employs the word "spiritual" in exactly the same sense.

This admission of "psychical" forces into the operations of intelligent animal life, must be held as corroborative of that proposition which declares that the evolution of man has a basis in spiritual principles, elements and forces.

Physical science also corroborates the higher science upon those propositions which declare that sex is the spiritual principle of affinity in operation, and that male and female represent the positive and aggressive, and the receptive and absorbing energies in Nature. The reader is asked to determine this for himself from a few disconnected statements concerning sex selection, and the characteristics of male and female nature.

Havelock Ellis says:

(1) "While the men among all primitive peoples are fitted for work involving violent and brief muscular effort, the women are usually better able than the men to undergo prolonged and more passive exertion."

(2) "The militant side of primitive culture belongs to man, the industrial to woman."

(3) "The characteristic implement of woman is not a weapon, but the 'ulo,' or the primitive industrial knife."

(4) "The militant element ruled throughout medieval Europe and that meant the predominance of men."

M. Letorneau, in his discussion of animal marriage, says:

(1) "Almost universally, whether she be large or small, the female is less ardent than the male, and in the amorous tragicomedy she plays from the beginning to the end a passive rôle. In the animal kingdom as well as mankind amazons are rare."

(2) "The female of mammals are always weaker than the male."

(3) "The female bird shows the natural reserve of her sex."

M. Letorneau sets forth what he terms the natural sex laws of courtship in lower life. On the masculine side he names this phenomenon, "the law of battle." On the feminine side he finds a law of coquetry.

The reader will be able to perceive that this so-called masculine law of battle is something more than a law of courtship. It is, in fact, the masculine law of being. The battle between male animals during courtship is but another phase of the inherent struggle for supremacy or conquest by force, which universally obtains in masculine nature. In the law of coquetry he will as easily discern the purely feminine method of accomplishment, viz., by self-surrender, or the promise of self-surrender.

The author says:

(1) "It is especially among the gallinaceæ that love inspires the males with warlike fury. In this order of birds nearly all the males are of bellicose temperament. Our barn-door cock

is the type of the gallinaceæ—vain, amorous, and courageous. Black cocks are also always ready for a fight, and their females quietly look on at their combats, and afterward reward the conqueror. We may observe analogous facts, only somewhat masked, in savage and even in civilized humanity."

(2) "Among fishes we begin already to observe another sexual law, at least as general as the law of coquetry, which Darwin has called the law of battle. The males dispute with each other for the females, and must triumph over their rivals before obtaining them. Thus, whilst the female sticklebacks are very pacific, their males are of warlike humor, and engage in furious combats in their honor."

(3) "The higher we ascend in the animal kingdom the more frequent and more violent become two desires in the males—the desire of appearing beautiful, and that of driving away rivals."

(4) "The law of battle prevails among aquatic as well as land animals."

(5) "The combats of the male stags are celebrated * * * Seals and male sperm-whales fight with equal fury, and so also do the males of the Greenland whale."

(6) "With birds, * * * the law of battle plays in important part in sexual selection. * * * The male Canadian geese engage in combats which last more than half an hour; the vanquished sometimes returns to the charge, and the fight always takes place in an enclosed field, in the middle of a circle formed by a band of the clan of which the rivals form a part."

* * * * * * *

(1) "But however short may be their sexual career, one fact has been so generally observed in regard to many of them (insects), that it may be considered as the expression of a law—the law of coquetry. With the greater number of species that are slightly intelligent, the female refuses at first to yield to amorous caresses. However brief, for example, may be the life of butterfles, their pairing is not accomplished without preliminaries; the males court the females during entire hours, and for a butterfly hours are years."

"We can easily imagine that the coquetry of the females is more common among vertebrates."

(2) "The courage and jealousy of the male (bird), his efforts to charm the female by his beauty and the sweetness of his song, and finally, the coquetry of the female, who retreats and thus throws oil on the fire."

* * * * * * *

Physical science corroborates the higher science as to the existence of a law of individual preference in sex selection. This is the law which physical science terms "inexplicable," but which the higher science explains as an expression of vibratory correspondence between two intelligent entities.

Letorneau says:

(1) "The animal as well as the morally developed man is capable of individual preferences. He does not yield blindly and passively to sexual love."

(2) "Many vertebrate animals are capable of a really exclusive and jealous passion."

(3) "Man alone should have the privilege of introducing free choice into love. It is not so, however."

(4) "According to observers and readers, it is the female who is especially susceptible of sentimental selection. The male, even the male of birds, is more ardent than the female, * * * and thus generally accepts any female. This is the rule, but it is not without exceptions; thus, the male pheasant shows a singular aversion for certain hens."

(5) Darwin finds that, "amongst the long-tailed ducks some females have evidently a particular charm for the males;" also that, "the pigeon of the dovecote shows a strong aversion to the species modified by breeders, which he regards as deteriorated. Female pigeons occasionally show strong dislike to certain males without apparent cause. At other times a female pigeon, suddenly forgetting the constancy of her species, abandons her old mate or legitimate spouse to fall violently in love with another male. In the same way peahens sometimes show a lovely attachment to a particular peacock."

(6) Letorneau says that "it is more especially females who

introduce individual fancy into sexual love. They are subject to singular and inexplicable aversions. * * * Very peculiar fancies arise in the brains of birds. Thus we see birds of distinct species pairing, and this even in the wild state. These illegitimate unions have been observed between geese and barnacle geese, and between black grouse and pheasants."

Darwin relates a case of this kind of passion suddenly appearing in a wild duck. Mr. Hewitt relating it says: "After breeding a couple of seasons with her own mallard, she at once shook him off on my placing a mail pintail in the water. It was evidently a case of love at first sight, for she swam around the new comer caressingly, though he appeared evidently alarmed and averse to her overtures of affection. From that hour she forgot her old partner. Winter passed by, and the next spring the pintail seemed to have become a convert to her blandishments, for they nested and produced seven or eight young."

If physical science had made no other discovery in the line of spiritual principles than the law of individual preference, it had in this one phenomenon sufficient grounds for modifying its own theories. Except for this fact of individual preference, there had been no such thing as choice in sex selection. In that case, the blind instincts of a purely physical passion would have prevailed. One mate would have satisfied the biological need as well as another, and promiscuity would have been the natural law of selection.

In this one phenomenon, then, an intelligent individual choice, is revealed that principle which establishes the sex relation as a spiritual relation, and raises sex love, even in animal life, above the biological need for reproduction. The introduction of an individual choice into the sex relation of intelligent creatures, quite transform the theory of life laid down by physical science.

Instead of a series of mechanical compulsions brought about by tyrannical instincts for reproduction, life is seen to be a series of *individual* selections brought about by an individual intelligence seeking its own self-content.

Natural Science holds that promiscuity is not a natural sex law. Physical science says:

(1) "Polygamy is common to mammals."

(2) "Mammals while less delicate than birds are already on a moral level incompatible with promiscuity."

(3) "Polyandry does not appear to have been practiced among animals. Polygamy is the commonest form. Monogamy is common and sometimes accompanied with so much devotion as to serve as an example to human monogamy."

The higher science holds that two animals as well as two humans may fulfill the universal law of vibratory correspondence. In such cases the love relation does not necessarily depend upon the degree of intelligence nor the evolutionary stage of animal life. It depends rather upon the degree of vibratory correspondence, that is, the degree of natural sympathy which obtains between any two individuals.

Letorneau says:

(1) "There is no strict relation between the degree of intellectual development and the form of sexual union."

(2) "We find among animals temporary unions, at the close of which the male ceases absolutely to care for the female; but we also find, especially among birds, numbers of lasting unions, for which the word marriage is not too exalted."

(3) "But if polygamy is frequent of mammals, it is far from being the conjugal régime universally adopted; monogamy is common and is sometimes accompanied by so much devotion that it would serve as an example to human monogamists."

(4) "Nearly all rapacious animals, even the stupid vultures, are monogamous. The conjugal union of the bald-headed eagle appears even to last till the death of one of the partners. This is indeed monogamic and indissoluble marriage, though without legal restraint. Golden eagles live in couples and remain attached to each other for years without even changing their domicile. But these instances, honorable as they are, have nothing exceptional in them; strong conjugal attachment is a sentiment common to many birds."

(5) "Examples of wandering fancy are for most part rare among birds, the majority of whom are monogamous, and even far superior to most men in the matter of conjugal fidelity."

(6) "Birds especially are models of fidelity, constancy and devoted attachment."

(7) "Among many animal species sexual union induces durable association having for its object the rearing of young. In nobility, delicacy and devotion, these unions do not yield precedence to many human unions."

At this point it would be of scientfic interest to know by what rule or process the author arrived at the conclusion that these durable associations have for their object "the rearing of young," rather than the satisfaction and content of the individuals composing those unions.

The author says further that there is the same diversity in the habits of the monkey as in those of the human in conjugal relations. He says:

(8) "Some are polygamous, some are monogamous. The wanderoo of India has only one female and is faithful unto death."

Natural Science finds that bird life represents the greatest physical and spiritual refinement and the highest vibratory condition of the animal kingdom. This being true, the intelligence of bird life would sense, enjoy and express those finer and higher conditions more readily and more harmoniously than any other department of animal intelligence. Partial proof of these statements are the superior love nature of birds, their gift of music, and the fineness and delicacy of their flesh.

Physical science corroborates these deductions when it says: "But it is particularly among birds that the sentiment, or rather the passion, of love breaks out with most force and even poetry."

"But among animals, as well as men, love has more than one string to his bow. It is especially so with birds, who are the most amorous of vertebrates. They use several æsthetic means of attracting the female, such as beauty of plumage and the art of showing it, and also sweetness of song. Strength seems often to be quite set aside and the eye and ear are alone appealed to by the love stricken males. * * * Birds often assemble in large numbers to compete in beauty before pairing. The tetras cuspido of Florida and the little grouse of Germany and Scandinavia do this. The latter have daily amorous assemblies, or cours d' amour, which are renewed every year in the month of May."

"Certain birds are not content with their natural ornaments, however brilliant these may be, but give the rein to their æsthetic desires in a way that might be termed human. * * * The palm is carried off by a bird of New Guinea, made known to us by M. O. Beccari. This bird of rare beauty, for it is a bird of Paradise, constructs a little conical hut to protect his amours, and in front of this he arranges a lawn, carpeted with moss, the greenness of which he relieves by scattering on it various bright colored objects, such as berries, grains, flowers, pebbles and shells. More than this, when the flowers are faded, he takes great care to replace them by fresh ones, so that the eye may be always agreeably flattered. These curious constructions are solid, lasting for several years, and probably serving for several birds. What we know of sexual unions among the lower human races suffices to show how much these birds excel men in sexual delicacy." * * * "Every one is aware that the melodious voice of many male birds furnishes them with a powerful means of seduction. Every spring our nightingales figure in true lyric tournaments," etc., etc.

In bird life, more particularly, is ample proof of a sex love which entirely transcends the physical affinity or physical passion. Here also we find that conjugal love is the earlier and far more enduring bond than parental love.

For example, Letorneau says:

"With the female Illinois parrot (Psittacus pertinax) widowhood and death are synonymous, a circumstance rare enough in human species, yet of which birds give us more than one example. When, after some years of conjugal life, a wheatear happens to die, his companion hardly survives him a month. The male and female of the panurus are always perched side by side. When they fall asleep, one of them, generally the male, tenderly spreads its wing over the other. The death of one, says Brehm, is fatal to its companion. The couples of golden woodpeckers, of doves, etc., live in a perfect union, and in case of widowhood experience a violent and lasting grief. The male of a climbing woodpecker, having seen his mate die, tapped day and night with his beak to recall the absent one; then at length, discouraged and hopeless, he became silent, but never recovered his gaiety."

"These examples of a fidelity that stands every test, and of the religion of memory, although much more frequent in the unions of birds than in those of human beings, are not, however, the unfailing rule."

What do all these facts suggest? That the struggle for nutrition is the inspiration of existence, or that love is—even in a bird—essentially a biological need? Do they not, rather, directly refute such a theory?

Here in this lower world of intelligence is substantial evidence of the spiritual relation in sex, such evidence as puts to confusion the theories of physical materialism. Here, in the lower kingdom of life, and the lower kingdom of intelligence, Nature establishes a bond which transcends every physical requirement, and endows even two birds with a love which obscures the procreative passion and ignores the claims of reproduction. By such widowhood and death it cannot be denied that bird love rises above the demands of physical nature. Thus, even a bird, in its ethical struggle for self-content, rises superior to the organs of digestion and reproduction. Indeed, by such fidelity and devotion, it defeats the claims of both.

Such a phenomenon is, indeed, "inexplicable," by all the rules of Darwinian philosophy. This is a fact, however, which must suggest that even the intelligence of a bird may be inspired by higher impulses and capacities than those which originate in the physical functions. With the demand for nutrition and the instinct for reproduction playing upon the intelligence of the bird, with nutrition at command, with other lovers at call—it is yet possible for even a bird to die of grief when the mate of its choice is gone.

If animal marriage can so far transcend the "requirements of the sex appetite," how vain to insist that human marriage has no other purpose than the regulation and satisfaction of an "imperious sex appetite."

Since two birds may live in conjugal loyalty a lifetime, how illogical to declare that such love and fidelity in human marriage is a result of sex appetite, or a habit induced by heredity, or a mere affection induced by the mutual care of progeny. Since physical science tells us that a widowed bird may die of grief, how opposed to Nature is the theory that such grief in a human lover is an efflorescence of sex lust, and disappointment in that passion.

The time approaches when physical materialism must explain what it now terms "inexplicable," and "unfathomable," or it must cease to dogmatize concerning the purposes of life and the nature and causes of love. It must reserve its judgments on these points, or offer scientific explanation that shall appeal to reason and intuition, to the commonest facts of life, to common observation and experience, and finally, to the personal impulse an aspiration of mankind.

Until such time, the higher science rests its case upon the judgment of a man eminent in the school of physical science.

It is Alfred Russel Wallace who says: "That theory is most scientfic which best explains the whole series of phenomena."

Mr. Drummond makes the strong point of his work through an oversight, or a misinterpretation of an actual fact, viz., he holds that the reproductive function is the source of love and of the moral order in human life. He assumes:

(1) That love comes into the world by reason of the enforced physical sacrifices of the female half of life.

(2) That any higher love between husband and wife is rather an affection or a friendship growing out of the maternal and paternal relations and obligations. In support of this position he says:

"Affection between male and female is a later, less fundamental, and, in its beginnings, less essential growth; and long prior to its existence, and largely the condition of it, is the even more beautiful development whose progress we now have to trace (maternal love)."

"The basis of this new development is indeed far removed from the mutual relations of sex with sex. * * * But the

Evolution of Love began long before these (the psychical relations of husband and wife) were established."

Mr. Drummond says further: "The idea that the existence of sex accounts for love is not true." And also: "Love for children is always an a priori and stronger thing than love between father and mother. * * * In lower nature, as a simple fact, male and female do not love one another; and in the lower reaches of Human Nature, husband and wife do not love one another."

Thus, from the moralist's point of view, sex plays but an incidental part in the evolution of love. This means that love between man and woman is not the cause, but a result, of reproduction; that the male half of life in lower nature is not susceptible to love. According to this philosophy, even man learns to love woman only through and by the care of, and contact with, his offspring.

To one familiar with the love relation and the conjugal habits of animals, as well as humans, this appears to be a very curious statement.

Physical science clearly establishes the direct contrary to be the truth. A single quotation from Letorneau serves to show the error into which Mr. Drummond is led by his zeal. This authority says: "As a matter of fact, both with birds and other animals, the paternal or maternal sentiment hardly lasts longer than the rearing time. When once the young are full grown, the parents no longer distinguish them from strangers of their own species, and it is thus even with monogamic species, when the conjugal tie is lifelong; the marriage alone endures, but the family is intermittent and renewed with every brood. We may remark that it is almost the same with certain human races of low development."

If this statement means anything, it is that parental love is a later development than is love between husband and wife.

Whenever and wherever the moralist is forced to consider sex at length, he unconsciously exalts the very principle in sex that his main argument undertakes to deny.

For instance, after hundreds of pages devoted to proving

that sex is nothing but a physical device for reproduction, Mr. Drummond is yet impelled to say of man and woman: "With the initial impulses of their sex strengthened by the different life-routine to which each led, these two forces ran their course through history, determining by their ceaseless reactions the order and progress of the world, or when wrongly balanced, its disorder and decay. According to evolutional philosophy there are three great marks or necessities of all true development —Aggregation, or the massing of things; Differentiation, or the varying of things; and Integration, or the reuniting of things into higher wholes. All these processes are brought about by sex more perfectly than by any other factor known. From a careful study of this one phenomenon, science could almost decide that Evolution was the object of Nature, and that Altruism was the object of Evolution."

This inconsistency with the main body of his argument, is, however, consistent with the general inconsistency of "The Ascent of man."

Again, for example, here is a large volume intended to prove that "man progresses, not by any innate energies of protoplasmic cells, but, solely by feeding and reinforcement from without." In the general summing up of this purely materialistic theory, the author is led to say: "If anything is to be implied it is not that spiritual energies are physical, but that physical energies are spiritual. * * * The roots of a tree may rise from what we call a physical world; the leaves may be bathed by physical atoms; even the energy of the tree may be solar energy, but the tree is itself. The tree is a Thought, a unity, a rational purposeful whole; the 'matter' is but the medium of their expression. Call it all—matter, energy, tree—a physical production, and have we yet touched its ultimate reality? Are we quite sure that what we call a physical world is, after all, a physical world?"

After writing hundreds of pages to show that "Love comes into the world at the point of the sword," the moralist enjoys that vision of the natural love principle which the writer has already quoted in full in another chapter. In this vision of the universal law of affinity, Mr. Drummond admits that, "Neither here nor anywhere have we any knowledge whatever of what is meant by Attraction and Affinity." What he says that is especially interesting in this connection, is that, "Here we are in the presence of that which is metaphysical, that which bars the way imperiously to materialistic interpretations of this world."

It will be remembered that Mr. Drummond started out to prove that, "Everything in the moral world has a physical basis." He is now, by a peculiar retrograde method, seeking to show that everything physical has a spiritual or metaphysical basis.

It will also be recalled, that he has formulated his moral philosophy, up to this point, without reference to the laws and principles which govern inorganic substance. He is finally, however, and almost as an after-thought, attracted to the principle of affinity which governs the inorganic kingdom.

What he says of attraction and affinity in the inorganic world, has been quoted fully in Chapter VIII. By referring back to this extract from his chapter on Involution, the reader will be able to form a fairly accurate judgment as to the logic of his work.

It will be seen that the moralist assumes:

(1) That the chemical activities of senseless atoms are governed by a metaphysical principle, while the ethical activities of rational human beings are governed by physical functions.

(2) That the union of two atoms illustrates a metaphysical affinity, while the marriage of two humans is based upon a physical passion.

(3) That centers of attraction in the mineral world are formed by reason of metaphysical affinities, while in the human world, centers of attraction are formed by physical compulsions and physical sacrifices.

(4) That there is an overshadowing principle of co-operation in inorganic Nature, but, "The training of humanity is under a compulsory act."

Nature, however, clearly demonstrates:

(1) That not only the chemical activities of atoms, but the ethical activities of men, are governed by a metaphysical or spiritual principle.

(2) That not only the union of two atoms, but the marriage of two human beings, represents a metaphysical or spiritual affinity.

(3) That centers of attraction in every kingdom, from atom to man, are expressions of a metaphysical or spiritual affinity which tends to equalize positive and receptive energy.

(4) That there is an overshadowing spiritual principle of co-operation or love in the development of human relations, as well as a metaphysical affinity in the aggregation of mineral atoms.

This new moral philosophy confesses that our solid earth was formed by metaphysical attractions and readjustments of individual atoms, which at last solidly combine "in the order of their affinities." At the same time it holds, that human morality has been forced upon the world by the physical coercions and compulsions of the individual female.

Nature, however, clearly demonstrates that, not only the solid earth, but a spiritually organized humanity is the response to the same metaphysical affinity.

While Mr. Drummond holds that chemical compounds represent super-physical affinities, he assumes that social harmonics are compelled by physical functions.

Nature, however, in its highest, as well as its lowest kingdom, demonstrates its spiritual basis. It demonstrates that social compacts, as well as chemical compounds, are expressions of the same general principle.

In brief, this new moral philosophy finds the inorganic world governed by metaphysical affinity, while the organic, sentient, intelligent and moral world is governed by physical compulsions. It claims metaphysical causation for chemical affinity, and physical causation for human affinity. In its last analysis, it declares that the bond which unites a positive atom to a negative atom is, primarily, metaphysical, while that which binds two intelligent souls is, primarily, a physical passion.

The reader can determine for himself the weight of such argument and speculation.

Love has a natural pathway from chemical affinity to psychical affinity. That pathway is an ever ascending scale of vibratory correspondences, known as chemical affinity, physical passion, spiritual sympathy, and intellectual companionship.

Deep down in the under world of senseless matter, the divided forces of Nature rush together, seeking equilibrium through vibratory correspondences. Each individual atom is engaged in this struggle for its own completion. Impelled by the intelligent spiritual principle of affinity, each seeks its polar opposite. All unions of atoms are, primarily, individual unions of atoms oppositely polarized. All effects are therefore, primarily, individual effects. Perfect vibratory correspondence between two atoms means an indissoluble union. All chemical compounds depend upon this primary necessity, effort, and purpose of individual atoms.

Gems (notably the diamond) illustrate the highest harmonics in the mineral kingdom. This means that the atoms of which a gem is composed, being so fine in particle and rapid in vibration, unite indissolubly with their polar affinities. The visible effect of this harmonic arrangement of refined physical matter is a material union remarkable for its molecular activity and for its solidity, durability, transparency and beauty.

A physical scientist aptly characterizes the diamond as being "alive," so rapidly do its particles move upon one another. The diamond is the highest expression of positive and receptive energy in the mineral world and represents an unconscious struggle of individual atoms for individual equilibrium. The gem may therefore, as a whole, be regarded as the offspring of electromagnetic energies, and an incident in the struggle made by the individuals composing the entire stone.

The diamond, by its close affinities, intense vibratory action, and indissoluble union, illustrates the true principle of marriage. By its crystallizations, reproduction is faintly foreshadowed.

Through vegetable life runs the same principle. Here, however, the energies of individual particles are reinforced and accelerated by the energies of another element, the vito-chemical life element. Attraction and affinity remain the same in principle, but are differentiated in essence and effect. Vito-chemical life generates living organisms which involuntarily draw and feed upon the universal elements. At the basis of all oganization, however, are the same efforts of individual particles for vibratory correspondence. Individual unions, thus become the center of all vegetable growth. All changes and effects are, therefore, primarily, individual changes and effects.

Flowers in the vegetable kingdom, like gems in the mineral, illustrate the highest possibilities of that kingdom, by way of material refinement and rapid vibration.

Thus, even in the vegetable world, reproduction is an efflorescence and incident in the individual struggle for equilibrium.

Animal life repeats the principle, processes and effects of the lower kingdoms, but with the reinforcements of another element, viz., the spiritual life element. While the same general principle continues to guide Nature's activities, the individualized intelligence supplements Nature by the individual effort for self-completion. Here, individual intelligence reinforces the involuntary chemical activities, by its own conscious effort for self-satisfaction. In this kingdom, the cleavage of positive and receptive energies is still further accentuated. There are now distinctly, male and female cells, organisms and entities. The sex principle is expressed with increased particularity.

Here, as below, Nature's divided forces rush together, seeking equilibrium between positive and receptive energies. That universal impulse is now, however, a matter of conscious individual impulse. Upon this impulse and effort of the individual all evolution depends. Reproduction and species both depend upon this initial impulse of individuals.

Impelled by the energies of all the elements they represent, animals rush together with a force and intensity which are well characterized as animal passion. That effort expresses itself upon the physical side as passion, upon the spiritual side as love, and in its ethical sense as content. The force and intensity of animal passion induce procreation. The seeds of life, generated by positive masculine force, are nourished and matured by the absorbing feminine energies. Thus, reproduction, even in the animal kingdom, is an efflorescence or incident in the greater struggle for self-completion. Animals live their lives, not in the struggle to reproduce, but in an effort to obtain and maintain those conditions and relations which are, in effect, self-content.

Even in animal life Nature's first command is not reproduction. On the contrary, here as elsewhere, Nature's first injunction is to seek harmony.

Animal passion, in its scientific aspect, represents that struggle of intelligence for individual expression and satisfaction, which finally perfects the physical body. This, indeed, appears to have been the supreme task set by Nature for the lower kingdom of intelligent life.

Human life is but a loftier life conducted upon the same general principle. Animal passion has finished its especial labors. It has perfected the physical instrument of intelligence. It has opened the doors to a higher life. The energies, powers and capacities of the highest soul element are super-added to that which has already been accomplished.

The sex principle is still more clearly defined. Sex specialization is immeasurably increased. It is improved in form and increased in complexity and capacity. It remains, however, the same in principle. The male has become the man, the female has become woman. To animal intelligence has been added psychical intelligence.

Here, as in the lower kingdom, the sex cleavage and sex affinity obtain. Men and women seek one another in terms of every vital element. To chemical affinity is added conscious animal passion. To animal passion is added the self-conscious needs of the soul. Though susceptible to, and influenced by, all of the elements of lower nature, the evolution of man and woman is really dominated by the superior energies of the intelligent soul.

Moved by the self-conscious will and desire of the soul, as well as by the impulses of the body, man and woman seek each other with the intensity and fervor of both a physical and psychical nature. In conformity to natural law, however, the correspondence sought is one which shall satisfy the highest element rather than the lowest. In this struggle of the soul for its own adjustment the impulse of men and women would be to unite, separate, and reunite, until that satisfaction should be found. Because of the soul's necessities individual choice is more sharply accentuated in human life. The physical relation, though a legitimate part of life, is below the level of the soul's necessities. In human life we have, therefore, in addition to the animal impulses, an intelligent soul seeking adjustment through physical conditions.

This effort for self-adjustment induces marriage in its many forms, and varying conditions. The force and intensity of this human will and desire govern procreation. The struggle for self-completion induces unions which generate physical life. Men and women seek one another in marriage with the single motive of an individual and ethical satisfaction, and not for the sake of reproduction nor the improvement of the family nor the preservation of species.

Children are therefore the efflorescence, or incidents, of the struggle for self-completion.

The completion of the individual upon the higher plane, as well as the lower, involves, primarily, the perfect relation of man and woman; that is to say, a human marriage which represents a perfect correspondence, physically, spiritually and psychically. Such a marriage would scientifically be defined as a perfect human harmonic.

Thus, a science which takes into account the spiritual side of Nature and the psychical element in man, finds a very definite and scientific pathway for love. This pathway is that supreme principle which impels every material entity, and every intelligent entity materially embodied, to seek correspondence in another like entity of opposite polarity.

Nature thus viewed, discloses a love principle which solidifies matter, evolves vegetation, generates animal life and inspires human intelligence to its higher intellectual and moral development. Love thus analyzed, rests upon an infinite series of harmonics, ranging all the way from the chemical affinity of mineral atoms to the psychical affinity of intelligent souls. The ascent of the individual is, therefore, through the simultaneous ascent of matter, life, intelligence and love.

The completion of an individual involves every principle, property, element and function of Nature. The ascent of the individual from the senseless atom to the self-conscious soul, may be considered in its several lines of development, viz.:

(1) As the ascent of matter and motion, meaning that vibratory process which refines matter and increases its vibratory action.

(2) As the ascent of life through those successive contacts of physical matter with the universal life elements.

(3) As the ascent of intelligence through gradual improvement of the physical and spiritual instruments intended for the uses of intelligence.

(4) As the ascent of love through increased harmonics between individual intelligences.

(5) The ascent of happiness through the increased effects which accrue to individual intelligence by reason of these correspondences and harmonics.

Thus it is, that the deductions of Natural Science form the basis of a new philosophy of life which is at once consistent with the universal law of motion and number, with the functions of physical nature, with the operations of intelligence, and with the activities of love.

Thus, evolution is something more than a preservation of species, or the improvement of a family. On the contrary, evolution is a universal scheme, primarily seeking the completion of the individual.

The object of Nature is to perfect men and women, and not to manufacture mothers. Individual existence is a ceaseless effort for an individual and ethical content. It is not merely contribution to species. Sex is a spiritual principle and not a mere physical function.

Love, which is "the greatest thing in the world," comes by a universal law of consent, and not "at the point of the sword."

All of Nature's vital energies are centered in individual man and woman. All of the activities and effects of those energies radiate from the individual man and woman.

This being true, the harmonics of life depend upon the balance and harmony attained between the individual man and woman.

Evolution is nothing more nor less than the spiritual principle of polarity or affinity in operation, and

"Love is the fulfilling of this law."

CHAPTER XVII.

NATURAL MARRIAGE

What constitutes natural human marriage is, as yet, an unsettled question in the mind of the average student.

The school of physical science faces certain facts of social development which are hard to reconcile with its preconceived theories as to the meaning and purpose of marriage.

It observes, first, that promiscuity in the sex relation is almost unknown in human society; secondly, that the masculine half of humanity inclines to polygamy either by law or in defiance of law; and thirdly, that prostitution is the universal accompaniment of all social life, primitive or highly developed. Finally it discovers that monogamous union is adopted by the most highly developed nations and individuals.

If the sex relation, as physical materialism claims, were a mere matter of sex appetite, there appears to be no scientific reason why promiscuity should not have been the true and natural sex relation.

The fact, however, that there is a universal tendency directly opposed to promiscuity, should suggest to the scientist that there is a universal principle involved. Physical science finds no better explanation for this phenomenon than a "caprice" or an "inexplicable" fancy which causes one individual to seek union with another particular individual. While it can but admit that this peculiarity of individual preference is a fact and factor in sex selection, physical science fails to follow the clue to a scientific solution.

To say that general promiscuity is prevented by caprice or even by individual fancy, is in no sense an explanation of why such caprice or fancy arises, nor why it is so universal and persistent in its demands.

On the contrary, this phenomenon clearly opens the way for those explanations which Natural Science has to offer. It plainly suggests that spiritual principle of affinity, or law of correspondence, by which the higher science explains "individual preference" as a universal law.

While physical science admits that promiscuity is not the natural relation of the human sex, it finds, however, that either polygamy or prostitution has characterized all times, all peoples, and all forms of government. It, therefore, and very logically, holds that a certain range of choice in the sex relation is both natural and desirable.

To one familiar with physical life and development only, but unfamiliar with spiritual laws and principles, this would appear to be an inevitable conclusion.

On the other hand, physical science admits that rational regulation of this natural sex relation has proved highly beneficial. It confesses that the highest types of societies and individuals are found under monogamous systems. It concedes that the most progressive nations adopt monogamy by law and public practice. It observes, further, that monogamous and indissoluble marriage is the ideal of the most highly cultured individuals.

This leaves physical science in a difficult position. It must now explain why an "unnatural" regulation of the sex relation produces the best results both in society and upon the individual. It must explain why the ideals of the most highly developed, universally point to "unnatural," monogamous marriage.

Physical science cannot answer these questions, inasmuch as the answer involves certain definite knowledge of the spiritual principle involved in marriage. Here, as elsewhere, the higher science supplements the modern school and offers to explain these apparently contradictory facts. It proceeds by answering directly that initial inquiry as to what constitutes natural marriage.

Nature declares that monogamy is the natural marriage relation, and that indissoluble union is the intent of natural law. The proof of this declaration rests, primarily, upon that universal spiritual principle which impels every entity to seek vibratory correspondence or self-adjustment in another like individual of opposite polarity.

Secondarily, the proof in question rests upon the commonest facts and experiences of human life, viz., the expectation, the hope and desire (of every normal man and woman), which point to some one individual as the ideal and inseparable lover and companion.

Physical materialism sees in the institution of marriage nothing but the "regulation of the sex appetite" in the interests of species. Nevertheless, in its study of the evolution of marriage it admits those facts which weaken that assumption and clearly sustain the deductions of Natural Science.

One important fact illustrating the natural spiritual law of selection is given by M. Letorneau in connection with the subject of promiscuity in sex relations.

The author says:* "We are warranted in believing that the very inferior stage of promiscuity has never been other than exceptional in humanity. If it has existed here and there, it is, that by the very reason of the relative superiority of his intelligence, man is less rigorously subject to general laws, and that he knows sometimes how to modify or infringe them; there is more room for caprice in his existence than in the life of the animals."

This "general law" referred to by this eminent French authority, is none other than the general law of affinity which guides the animal, and ordinarily the human, in sex selection. This "general law," as already explained, is none other than the law of individual preference. This law, though permitting an experimental range of choice, yet discourages promiscuity.

Physical science is correct in assuming that man may defeat this general law by reason of a more independent will and individual powers of execution. Physical disease and deformity and

---

*"The Evolution of Marriage," p. 38.

mental and moral degeneracy testify to man's ability to defy and contravene the universal spiritual laws of Nature.

The fact that man is not generally promiscuous in sex relations leads physical science to rightly regard such practice as unnatural. It is, however, led to other conclusions concerning polygamy. From history and observation the physical scientist concludes that the natural sex relation is polygamous, that is, upon the masculine side. In support of this theory it cites the almost universal practice in savage and semi-barbarous life. It refers to the various forms of slavery, to the legal concubinate, and to divers forms of religious polygamy. It reinforces this evidence by the indisputable fact of prostitution in all countries, among all grades of development, in all races and under all religions.

The author, in referring to prostitution, which is the same in principle as polygamy, says:* "To sum up, the origin of prostitution goes back to the most primitive societies; it is anterior to all forms of marriage, and it has persisted down to our own day in every country, and whatever might be race, religion, form of government, or conjugal régime prevailing. Taken by itself, it would suffice to prove that monogamy is a type of marriage to which mankind has found it very difficult to bend itself."

These facts it were idle to dispute.

The conclusions of physical science, based upon these facts, are perfectly logical—from the physical scientist's point of view.

Physical science concludes from these undeniable facts that polygamy, or a certain range of choice, is the natural sex relation. It makes other discoveries, however. It finds that monogamy is adopted by the most highly developed peoples; and further, that the ideal of marriage entertained by the finest individual types, is universally a monogamous and indissoluble union. From this, physical science is forced to concede that monogamy accompanies the highest national and individual development.

As will be seen, the two sets of facts and several conclusions

---

*"The Evolution of Marriage," p. 160.

lead physical science into a paradox. It first declares that polygamous sex relations must be natural. Next, it admits that monogamy accompanies the highest national and individual development. This leaves the physical scientist in the position of declaring that natural law does not produce the best results for man; or the reverse, that man reaches the highest development under direct violation of natural law.

It will be recalled that physical science holds that the institution of marriage has no other meaning than "regulation of the sex appetite." It finds, nevertheless, that the best results as to reproduction are effected by love unions, or union by individual choice. On the strength of this fact a third proposition is added.

The physical scientist has already declared:

(1) That polygamy appears to be a natural relation.

(2) That monogamy accompanies the highest known development.

He now adds his third proposition, which, to say the least, is inconsistent with the second. Satisfied that love unions produce the finest progeny, the scientist therefore concludes, not merely that love marriages should be encouraged, but he declares that the individual should be permitted to form as many love marriages as he chooses. This conclusion of the scientist is clearly stated in his chapter on "Marriage of the Future." The author says in substance: "Marriage if monogamous, should be made and dissolved at pleasure."

Thus quoting this bald proposition, makes it appear more repulsive than when presented in the author's attractive style. The theories of physical science, however, like other monstrosities, are always rendered more repulsive when stripped of an attractive garb.

The position taken by this eminent French scholar is not exceptional. His bold proposition quite accords with the moral philosophy of Darwinism and the entire school of modern physical science.

This, however, is a position which the great common sense of civilization will condemn. It is an extreme position, possible only to minds that have become so engrossed with the functions of physical nature as to lose sight of those higher laws and principles which govern intelligent and moral nature. This is a position which reduces man to the animal, levels love to a matter of lust, and eliminates the question of moral responsibility in the family relation.

The natural corollary to this singular proposition is frankly set forth when the writer admits that his scientific system would mean that children become wards of the state, cared for by public officials at public expense. This "improved" system, it will be seen, eliminates the parents and natural guardians. It discards, as unnecessary, parental and family love which have so long been regarded as the foundation of the social order.

This, however, is a logical outcome of Darwinian doctrine. Here we have, not merely a marital, but a social, educational and ethical system outlined upon the theory that "man is a mammal like any other with a better cerebral development than a horse or a dog."

Thus, in the very face of a system which admittedly accompanies the highest development on earth, it is suggested that we substitute a practically free selection; plainly speaking, a successive polygamy sanctioned by law. Here we have it seriously suggested that parents be relieved of their natural and moral responsibility for their own children, and that children be deprived of parental love and personal influence whenever those parents desire to form other ties. We have also a marriage system proposed that shall be governed by the caprice or passion or self-interest of the individual.

Physical science rightfully defines promiscuity as an unnatural practice. It states facts when it claims that polygamy or prostitution has been or is the accompaniment of all grades of social development. It is also correct in assuming that strict monogamy characterizes the highest civilization, and that it is best for the whole people. It is right again when it finds that the best results as to reproduction depend upon love unions.

Herbert Spencer was partly right when he said: "A time will come when union by affection will be considered the most important, and union in the name of the law the least important, and men will hold in reprobation those conjugal unions in which union by affection is dissolved."

Mr. Spencer, however, might better have said, the time will come when husbands and wives who have ceased to respect and love each other will abstain from all intimate relations; or, the time will come when the mismated will cheerfully preserve the legal form of marriage so long as mutual obligation to their own children demands such self-denial.

The evolution of marriage, up to our present system of strict monogamy, represents the evolution of the rational intelligence and the moral nature of man. If we admit that the present Christian nations of the earth represent the highest stage of evolution, we must also admit that its marriage system has produced the best results. The progress of civilization unquestionably proves that the rational mind and the spiritual intuitions of man set steadily toward monogamous and indissoluble union.

Such a marriage system has been made possible only by reason of the fact that man recognizes in himself something more than an animal. It means that he recognizes and accepts the responsibilities imposed upon a rational and moral being. It means that little by little he rises above the animal side of his nature and conforms to regulations which satisfy the higher nature of the intelligent soul.

Nor could any evolution of marriage have been possible, except for the fact that man realized such restrictions as beneficial. Even the very individuals who secretly indulge and traffic in animal propensities would not, if they could, repeal the laws which guard the physical, mental, moral, material and social well-being of the race.

When physical science talks of the "naturalness" of polygamy and prostitution, it considers merely the animal elements and impulses which are a part of human nature. It fails to recall that the evolution of rational intelligence and moral perceptions has been universally accompanied by a more and more rigid system of monogamous union. It ignores the fact that the highest nature of man universally inclines to such a system. No man, and surely no woman, well balanced morally as well as mentally, could condemn monogamy in its principle. Even the libertine and prostitute would admit it to be an essential in the development of government, society and the home.

The average individual condemns it only when he has mistakenly assumed its obligations and is cheated of his personal happiness. Even the average of men in civilized life, those given to personal indulgences, would be the first to resent that return to barbarism which is implied by, "marriage made and dissolved at pleasure." It is safe to say that the great majority who enjoy the benefits of civilization and Christian development, would protest against any scientific system that reduces marriage to a question of individual caprice, cupidity or passion, and eliminates the responsibility of parents, consigning children to the care of the state.

The rational and moral evolution of the average man has been carried too high for serious consideration of any system which would wipe out all that makes life worth living, viz., the mutual love and loyalty of men and women in the marriage relation, the moral responsibility to children, and those ideals which bind men and women and children to the home.

Nature and history sustain physical science up to a certain point. There are, however, other facts in Nature which contravene those theories of physical science concerning the practices of polygamy and prostitution. These other facts, at present unknown to the physical scientist, show, on the contrary, that such practices are not the intent of Nature. These other facts go to show that such a marriage system as outlined by physical science, if universally applied, would mean the physical, intellectual and moral degradation of humanity.

Promiscuity is unnatural in that it directly contravenes the universal law of affinity. Polygamy and prostitution represent, not fulfillments of the natural law of marriage, but *the effort to fulfill it*.

Every entity, from atom to man, is an individual seeking vibratory correspondence in another individual of opposite polarity. Every individual seeks a correspondence in each and all of the life elements of which it is composed. The requirements of each entity are, therefore, simple or complex, according to its place in the scale of development; for example, the mineral atom whose energies are electro-magnetic only, finds an affinity or mate more readily than a man who represents the combined elements of electro-magnetism, vito-chemical life, spiritual life, and the life of the soul.

Every human being, as a distinct individual in Nature, is a vibratory law unto himself. That is to say, he has a degree of material refinement and a rate of vibratory action in both his physical and spiritual organisms, which are distinctly his own. He has also a psychical nature, that is, a quality of intelligence and a code of morals, which are peculiar to himself.

These, indeed, are the peculiar conditions, qualities and properties which constitute individuality.

Nature designs that this individual shall seek his complementary half, or his polar opposite, in another individual who must be in such correspondence with himself that the two shall represent the completed individuals.

This natural law of vibration or affinity is the pathway along which Nature guides the individual in his sex selections. Individual preference, in such selection, is the only sign by which Nature informs man of his obedience to law. The scope and power of this principle cannot be appreciated until it is once recognized as the universal guide to man in his efforts for self-completion and happiness.

Such, however, is the law which forbids promiscuity, suffers polygamy and leads on to monogamous and indissoluble union where all of its conditions are fulfilled.

Polygamy and prostitution are deplorable phases of the struggle for self-completion and happiness. They represent the ignorant efforts of the undeveloped, the gross and the vicious, for self-adjustment. Both forms of sex relation are simply due to undeveloped reason and a low grade of morality. In the higher science these practices are simply regarded as immature stages in human development. They are practices which may be termed experimental rather than unnatural. They are the negative means of education. That is to say, they teach man through experience that he does not find permanent satisfaction in such relations, and are practices which retard true development so long as maintained.

Polygamy and prostitution must, therefore, be classified as schools of experiment in which human intelligence learns the penalties of false sex relations.

These experiments, however, appear to have been the necessary trial of the law which ignorance must make in the midst of Nature's subtle principles and forces. These gross practices, in their evil results, are a part of human education. They are, therefore, not laws of Nature, but they are lessons involved in learning the law.

Thus, polygamy and prostitution, whether immediately governed by passion, ambition, avarice, vanity or religious superstition, must be interpreted as a part of the long struggle for happiness.

No matter how coarse and repulsive those practices appear to the refined mind, science detects under all revolting detail, the universal search of the individual soul for its natural affinity and mate, for its own completion and happiness. This search, not for the ideal, but the real, is a long one in point of both time and experience. It represents an almost infinite series of individual steps. It involves an almost infinite number of experiments and mistakes. It effects an almost infinite series of improvements.

This is proved by the stupendous fact of evolution itself, for evolution is simply the history of that search and struggle.

The rational powers of the soul element introduce new factors into the marriage relation. Students of Nature are struck with the superior order, harmony and consistency which appear to characterize the animal sex relation.

This is accounted for by the fact that the intelligent animal responds intuitionally (or instinctively) to the spiritual principle of affinity. He therefore responds to each new attraction with a readiness which makes the act appear as almost automatic. In animal life are none of those restraints or restrictions which prevent response to every stronger attraction. The individual animal, therefore, escapes the temperamental friction which comes in human life through social and legal restraints.

In human life, however, marriage is a relation which calls for the exercise of an individual and independent reason, as well as spiritual intuition. To intelligent intuition is now added an individual and rational will with the independent power of execution.

To these higher and independent capacities we must attribute the errors and confusion brought into the orderly operations of the natural law of affinity. Man feels the same impulses as the animal, but he indulges or restrains them at will. Man, however, has impulses which animals do not feel. These also he may gratify or deny at will. Man, like the animal, is forever seeking adjustment, but he does not seek it in exactly the same way. The animal has only his intuitions to guide him. He does not make independent rational experiments. Man, however, though continually admonished by intuition, has yet the individual power to reject those intuitions and to follow a path marked out by his own individual judgment.

Thus, it appears that animal intelligence employs but one method in the struggle for self-completion, while human intelligence employs two. This means, therefore, that the natural law governing human marriage is a higher law than the natural law governing animal marriage.

In the lower reaches of human life the individual acts of men appear, and are, almost an unbroken series of blunders and perversions as to the general laws of Nature. Nevertheless, man is pursuing the only path possible to an independent, rationalized intelligence.

The primitive man resembles the child emerging from the unreasoning state of infancy. Indeed, child life as clearly represents the evolution of individual intelligence as the prenatal development of the child represents the processes of physical evolution.

Human infancy is clearly analogous to the animal stage of intelligence. It is characterized by the same passions of love, fear, anger, jealousy, etc., which move the animal. As with the animal, however, so the infant lacks the rational powers and the moral capacities which give dignity and value to adult human intelligence.

With the dawning of its reason the child clearly represents that evolutionary stage where the prepared animal organism inspired the higher element and became the rational, living soul. During the childhood, whether of a man or a race, life is a succession of experiments, mistakes and penalties. A race, just as a man, must learn the nature and effects of law and the transgression of law, by independent experiment.

A child generally learns the nature and effects of fire through willful meddling with it. To the burned child, therefore, fire appears to be an evil. The man, however, knows that it is an essential to life and when rightly understood and controlled a beneficent provision in the natural order. It would be difficult to teach the child this truth. Unable to exercise reason, and smarting from the burn, he would naturally feel that fire is hostile to comfort and the pursuit of happiness.

This is the seemingly hard path of human development. It is, however, the only path for a child or a man or a race of men. The higher science declares this path to be, after all, only seemingly hard. This is said with confidence, for the end and the fruit of all this experiment and suffering are compensation to the soul, full and complete.

The Natural Scientist does not pretend that he has discovered all of the compensations which Nature has in store for man. He has, however, so far investigated Nature's plan and purpose, as to feel justified in saying that individual completion and happiness here and hereafter, come within the scope and intent of the law. He claims further, that the primary necessity in such an estate is an individual love relation which fulfills the natural law of selection, or marriage.

In such a relation science explains to the individual that he or she finds that correspondence, sympathy and companionship in the several planes of being which meet and satisfy the individual demands of the soul.

This is the vibratory correspondence Nature is forever seeking to establish. This is the harmonic relation which constitutes the highest ideal of the soul.

It is important that every individual should know that every right endeavor is a step nearer to Nature's purpose. The great general task of Nature is lightened by every individual process which refines the physical body, develops reason, and induces morality. To aid the individual in his search for this true relationship, science says to him:

"The fact that you are an individual, differentiated from all others of your sex, must suggest to you that Nature can furnish but one perfect complementary of the other sex. No two men nor women ever had or ever will have the same physical, spiritual and psychical constitution, quality or capacity. No two of the same sex are alike in their necessities. Therefore the individual who fulfills every requirement of your own nature must belong to you and to none other. It is only such an individual who can effect the true harmonic in your life. Furthermore, in this relation, as in none other possible, you will find continual rest, companionship and happiness. In this perfect relation only you may hope to escape satiety."

Here, then, is that marvelous principle of completion through vibratory correspondences, which forbids promiscuity, governs repulsion, and necessitates divorce, thus leading on to new experiments and final accomplishments.

Universal experience is the proof of this law.

Every man and woman can recall his or her own individual fancies and unaccountable attractions and repulsions for those of the other sex. Each one knows from observation or from a personal experience that there is a principle of sex attraction and selection which defies the arbitrary rulings of the civil law, the conventions of society, and even what appears to be one's own physical, material and moral interests. Here is the principle which so frequently runs counter to social and commercial advantages that have been so "reasonably" planned for personal advantage. This "inexplicable caprice" in sex selection is, more than anything else, the factor which colors the history of nations and shapes the individual destinies of men and women.

Though Nature furnishes human intelligence the true principle of selection, the individual has not that independent knowledge of the law which enables him to reach an a priori judgment. Men and women, eager for happiness and full of their own preconceived ideals, rush into legal or illegal marriage with the first individual who seems to meet the requirements. As a result, the large proportion of marriages represents only partial correspondences instead of the perfect harmonic.

Marriage, therefore, as a whole, at the present time, occupies that broad middle ground between complete discord and perfect harmony. This means that few married pairs at present either hate or absolutely love each other. It means that in the average marriage, affection and antagonism alternately play. Thus, the great body of married people live in an individual relation which is as far removed from actual happiness as it is from actual misery.

This will not be disputed by any student of human nature. It will be secretly conceded by most of the married.

Consciously or unconsciously, which means rationally or intuitionally, every man and woman is diligently seeking this true harmonic. Each one hopes to escape the discords which are so apparent in other lives.

While every soul cherishes its own ideal of a perfect love relation, there are, however, very few who really believe in a natural law of realization. To seriously claim that there is such a law, is to lay one's self open to the criticisms of ignorance, and possibly the ridicule of scientific skepticism.

Such, nevertheless, is the law.

The individual who has the courage to give out any unusual knowledge he has come to possess, must take his chances with the public temper and intelligence.

There is a natural law of perfect marriage, and all social inharmony arises through transgression of this law. All of our conjugal infidelities, deceptions, discords and sorrows represent either the innocent or ignorant or willful violation of the law.

Here and there are individual pairs who appear to be fulfilling the natural law of selection as well as the legal contract of marriage. As yet, however, such unions are exceedingly rare. When it is conceived that a perfect marriage relation is a part of the evolutionary scheme, the mind has some conception of the task which Nature and the law have set out to accomplish. This will appear to the skeptic as even a greater task than the gradual evolution of a man from a mollusk.

Advanced science, however, which has been studying and experimenting for ages along these lines, declares that the climax of the individual development rests, primarily, upon this perfect relation rationally effected.

A happy love relation is intensified by rational knowledge of the principles involved, just as the appreciation of music increases with the rational understanding of the laws of harmony. It is not denied that love and happiness may be experienced and enjoyed by the immature and the ignorant. Rational love and rational happiness, however, are joys which Nature reserves for the higher orders of human beings.

The general average of individual happiness will be immeasurably increased when the individual is willing to expend the same intellectual energy in the selection of a life-long companion that he does in the study of the arts and sciences, or even in the matter of horticulture and stock breeding.

The natural tendencies of evolutionary processes may be quickened by the individual and intelligent efforts of man himself.

Physical science demonstrates how human intelligence, when brought to bear, may improve animal and vegetable species. It clearly proves that an intelligent breeding and training of animals improves them physically and increases their intelligence. It proves also that intelligent grafting and pruning improve vegetation in form, in luxuriance of foliage, in beauty of flowering and in the quality of fruit.

If intelligent employment of natural laws thus accelerates the development of the animal and the plant, how much more of satisfaction and benefit the individual will receive through the same character of self-improvement.

Thus, the general purposes of intelligent Nature may be rapidly promoted by the willing co-operation of individual intelligence. Every effort which refines the physical body, quickens the spiritual faculties, increases knowledge, and uplifts the moral nature, is a distinct step toward self-completion and happiness. By such efforts and such steps Nature guides the individual into alliances and relations and conditions of increasing personal happiness.

To even the most unobservant, it must appear that the individual who is himself refined and learned, who has keen sympathies, noble aspirations, and high moral principles, is the individual who has the greater chances for a healthy, successful and happy human life.

It must be admitted that such a man or such a woman is best equipped to make the true selection which is the basis of self-completion and happiness.

The message of Nature to the individual, when rightly understood, is therefore one of encouragement and hope. To every living soul it says explicitly and directly: "There is a universal law of individual development and fulfillment. There is a natural law of marriage and a natural law of happiness. The universal ideal of love and of individual companionship between man and woman is neither a dream nor a delusion. That perfect ideal is the soul's perception of a natural relation. It is the soul's prophecy of an evolutionary possibility."

CHAPTER XVIII.

LEGAL MARRIAGE.

By Legal Marriage is meant marriage in conformity to civil laws and legislative enactments of man, from time to time.

The purpose of this chapter is two-fold, viz.:

(1) To briefly trace the laws of marriage as they are recorded in the past customs, and upon the statute books of civilized nations.

(2) To show that such regulation is a necessary and natural accompaniment of the higher evolution of man.

The limitations of this work forbid more than the briefest outline of the history of marriage. That outline, however, will enable the reader to trace for himself the rise of the marriage relation from its lowest rational condition to the nearest approach to the perfect relation that has been made under civil codes.

In presenting this brief review of marriage customs and codes, the writer relies largely upon an authority of the modern school who has already been referred to. M. Letorneau, in the "Evolution of Marriage," has undoubtedly presented that subject in fuller detail and with greater accuracy than any other modern student and investigator. The very first question with which that eminent writer deals is the question of promiscuity, as a primitive sex practice. As will be readily seen, almost insurmountable difficulties surround this subject. Exact data concerning prehistoric conditions are out of the question. The best that science may do is to compare traditional report and observe the actual sex conditions among living tribes of low development.

In drawing his final conclusions from meager historical fact and the varied present customs of savages, M. Letorneau takes into consideration another range of fact not usually considered in this connection. He holds that, at its base, human and animal marriage are the same in principle. With animal sex relations as a basis for observation, he draws the following general conclusions as to primitive human marriage.*

"Do we thus mean to say that there is no example of pro-
"miscuity in human societies, primitive or not? Far from it.
"It would be impossible to affirm this without neglecting a large
"number of facts observed in antiquity or observable in our own
"day. But we are warranted in believing that a very inferior
"stage of promiscuity has never been other than exceptional in
"humanity."

The scientist next makes a statement which the reader is called to particularly note, when he adds:† "If it has existed "here and there, it is that by the very reason of the relative "superiority of his intelligence, man is less rigorously subject to "general laws, and that he knows sometimes how to modify or "infringe them; there is more room for caprice in his existence "than in the life of animals."

The author of this statement did not realize its deep significance. He did not know that these words are clearly corroborative of Natural Science which declares that sex is an intelligent spiritual principle. It will be recalled, that M. Letorneau notes a law of individual preference when he is dealing with animal marriage. This law, he claims, discourages promiscuity and establishes polygamy and monogamy as the common form of marriage, even in the animal kingdom. In this instance his general law is the same law of individual preference. He does not, however, attempt to explain this general law, nor to identify it with the law he had previously observed. In fact, neither here nor elsewhere does he demonstrate the principle involved in that general law of individual preference which maintains harmony in the animal sex relations.

---

*"The Evolution of Marriage," p. 38.
†Same, p. 38.

The scientist is right. There is such a law. That general law which physical science perceives but does not identify, is the universal law of polarity or affinity. Again he is right when he attributes man's infringement of that law to his superior, independent, rational powers. Man possesses, not merely the capacity to form an individual choice, but he possesses the rational intelligence to execute his Will and Desires. Thus, once more physical science supports the position of the higher science. We may, therefore, from the standpoint of both schools of science, conclude that promiscuity has never been generally practiced in the human family.

This brings us to the consideration of those rational steps by and through which man himself has sought to properly establish the sex relation.

The limitations of this work forbid more than the bare outline of the history of legal marriage. In that meager outline no more can be done than to indicate the several distinct steps that have been taken between the chaotic conditions of primitive marriage, and the civil code under which the best developed live. The legal institution of marriage gradually developed out of what appears to us as mental and moral chaos. Human society, in its primitive stages, displays apparently nothing more than the fierce ambition of the male half and the extreme stupidity of the female half. In reality, however, it is the effort by which undeveloped reason seeks to attain an individual gain or satisfaction.

Out of these natural conditions of force on one side and inertia on the other, but one result has been possible, viz., masculine domination by force and feminine subjection through weakness.

The men owned the women and children. Women were distributed, not married, to suit the gross caprice or fancy of the men of the tribe. They were the property of the tribe, used and misused to satisfy the savage passion of the stronger half of the community.

In tracing the origin of the legal system we cannot go farther back into the mental and moral twilight of humanity than is illustrated in

"Marriage by Capture"

In this brutal custom it is safe to say we strike the foundation of our own highly developed legal system and codes. Marriage by capture refers to the forcible capture and marriage of the women of one tribe by the men of another. This form of sex relation cannot be correctly designated as marriage. It represents more correctly the results of physical combat and the physical supremacy of man over woman and over other men. It marks the lowest point of human nature, and the smallest degree of natural intelligence. On the other hand, it represents the free play of man's naturally fierce passions and ambitions. Marriage by capture involves no greater exercise of intelligence than to effect capture and hold the captive against her will and against other rivals.

The next general step in the rational development of marriage is scarcely an appreciable one to the moral sense of a civilized being. It is,

"Marriage by Purchase and Servitude."

As the terms indicate, this refers to that condition of barbarism where women have become a matter of commerce and barter. At this stage the men not merely own and dominate women and children, but they have risen to an appreciation of their commercial value. The husband owns his slave wife or wives. A man may kill his wives or sell them to other men. He owns his daughters and sells them in marriage. In point of morality there is nothing to mark the distinction between marriage by purchase and marriage by capture, except that we here find the first indications of the recognition of the property rights of others. Marriage by purchase and servitude involves a certain exercise of the rational powers. It is one step in advance of pure lawlessness. It necessitates a crude set of customs or laws. It embraces regulations, confers rights and privileges, and imposes punishments for the infringement of the rights of others.

There is a considerable degree of difference between the period where each man maintains his own supposed rights by his individual strength, and the point where individual rights are recognized by the community or tribe. Marriage by purchase and servitude, with all of its brutalities, involves the exercise of reason and the first faint perception of the rights of others. As may be imagined, this complete subjugation of woman affords her little opportunity for advancement. The sex relation becomes a matter of barter and sale. It is subjected to every species of perversion that the fierce passions and low intelligence of man can devise. This engenders a gross polygamy, which prevails until man attains the next step in evolution known as,

The Concubinate.

This is the general form under which pure savagery emerges into semi-barbarism. This marks the period when the human mind has risen to a rational conception of a civil code of laws.

M. Letorneau refers to the concubinate as that form of marriage which fills the wide space between animal love and the noblest monogamic unions. Referring to legal concubinage, he says:* "The concubinate, admitted and practiced, as we shall "see, in so many countries, is a sort of free marriage, tolerated by "custom, recognized by law, and co-existing by the side of mo- "nogamic marriage the rigor of which it palliates."

This, in fact, is a blending of polygamy and monogamy, which at the same time gratifies the sex principle of selection and contributes to the material wealth of the individual man. The concubinate will be found to cover the sex relation of a larger number of people than any other form of marriage.

We come now to the highest phase of the evolution of marriage. This is the form which characterizes the highest civilizations on earth.

Monogamy.

Here, at last, we have a legal system which, in theory at least, represents the perfect sex relation. Here we have in crystallized form a rational conception of marriage which accords with the spiritual and psychical principles involved in true marriage.

The limitations of this chapter exclude reference to those

---

*Page 160.

countless unnatural marriage customs which have, from time to time, distinguished small communities or groups of people. In fact, the varied forms of marriage which lie outside of the four grades named, may be defined as local, rather than general, in their nature. They represent immediate and local causes, rather than the great general steps by which the rational and moral nature of man gradually ascended to higher levels. Geographical, political and religious conditions give rise to countless vagaries in marriage. For example, polyandry has prevailed where men were largely in excess of women. The levarate is sustained by the codes of both Moses and Manu.

It is also true that prostitution has been and is the universal accompaniment of all legal forms of marriage. It is an almost open practice among savages. It flourishes under the concubinate. It is the licensed crime of many civilized peoples. It is an unlicensed indulgence, not unheard of, in what we define as "Society."

This, then, is the brief outline of legal marriage and of the sex relation in the past and present.

What part does legal marriage play in the evolution of the higher sex relation? What is the attitude of Natural Science toward the already established codes of civilization?

These are the questions which the reader is entitled to put to the writer. They are questions he has the right to ask. They are questions which this philosophy is prepared to answer. Indeed, the whole purpose of this chapter, as indicated at the beginning is to answer these questions.

Natural Science, equipped with knowledge of the spiritual and psychical relations, as well as the physical functions of sex speaks with assurance upon the question of marriage. It holds that legal marriage, as developed by man, is in accordance with the spiritual principles of evolution. Legal marriage represents the effort made by the independent, rational mind of man, to adjust the sex relation. The laws of man upon this subject are only the expressions of his rational and moral self-development from lower to higher planes.

The history of legal marriage is the history of human experiment in the midst of Nature's hidden principles and subtle forces. It is the history of rational intelligence working out spiritual and psychical principles in Nature. The self-evolution of reason and morality is necessarily by and through such experiment. Human reason approaches a knowledge of, and compliance with, natural laws by a long and devious route. Experience by experiment, and wisdom by experience, constitute the only path to individual self-completion and rational happiness. The history of legal marriage is, indeed, a long record of ignorant experiment and consequent penalty. All of this has been necessary to the final rational comprehension of the true sex principle.

The history of legal marriage is vastly more than a history of experiment and suffering. It is also the history of individual achievement; a history of the higher rational and moral evolution of man. It is man alone who introduces a rational selection as a legitimate part of natural selection. This means that men and women unite, not only as the animals, involuntarily and intuitionally, but also through voluntary and rational processes. Legal marriage represents the struggle for self-completion and the struggle for happiness through rational methods. Here is presented the phenomenon of individual and independent reason rising out of an intuitional condition of intelligence. Here is the record of a rational marriage relation evolving from irrational and chaotic impulses and passions.

The phenomena of legal marriage support another proposition of Natural Science, viz., that man is the aggressive and woman the receptive factor in the evolution of man. From the beginning woman has been the receptive rational, as well as the receptive physical factor in legal marriage. She has exerted influence, it is true, continually and to an immeasurable degree. So far, however, as the rational conception and the rational formulation of legal principles are concerned, she is not in evidence. As far as legal enactments are concerned, woman, from first to last, has occupied the position of a non-resistant subject.

The history of legal marriage is the history of masculine domination and feminine acquiescence. From the beginning woman has submitted to and supported laws in which she had no voice.

These were laws which disposed of her personal liberty, her body, her children, he will and her property. The legal marriage system, as well as all civil and ecclesiastical codes, is the direct outgrowth of the masculine mind and not of the feminine. It is the aggressive masculine intelligence which inclines to organization as a means of individual benefit. Civil law represents the aggressive mind seeking self-satisfaction through forceful conquest and acquirement. In the beginning legal marriage embodies neither justice nor morality. Instead, it represents the crude efforts of masculine ambition seeking to gratify itself. It represents the operation of undeveloped reason, supported by a strong will and a strong body.

At the beginning woman had no voice. For ages she was a subject only. So long as man remains the savage, woman remains the slave. If man at the beginning was fierce, woman was stupid. Physical force subjected physical cowardice. The strong will dominated the weaker one. Man assumed control, woman acquiesced. Man was a tyrant, woman a slave.

Because of these fundamental differences in masculine and feminine nature, man was assigned the larger share in the rational development and material progress of the world. To woman, on the contrary, is particularly due the credit of conserving the love relation and promoting ethical content. True to his nature, man primarily bases his legal code upon property considerations and the principle of inheritance. In this he cooperates with other men. Woman, on the contrary, more closely bound to the personal and family relation, attempts to conserve her own interests by personal means and influence. Men concern themselves with general affairs, women with particular relations. The one legislates for himself and the community, the other strives for herself and her family.

Thus, the masculine mind seeks self-completion largely through rational processes, while woman is content to rely upon her spiritual intuitions. Legal marriage at every stage of evolution bears the impress of the masculine mind and masculine nature, rather than that of the feminine. It is, however, to the more spiritual nature of woman that we must look for the encouragement of the spiritual and psychical relations.

Thus, legal marriage represents the co-operations of man and of woman, gradually moving toward the perfect ideal, viz., monogamous and indissoluble union, rationally contracted and legally sanctioned.

This brings us to a consideration of those particular causes which have impelled man to continually change and modify his own laws. When we say that legal marriage represents the rational development of man, we have only named the method by which he is able to improve upon primitive sex relations. This does not explain the cause of improvement. It does not explain why that rational method carries him from the grossest forms of polygamy to the noblest systems of monogamy. Here the inquirer must turn to the higher science.

The secret cause of an improved marriage system is the same cause that binds two birds in conjugal content for life. Here we encounter the same principle which has already been explained as the law of individual preference.

This is the one principle which this work undertakes to elucidate, viz., The Universal Law of Affinity.

The evolution of the legal marriage system, is due, neither to the organs of digestion, generation, nor reproduction. Instead, it is based in the universal spiritual and psychical law of affinity which impels the individual man to seek self-completion and happiness. Legal marriage does not represent a "struggle for nutrition in the midst of a hostile environment." On the contrary, it represents a struggle for material benefit and rational happiness in the midst of unknown principles and forces. This struggle for happiness is carried out by man, not merely through rational methods, but by the intuitional obedience to the universal law of affinity. The legal code is improved not merely by rational judgments, but through natural impulses or affinities, supported and carried out by rational judgments.

The spiritual law of affinity and the individual psychical struggle for happiness, have been the corrective agents in the development of legal marriage. Physical science is right when it refers the depravities of human sex to the independent reason of man. These independent powers of will, of desire, and of execution, enable man to pervert, if not to abrogate, every natural law of being. It is significant, however, to note that even with the power to do so, man does not incline to general promiscuity in the sex relation. Here the universal principle of affinity is more potent than the caprice of undeveloped reason.

While legal marriage represents the rational nature of man, it is the principle of affinity that acts as a check upon his ambitions, his greed and his vanity. Rational judgment restrains the impulses and raises marriage from an irrational to a rational act. On the other hand, it is the spiritual law of affinity and the psychical law of individual preference which sway the rational judgments and prevent marriage from descending to an act of cold calculation.

In marriage, as in every other relation, the rational mind of man has sought regulation for the purpose of increasing his own satisfaction. In this, however, as in every other experiment, the immutable principle of affinity has operated to correct him and to revise his judgments. It sweeps away man-made regulations, from time to time, and vindicates itself in more equitable laws and codes. Even in savagery the natural law of selection by affinity prevails over arbitrary customs. The individual struggle for happiness continually overrides the barbarous marital codes. The love of personal liberty and the desire to give one's self freely in accordance with the individual will and desire, are as strong in the savage as in the civilized. Slave women, incapable of physical resistance, and too timid for open defiance, will yet elude their purchasers, risking life, and give themselves to lovers of their own choosing.

The brutal penalties imposed by the master upon his unfaithful slave wife, fails to compel loyalty, just as the legal penalties of our own code sometimes fail to compel the loyalty of wives to husbands. If sex involved no other gratifications than physical ones, no married woman, whether savage or civilized, would invite the punishments that have always fallen most heavily upon women.

The desire to follow individual preference in the sex relation is the natural pathway toward true marriage. At the same time, it appears as the incorrigible factor in legal marriage. A careful review of the conjugal customs and characteristics of savagery show the same individual struggle for happiness that moves the highly developed. The amours of the Hottentot will not bear discussion by refined people, and yet, back of those gross practices is the same motive which impels the cultured gentleman to seek the lady of his choice, viz., an overwhelming and overmastering desire for an individual satisfaction and happiness.

From the lowest to the highest stage of development the natural law of affinity continually operates to improve the legal system and to establish it upon the basis of individual love, as well as upon rational judgment. Another factor has been in operation toward the improvement of marriage which has not been considered by itself. We have seen that it is the masculine mind which particularly governs the rational development of marriage. On the other hand, it is necessary to point to the feminine nature and the feminine quality of intelligence which more particularly operate to develop the love relation. Man, as already shown, represents conquest by force, psychically as well as physically. Woman, on the contrary, represents achievement by non-resistance, psychically as well as physically. From the beginning, the masculine mind has been the organizing factor of society. It does not matter what the motive, nor how those organizations were maintained. The important fact in this connection is that the gradual rise of the familial clan, community and state, is due to masculine will sustained by masculine physical force.

From the beginning woman has been the receptive factor, not only in legal marriage, but in all social organization. True to her non-resistant nature, she has occupied herself with her maternal duties and in the close personal relations of life. Man, from the beginning, seeks to gratify his nature through organization and rational compacts with other men. Woman, on the contrary, has depended almost entirely upon her personal influence in the personal relations. Though deprived of legal power in person, in will and in estate, woman has not been powerless,

The earliest lesson she learned was the potency of her love nature, the strength of non-resistance. From the savage slave woman to the noblest lady of the land, woman employs the love relations to effect personal achievements. In this woman trades. Through this she matches masculine ferocity and defies masculine codes of law. In this she influences marriage as a spiritual compact, rather than as a physical and commercial one.

From the beginning, man and woman have sought self-completion in accordance with the aggressive nature of the one and the receptive nature of the other. Man has looked mainly to legal processes, while woman has relied wholly upon the personal relations. It was ages before the forceful nature of man made appreciable impress upon woman. It was ages before the pacific nature of woman modified the ferocity of man. Woman has been long in acquiring force of will and rational power. Man has been long in acquiring the powers of non-resistance and developing the intuitive faculties.

Legal marriage represents the struggle for equilibrium as between these differing principles in Nature. The monogamous marriage system of the highly developed nations represents that struggle and the near approach to the true balance between man and woman. For the first time in the history of the world a marriage system seems imminent which will fulfill the demands of Nature and of true philosophy, viz., monogamous union, freely contracted and indissolubly maintained by individual love, and by sanction of the law.

Physical science, in considering the evolution of marriage, says: "The greatness of a nation is measured by the position its women occupy." Natural Science could not state that truth more strongly. For corroboration of this the reader is referred to the history of nations. He is asked to study the existing governments of the earth. In this study he will satisfy himself that all the inferior nations of the earth hold their women in subjection or bound to unnatural customs. Notably among the nations where the perversion of the sex principle has arrested development, are China, Japan and Turkey.

Man-made laws, whether in the name of government or religion, have imposed terrible penalties upon women. They have, as well, imposed penalties upon men and upon nations. Not only woman, but man and the nation, pay the penalty of ignorant rulings and unnatural sex laws. Just in proportion as man perverts the natural, equal and reciprocal sex relation, in that exact proportion he is degraded and his nation is weakened. Japan, the most ambitious of all the eastern nations, cannot hope for great development until her women have been liberated from servile obedience to men, and both men and women have been liberated from the degrading effects of ancestry worship.

What is the attitude of the higher science and the higher philosophy toward the established legal marriage codes of civilization?

This is the final question which the reader will ask in connection with this subject. The position of both science and philosophy in this matter is an unequivocal one. It is important that this position should be explicitly stated by the writer and clearly comprehended by the reader .

Natural Science and philosophy based upon that science, stand squarely for the preservation of the legal institution of monogamic marriage. This position is based upon several considerations, viz.:

(1) Law and order are a necessary part of man's rational development.

(2) Legal marriage is a natural and legitimate accompaniment of the rational and moral development of the sex relation.

(3) Legal marriage conserves the earthly and material interests of the individual man and woman. It conserves the material and moral interests of children, of society and of the nation.

(4) The legal system of monogamic union represents the true spiritual and psychical relation of man and woman.

This position must be understood to mean:

(1) Rejection of the free marriage system as proposed by physical science.

(2) Opposition to everything that would overthrow the mo-

nogamic system, and relieve individuals of their personal responsibility to children and to society.

(3) Uncompromising hostility toward that moral laxity which would substitute either "free-love" or a licensed polygamy for our own rational, moral and natural system of monogamic union.

It is confidently hoped that such explicit statement will never be confounded with the countless matrimonial vagaries of would-be reformers. It is hoped that the philosophy laid down in this work may not be misinterpreted by those who, through general prejudices, distrust any individual or any school that seeks to improve the marriage relation.

What is said does not imply that our monogamic system is the ideal fulfilled. It does mean, however, that in outward form and by intent, our legal code represents the true relation of man and woman. Our legal code presupposes that the marriage of two individuals represents a free and natural selection with a perfect community of interests, material, intellectual and moral. Acting upon this supposed declaration on the part of the individuals, it unites them (by intention) for life.

Whatever may be the individual disappointment and suffering under our legal system, it is not through fault of the civil code. Instead, it is because of the mis-mated natures which the code continues to bind when love, respect and common interests between the pair, are gone. These penalties, however, are not due to legal restrictions. They are the result of ignorant, hasty and mercenary marriage. They are the results of legal marriage which contravenes natural marriage, and not the effects of a faulty legal system.

When parents, teachers, society and the law, labor to *prevent* unnatural marriage, there will be found less and less friction *in* legal marriage. The root of all matrimonial unhappiness lies outside and prior to legal marriage itself.

The man and woman who love each other rationally, who marry rationally, and whose mutual relations are based upon spiritual and psychical principles, will not condemn the monogamic code. To such as these the obligations, restrictions and requirements of legal marriage are accepted as the highest earthly privileges.

Legal marriage, as well as all other forms of sex union, represents the struggle for happiness. Every man-made custom and code which sought to improve the marriage relation, represents the rational effort to effect changes which would increase human happiness. The fact that the legal code is an infinite series of legal reversals, corrections and amendments, shows that the true relation has not yet been attained. Man confesses this failure by his own reversals, and by every new amendment he offers. Every experiment shows that man has labored to achieve an individually happy marriage relation.

The best legal minds among the best developed races agree that monogamous marriage is accompanied by the best results, physically, morally, socially and nationally. They are satisfied that monogamous and indissoluble union is the true type of marriage. No great jurist, however, would claim that our American code is flawless, nor that our civil regulation of marriage is absolutely equitable. In fact, the continual modifications of our present statutes is a confession that the legal mind has not reached its own ideal of a perfect marriage system.

It is true, however, and a fact which every student of science regards with satisfaction, that our monogamic civil code is a rational approach to the true relation. Natural Science, therefore, stands for the legal code, as well as for the natural union of man and woman. It stands just as firmly for the moral obligations of the legal system as it does for the spiritual and psychical affinities of the natural relation. True philosophy, therefore, declares:

(1) Whatever tends to prevent natural marriage, or union by love, is detrimental to the individual, to offspring and to the race.

(2) Whatever tends to weaken the monogamic legal system is detrimental to the material and moral interests of the individual, of children, of society and of the nation.

The civil code looks only to the material, physical and social interests of men, women and children. Nature looks only to a completed individual through vibratory correspondences, physical and spiritual. True philosophy, based upon Nature, looks to individual happiness through and by this perfect physical, spiritual and psychical affinity, sanctioned and guarded by all legal ceremonies and provisions.

CHAPTER XIX.

DIVORCE.

The question of divorce must be regarded as purely incidental to this discussion of marriage.

The primary object, in this connection, is to state the law of attraction. It is not to discuss the negation of that law.

Modern science has accustomed us to speak of the law of "attraction and repulsion." Strictly speaking, there is no such thing as a law of repulsion between individual entities. There is but one universal law of evolution, viz., the law of attraction. The law of attraction is the one and only principle of positive, generative, formative and creative energy. When attraction ceases between any two entities, whether those entities be atoms or organisms or individuals, it is because stronger attractions draw them apart. It does not matter whether that stronger attraction is another atom or organism or individual, or whether it is simply the attraction of a more congenial environment. The action of heat even is no exception.

The chemist demonstrates this very clearly in the mineral world. For illustration, oxygen and hydrogen do not separate until Nature furnishes a substance which possesses closer vibratory affinities with the one or the other.

Repulsion, therefore, in its proper scientific sense, is simply the expression of a closer affinity or a new attraction elsewhere. The effort of the mineral atom is not to separate itself from its fellow atom. The effort is merely to unite with another atom in closer vibratory correspondence with itself.

Legal divorce represents the same general principle. Men and women seeking self-adjustment are misled into unions which furnish but a temporary and imperfect correspondence. The struggle for self-completion or happiness soon impels them to separation. The natural law of affinity continually operates to render the bond intolerable. Those stronger attractions which impel separation may be a place, a person, an ambition, or it may be merely the natural love of liberty for the pursuit of happiness.

Unnatural human marriage engenders strife, resentment and mutual dislike. In this particular human marriage is unique in the kingdoms of Nature. Two animals mate and divorce themselves without engendering mutual dislike and hate. This is because they instantly respond to the natural law of attraction which draws them elsewhere. They instantly obey the dictates of those lower elements which go to make up animal nature.

In human marriage, however, the responsibilities and obligations growing out of a higher element in Nature have created legal and moral barriers to unrestrained divorce. In human society the individual is not permitted to follow the dictates of new attractions. He is bound to an inharmonious mate until natural inharmony is deepened into a definite and aggressive dislike.

The phenomenon of "repulsion," like the law of attraction, is accentuated in each higher kingdom. The so-called "repulsion" which occurs between two atoms is a different thing from that which obtains between two rational beings forcibly bound in an intimate relation. Chemical atoms are continually seeking closer affinities. Chemical "repulsion" is, therefore, but an unconcious incident in the operation of the law of attraction. Even in animal life separation appears as merely a part of the process by which the animal forms a more desirable association. It is an act apparently without individual hostility or moral significance.

In human marriage "repulsion" is intensified, first, by the energies of a higher element, and next, by the legal restraint imposed after natural separation has occurred. The introduction of this highest soul element into marriage intensifies both attraction and "repulsion" and adds moral responsibility to both the act of separation and that of legal divorce.

Thus, the very element which transforms marriage into a rational and moral relation is the same element which imposes legal restraint and causes unhappiness where the marriage is inharmonious. The animal frees himself so easily from an undesired relation that he suffers nothing from his experiment. In legal marriage, however, the natural inharmony of two natures is aggravated by restraint into an active and individual hostility. While the animal is free to follow the natural law of affinity, the individual man or woman is bound to an object which directly bars the way to the pursuit of happiness in another direction.

Though the law of attraction operates in human life with increased energy, man himself has erected barriers which restrict the free and public expression of that law. Man responds to the general law of affinity as readily as the animal, but he also recognizes the responsibilities which his own higher human nature has imposed. Because of such recognition he undertakes to regulate those general laws for the best good of other rational beings like himself. In this effort he formulates laws, marks out duties and raises barriers which would seem to interfere with Nature.

It must be remembered, however, that the rational operations of the human mind are just as natural as those general laws to which the lower entities in Nature so readily yield.

For this reason legal marriage and legal divorce are just as natural as are the free selections and separations of animals. While the dissolution of marriage in lower nature is either an unconscious or an intuitional act, legal divorce in human life is a voluntary and rational act directly chargeable to the contracting parties.

It is true that both human and animal divorce are acts incidental to the same universal principle, viz., that principle which impels every entity to seek vibratory correspondence in another like entity of opposite polarity. Nevertheless, human and animal divorce must be considered from points of view as widely different as are the controlling elements in the life of man and that of animals.

It is true that the general purpose of animal and human marriage is the same, viz., the completion of the individual. It is also true that the individual purpose is identical, viz., self-completion. At the same time, the higher intelligence and the moral nature of man furnish an element in both marriage and divorce which calls for particular laws and particular regulations and restraints. Though the general purposes of individual life, whether of man or animal, are always the same viz., self-adjustment; the effects, however, of individual acts, are as widely different in the two kingdoms as are man and animal in physical appearance, intellectual power and moral capacity.

Legal divorce is the concession made by the general intelligence of society to individuals who are mismated. Legal divorce is admission of the fact that a natural separation has already occurred between the legally bound. The difficulties with which legal divorce is attended go to show that the law is considering the obligations of marriage rather than the desires of the individual for liberty.

Legal divorce is one of the expressions of the rational soul which has risen to the consideration and control of its own affairs. This is in conformity to Nature, for each great kingdom of Nature is directly governed by its own highest element. Human affairs, therefore, bear the impress of the soul element and must be measured and regulated and judged from the human instead of the animal plane. A man is more than an atom or a plant or an animal. He is all these and more. He combines the elements and energies of all lower entities, but in addition enjoys the potencies, powers and responsibilities of a distinctly higher element.

Man, therefore, mates and divorces in terms of this highest nature as well as in those of the lower. Thus, human marriage and divorce produce effects which are far more varied and more important than the matings and separations of animals.

Man is a mammal and something more. He is a living soul, endowed with self-consciousness, the consciousness of other selves, with reason and memory, with an individual will and desire and the powers of their execution.

Self-consciousness, together with memory, involves the consciousness of other selves. This consciousness of others, together with memory and reason, gives rise to what we know as the sense of an *individual moral responsibility*.

Thus, Nature imposes upon man alone that knowledge of his own acts and their effects, which constitutes him a morally responsible being. Self-consciousness, consciousness of other selves, and the sense of individual responsibility, are the essential phenomena of the soul element. No living soul, normally conditioned, escapes the consequences of this highest element in himself. No normal man escapes the knowledge of his own acts, their results, and his own responsibility for them. In this one element we find the causes of differentiation between man and animal. Here are the factors which confer greater happiness and at the same time the capacity for greater suffering. Here are the higher powers which impose individual responsibility and invoke penalties when such responsibility is evaded.

Civil law is distinctly the outgrowth of these higher human capacities, and every code of laws stands for the general recognition of an individual responsibility. It is only the lowest of human beings who would be a law unto himself. The very first step man takes in the direction of law and order is a surrender of some of the things he has heretofore held as individual rights. He recognizes the fact that individual concession means the general betterment of the community. He sees himself as a part of that community, enjoying certain other privileges which compensate for his concessions.

This, in fact, is law and the intent of law, viz., to secure the general good through individual concession and individual restraint.

The law of marriage is based upon the effort to regulate the sex relation for the best interests of society. The law of divorce is based upon exactly the same intent. It is, therefore, a law looking, not to the happiness of the individual, but to the best material and moral interests of society. Divorce, therefore, must be viewed from the point of individual responsibility to society and not from the point of an individual personal happiness.

Civil law implies that the good of society demands, primarily, the proper care and rearing of its children. It therefore assumes that the natural parents are the proper legal custodians and protectors of their own children.

Just here, in this relation of parents to children, arises that moral responsibility which must take precedence of all questions as to the happiness or unhappiness of parents. Here is involved an issue that does not obtain in animal life, but is of vast importance in human society. It may be true that the individual man and woman were impelled to union by a temporary or imperfect correspondence of impulse such as moves the animal to union. There may also arise the same "repulsion" which would immediately separate animals. This impulse for separation is as natural to humans as to animals when the natural principle of harmony it outraged.

Here, however, but not as with the animals, a higher element asserts itself, new considerations enter in and a new principle is evoked for the government of Nature's higher marriages. Animal marriage and separation represent the intuitional, and therefore irrational and irresponsible obedience to a general law of affinity. Human legal marriage may include the same intuitional, irrational and irresponsible impulses, but in addition it represents an independent, rational contract, imposing present and future obligations. This legal contract implies, not merely the mutual loyalty of the two who unite, but it also implies an obligation to the children of that union and to society in general.

It will be seen, therefore, that while animal mating and separation are purely egoistic in their nature, the intent of legal marriage and legal divorce is purely altruistic. Legal divorce is regulated for the good of society alone, and the law holds that the best good of society rests with the proper care and training of children. Because of this fact it finally comes to mean that the question of divorce must be regulated for the good of the child, and not by later impulses, passions or desires of the parents.

A human infant is something more than a digestive apparatus. A child is something more than a mammal with a better cerebral development than a horse or a dog. If a child were nothing more than a mammal with a better brain than a young puppy or monkey the law of divorce might be simplified. If human parentage did not involve other responsibilities toward offspring than nutrition and physical comfort, our complicated divorce system might well be amended. If a child were merely a mammal like any other, the question of nutrition and physical well-being would, at best, cover the obligation of parents. If such were the case men and women could mate and separate as do the animals. If such were the fact children might well be assigned to the care of public officers and "reared by the state."

A child, however, is something vastly more than a mammal with physical functions and cerebral activities. It is, instead, a living soul destined to live here and hereafter. It is a spiritual being with infinite possibilities for good or for evil, for development or for degeneracy. More than this, this human infant has been brought into this world by the self-conscious, voluntary, and therefore responsible act of its parents. It appears in response to a natural law that has been evoked by two intelligent beings fully acquainted with the consequences of the law. This fixes an obligation which does and should take precedence, in both law and conscience, of every personal desire and demand.

This recognition of moral responsibility to offspring separates human from animal intelligence and raises human life immeasurably beyond the life of the animal.

Parental responsibility is read from the book of Nature with equal clearness by science, by law and by religion. Modern physical science demonstrates physical responsibility in heredity. The proved facts of heredity show that a large proportion of children born are the victims of parental deformity and disease. The Law, which represents the rational intelligence of man, holds that parents are responsible for the physical, material and intellectual well-being of their own children. Religion, which represents the spiritual intuitions and moral principles in man, declares that parents are as responsible for the moral training of their children as for their physical and material comfort or their intellectual development.

With the several findings of science, law and religion, Natural Science and its correlated philosophy agree. Both science and

Nature declare that every human infant has certain natural and inalienable rights, viz.:

(1) A perfect physical body and a normal brain.

(2) Material care and provision until old enough to be self-sustaining.

(3) Intellectual and moral training under the daily, personal and loving supervision of both parents.

When the average intelligence rises to a clear perception of the moral obligation to children the demand for divorce will decrease in proportion. That is to say, when the average man and woman recognize their full moral obligation to their own children they will seek to fulfill rather than to evade that duty. This they will do irrespective of present personal desires and impulses. This moral obligation will preserve the marriage tie, in form at least, even when the relation falls short of the ideal.

It is safe to say that this mutual obligation to children, rather than mutual love, binds a majority of married pairs even at the present time. In thousands of households the physical relation of husband and wife is dissolved, while the relation of father and mother safeguards the children in their material and moral rights.

There are, of course, flagrant breeches of loyalty and of decency and of marital duties which justify and necessitate divorce. For such cases the law to-day amply provides.

More than this, certain of our own states have gone so far as to recognize that there is a spiritual as well as a physical and financial relation in marriage. When "Incompatibility" was added to the statutory grounds for divorce, the legal mind had come to recognize a higher principle in marriage than either physical fitness or chattel rights. "Incompatibility of temper," means neither physical disloyalty, criminal conduct, nor failure to meet the material obligations.

Instead, it means temperamental friction and conflict of natures in the higher intellectual and moral life. "Incompatibility" stands for discordant vibratory conditions and for an inharmonious alliance between two intelligent souls. This is one of the legal grounds of divorce which could be avoided through the mutual intelligence and self-control of any two individuals. No matter whether they love, or do not love, in the sense of husband and wife, any two people may adjust themselves to a mutually respectful and mutually friendly relation. This it is their duty to do when the best interests of their children are involved.

Except for the mutual obligation to children, the law might profitably divorce mismated pairs upon their mutual request. The breaking of a burdensome contract between a childless couple would in no sense be detrimental to society. On the contrary, it would benefit the individuals, giving them fresh opportunities for self-development.

Neither science, law, religion nor society is conserved by the forced intimate association of any one man and woman who are without children. In such cases the "collusion" of such a pair instead of being a legal offense, should be the one proper reason and condition of divorce.

The civil marriage codes of the superior nations are in perfect accord with Nature's intent. That is, they grant every subject man and woman a prior right to free selection in marriage. At this point the law joins with Nature to furnish every soul the opportunity to secure its individual happiness. After that choice is made the law, again reflecting the higher nature of man, binds those voluntary partners to a full responsibility for all of the consequences of that contract.

When once that choice is made and the contract signed the individual man and woman have then passed from an egoistic to an altruistic obligation. Their relation is no longer an individual question. It is one that affects society in general.

There are some palliatives for our present marital inharmonies. None of these, however, is a more lax or more stringent divorce law. Criticism and denunciation of our divorce system are both illogical and unjust, so long as there is almost no legal restraint put upon the marriage of youth, upon ignorance and inexperience, or upon poverty, deformity, disease and crime.

Here, at the root of the evil, and not in the flower and fruit, should legislation strike. With our almost unrestrained marriage system and the almost total ignorance of the true principles involved in marriage, our liberal divorce laws are simply preventive of still more flagrant offenses.

It were far better for the nation, society, the family and the individual that our legislators turn to the framing of better marriage laws, rather than to restrictions or amendments of the divorce law.

A prominent reformer of New York City is reported as having said that fully one-third of the married population of New York is disloyal to the physical obligation. If this be true of the purely physical relation, who would undertake to calculate the intellectual and moral "incompatibility" of the other two-thirds? The force of this suggestion will not be lost upon any one who has studied married life, who has been the confidant of married people, or who has had much personal experience.

Even with our present lax marriage laws, there are yet other preventives and other remedies for marital inharmony, viz.:

(1) The study and intelligent comprehension of the universal principle of affinity which is involved in the true relation.

(2) The honest and intelligent effort of young men and young women to marry in conformity to that principle.

(3) The honest endeavor of the married to fulfill to the uttermost all self-imposed obligations, especially those which relate to the personal supervision and training of their children.

It is impossible to discuss legal divorce without reference to the Ecclesiastical Codes, which are felt by a large number of people to be even more binding than the legal contract. Ecclesiastical marriage has been universally more oppressive than the civil codes. This is especially true of the Catholic Church. The church, Protestant and Catholic, recognize the spiritual relation in marriage as well as a physical one. The church recognizes monogamous and indissoluble union as the spiritual ideal.

In this the church is right.

It errs, however, when it sets the seal of spirituality upon all of the marriages it sanctions. Without regard to either physical, intellectual or moral fitness, it joins in an indissoluble union every shade and grade of humanity. Without knowledge of, or regard for, the conditions and motives which lead to marriage, it pronounces as spiritual every union it is asked to solemnize.

Ecclesiastical codes, past and present, have sanctioned and continue to bless every shade of unnatural, immoral and debasing union. It imposes the mockery of a "divine benediction" upon marriages contracted from the most unworthy motives in all human nature. It holds as "spiritually indissoluble" marriages which mock religion and stultify the individual honor of the men and women concerned.

Such are the marriages contracted from political, financial or social considerations; or in fact, from countless other motives than mutual respect and love.

Having declared all marriage a spiritual and indissoluble bond, the Catholic Church is particularly oppressive in maintaining those relations. Through and by the threatened penalty of excommunication, it holds countless men and women in legal bondage, even after crime and brutality have made physical separation necessary.

Thus, while the Catholic Church recognizes a spiritual principle in marriage, it fails to recognize the mistakes of men and women in entering into that relation, and makes no provision for such mistakes.

"No divorce" is as iniquitous as too easy divorce. There are marriage relations which it is both immoral and dangerous to sustain. A separation that is not legalized is a blight to any life. "No divorce" is more conducive to general immorality than is an easy divorce system.

Between the good of society and the natural rights of the individual the law should endeavor to deal equitably. The divorce laws of our many states are fair representatives of the various stages of development and the sectional differences which obtain in the several legal jurisdictions.

It is a commentary upon the "no divorce" system of certain southern states to note the fact that a large proportion of its colored population is mulatto. Of course, it is known that whites and blacks cannot legally marry in the south. The general morality prevailing in certain of our northwestern states noted for lax divorce laws, will compare favorably with the general morality of those southern states in question.

From all that has been said, it must appear that the intent of law, as well as of nature, is monogamous and indissoluble union. It must be seen that divorce is the legal recognition of failure of Nature's purpose. Legal divorce, therefore, testifies to the violation of natural law, as well as the violation of a legal contract. It must be recognized as an expedient and a compromise. It must be accepted as one of those penalties which man is perpetually paying through ignorant experiment in natural law.

Nature's effort is to effect indissoluble union. Divorce represents the protest of the individual against unnatural relations which he has ingnorantly assumed. The demand for divorce is simply a demand for individual liberty in the pursuit of happiness.

Divorce, in itself, is not a factor in development. It is no part of either intellectual or moral progress. One may suffer smallpox from having unwittingly exposed himself. That experience, however, is valueless except by way of warning to avoid contagious disease. Smallpox weakens and mars the physical body, and, temporarily at least, interferes with all of the interests and activities of life.

There are certain classes of experiences which tend to weaken or to harden and demoralize, rather than to develop and improve men and women. The conditions which necessitate divorce are among such experiences. For this reason, divorce, like smallpox, is to be avoided if possible. As with smallpox, divorce is also more easily escaped by prior avoidance of exposure.

What the world needs to-day are better marriage laws. To these, and not to more stringent divorce legislation, must we look for improvement in the marriage relation.

The first object of law should be the regulation of marriage with the view of decreasing the number of false and mistaken marriages. Such legislation is the true province of law, since it directly benefits the nation, society, the family and the individual. While it is true that a keener sense of moral responsibility would decrease the number of divorces, it would not lessen the number of young men and women who are permitted to ignorantly bind themselves in unnatural and loveless marriage.

Such legal enactments, however, would not effect this improvement if prematurely forced upon society. The mere enactment of a statutory law does not necessarily compel the improvement which it implies. The improvement of human relations comes, in reality, through the gradual improvement and self-development of individual men and women. Human beings are restrained, but they are not made better nor wiser nor happier, by mere acts of law.

Nor is the proper development of law possible until the average intelligence and the average morality demand revision and amendment. Neither marriage nor divorce can be legally improved to any extent until public sentiment warrants reform.

This being true, science and philosophy look, primarily, for such reform, to the individual men and women who are the creators of public opinion and public morality. The prevention of false relations and consequent suffering and divorce rests upon marriage which conforms to Nature's law. That is to say, it rests upon marriage which fulfills the universal, spiritual law of affinity or love.

Our present stage of intellectual and moral development promises these improved relations for our younger generations. The slow processes of evolution have already brought the highest type of men and women into an approach to the true sex equilibrium. To the most advanced of both sexes marriage has come to be recognized as a spiritual relationship which must be contracted and guarded and fulfilled intelligently under the laws of both Nature and man.

To such as these an intelligent understanding of this philosophy is a necessity. Such as these would soon be prepared either to teach or to exemplify the law.

It is the part of science to state the principle involved in natural marriage. It is the part of philosophy to explain the effects. It is the duty of civil law, of society, and of the individual, to consider the findings of science and to accept the suggestions of philosophy.

Science lays down a principle of natural marriage and explains the conditions which produce discord and compel divorce. It rests with the general intelligence to accept or reject these deductions. It rests with the individual whether he will seek to so inform himself that he may personally know and prove the law.

It remains to be seen whether legislators will consider these principles in law, whether educators will incorporate them in educational systems, or whether parents and guardians will fortify and guard the young with a knowledge of these truths.

There is but one natural and just preventive of legal divorce. That is marriage which fulfills the law of love. There is but one path to perfect marriage, viz., through a rational knowledge of the principle governing true marriage and through the honest endeavor of free men and women to seek union in accordance with that perfect law.

CHAPTER XX.

TRUE MARRIAGE.

The Mathematics.

Perfect marriage upon the physical plane is a scientific possibility.

The principle involved is the universal principle of affinity.

The process involved is the effort for vibratory correspondence between two individuals of opposite polarity.

The effects of true marriage are three-fold in character, viz., mathematical, harmonic and ethical.

The mathematics of marriage have to do with the rates of vibratory activity in the individuals only.

The harmonics of marriage have to do with the ratios which exist between the rates of vibratory activity of two individuals of opposite polarity.

The ethics of marriage have to do only with the effects produced upon the individual intelligences of a man and a woman, by harmonic or inharmonic ratios between their individual rates of vibratory activities.

Thus, true marriage is vibratory in principle, harmonic in relation and ethical in effects.

The principle, process and effects of true human marriage are analogous to the principle, process and effects of perfect union between entities in all of the lower kingdoms of Nature.

Human marriage includes all of the affinities and effects which characterize the union of two atoms, two cells or two animals. The higher marriage, however, embraces affinities and effects superior to those of the lower unions. Human marriage adds those closer affinities, sweeter harmonies and richer emotions and ethical effects which flow from the union of two rational souls.

The difficulties which stand in the way of true marriage, rationally and scientifically contracted, are great, but they are not insurmountable. The obstacles are many, but they are not endless. There is a natural law of perfect marriage, and knowledge of that law is obtainable. Its fulfillment is a human possibility.

The immediate duty of science is to set forth the principle and process, and to classify the facts involved. The immediate duty of philosophy is to set forth the effects of that principle and process, and to apply them to the individual life of man and woman.

All that has been previously declared as to the universal principle of affinity, applies to the intelligent soul of man and woman. All that has been elucidated as to the process involved in the union of positive and receptive entities, applies to the union of two rational beings. All that has been shown by way of effects, mathematical, harmonic and ethical, applies to the physical, spiritual and psychical union of man and woman.

Thus, the universal principle of polarity or affinity, variously defined as the Law of Motion and Number, the Law of Vibration, the Natural Law of Selection, is also the Law of True Marriage. One and the same universal principle governs the involuntary affinities of mineral and vegetable substance, the passions of animals, and the loves of men. As Nature advances, however, the processes involved in marriage rise in importance. The universal but involuntary struggle of mere substance for vibratory correspondence, is finally replaced by an individual self-conscious struggle of the rational soul for happiness. The same is true as to the effects of marriage in the higher kingdom. They are immeasurably advanced in value to the individual. Indeed, they are so advanced as to make it appear that man and the atom have nothing in common.

For illustration, the chemical affinity of two mineral atoms results merely in a material cohesion which may be readily overcome. The vibratory correspondence of two mineral atoms establishes nothing more than equilibrium between the electromagnetic energies of those atoms. This union represents merely a correspondence of vibratory action in the physical and spiritual substance of which the atoms are composed.

A man and a woman, however, by true marriage, attain something more than a vibratory correspondence in their material organisms. They arrive at an individual self-conscious relation of mutual sympathy. They experience, not only a sensation called passion, but an intelligent emotion called love. They realize an individual effect which they define as happiness. They establish an equilibrium of forces psychically, as well as physically and spiritually.

After ages of research and experiment, science lays down as facts of Nature, certain propositions which directly contravene physical materialism. In these propositions, it offers to man a new inspiration to action and a new ideal of life.

Natural Science declares:

(1) Love is the supreme activity of the intelligent soul.

(2) Rational happiness is the highest attainment of the soul.

Science goes further. It declares that the power to love and the attainment of rational happiness rest, primarily, upon a possible, perfect marriage relation.

From what has already been said up to this point, it will be understood, that man upon this earth plane has a three-fold nature. In other words, man is a triune being, made up of three distinct elements, as follows:

(1) A physical body composed of physical matter, coarse in particle and slow in vibratory activity.

(2) A spiritual body composed of spiritual matter, fine in particle and rapid in vibratory activity.

(3) An intelligent soul which operates both of these bodies and manifests itself through them.

The physical body possesses physical sensory organs, by means of which, what we term physical sensations, are conveyed to the intelligent soul.

The spiritual body possesses spiritual sensory organs, by means of which, what we term spiritual sensations, are conveyed to the intelligent soul.

Thus, the intelligent soul of man is equipped with the instruments necessary for communication with both the physical and the spiritual worlds of matter. If it does not always so communicate, science has at least demonstrated that it possesses the necessary instruments and may do so under proper conditions. The activities of the soul, wherever they are normally manifested, whether in the world of physical matter or in that of spiritual matter, are intelligent in their operations.

With these facts clearly in mind, it will be understood that whenever reference is made to the physical nature of man, the term "physical" includes the physical body with all of its sensory organs, sensations, powers, properties, activities and functions.

Whenever reference is made to the spiritual nature of man, the term "spiritual" includes the spiritual body, with all of its sensory organs, sensations, powers, properties, activities and functions.

Whenever reference is made to the psychical nature of man, the term "psychical" includes rational intelligence and all that is commonly understood by the term "Ego" or "Soul."

At this point it will be of interest to note the fact that from time immemorial, the equilateral triangle has been the acknowledged metaphysical symbol of man, in his three-fold or triune nature.

The equilateral triangle (fig. 1) is a plane figure, bounded by three equal sides and having three equal angles. It is doubtful if the metaphysical symbolism of the equilateral triangle is fully understood or appreciated by our western people in these modern times. It may not be profitless, therefore, to study its significance a little more carefully.

Let the base line, MN, represent the physical side of man, MA, the spiritual, and, NA, the psychical. It will be observed:

(1) That the three lines are the same in length. This means that the three natures of a normally developed man are equally developed. From this it will be understood that the equilateral triangle is a proper symbol of only the normally developed, or

perfectly balanced man.

(2) The three angles at M, A, and N are equal. This means that in the normally developed man, the soul, NA, sustains an equal relationship to the spirit, MA, and the body, MN.

Thus, there is perfect harmony at every point of the individual who is equally balanced.

The reader is asked to particularly note figure 2, which is also an equilateral triangle. It is always understood that the

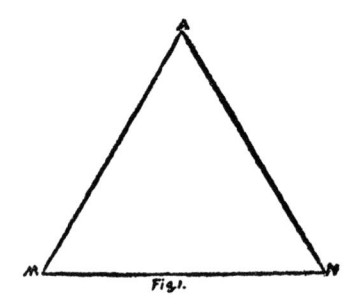

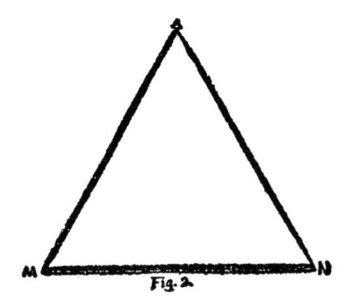

Fig. 1.

Fig. 2.

lines of a geometric figure have but one dimension, viz., length. In figure 1—the figure universally used as the metaphysical symbol—the lines are supposed to be geometric lines, possessing but the one dimension of length.

Strictly speaking, such a figure does not properly symbolize the triune nature of man. For this reason, figure 2 has been substituted. Here is the same equilateral triangle, but with shaded lines, the base line, MN, being a heavy line, the oblique line, MA, a medium line, and the oblique, NA, a hair line. This figure has been chosen with special reference to the scientific explanation necessary in this connection.

The physical body of man is the coarsest of the three natures composing the trinity. It is, therefore, properly symbolized by the heavy base line, MN. For similar reasons, the finer spiritual organism is best represented by the medium line, MA, while the intelligent soul or Ego is properly represented by the hair line, NA.

A clearer idea of the symbol may be obtained if we conceive:

(1) That the base line, MN, is composed of a tow string one-fourth of an inch in diameter.

(2) That the oblique line, MA, is a silk cord one-sixteenth of an inch in diameter.

(3) That the oblique line, NA, is the finest possible silver thread, a thread so fine as to be no more than visible.

Having in mind a triangle so constructed, the reader may obtain a clear conception of the actual relationship of the body, spirit and soul. Let us now suppose that the three sides of the triangle are set in vibratory motion. It will be understood that the coarse string, MN, will vibrate slowly, while the more slender cord, MA, will vibrate more rapidly. The third, NA, will move at a still higher rate of vibratory action.

From this it will be seen that while the length of the sides is the same, they vibrate at entirely different ratios, corresponding to the thickness and the tension of the rope of tow, the cord of silk and the thread of silver. The base line, MN, vibrates slowly, being large in size and low in tension. The line, MA, vibrates more rapidly, being smaller in size and higher in tension. The same is true of the line, NA. This line is almost inconceivably smaller and higher in tension. In consequence, its vibratory action is immeasurably increased.

Scientifically, the triune nature of man corresponds to the three sides of a triangle as here represented. They naturally sustain to each other this vibratory relation. When man is normally or equally developed in each of his three natures, their vibratory action sustains a perfect harmonic relation to one another.

For illustration, suppose the vibratory action of the base line, MN, were represented by the number 3. Any multiple of this number would constitute a harmonic. This being true, the line, MA, should represent some multiple of 3, as 9, 27, 81, etc., while the line, NA, should be keyed to a harmonic relation with both of the other lines.

Thus, it must be understood that man, whether in the grossest or most highly developed condition, has three natures, unlike in degrees of refinement and vibratory action; or, to return to our figure, man is always a triangle and that triangle always has three sides unequal in point of refinement and vibratory action.

Science determines that the vibratory action of the two material bodies is governed by the intelligent Ego, the third and highest entity. This means, therefore, that the soul is the important factor in the higher or self-evolution of man. To briefly state this principle, it is only necessary to say, that all of the individual and voluntary activities of man are set in motion by the intelligent Ego, the soul. This is in conformity to the universal law which leaves the government of each kingdom to the energies of its highest element.

The soul is the governing entity in the triune nature of man. While it is true that the involuntary affinities of lower nature have their influence, they do not control the life of man. The vegetable kingdom is controlled, primarily, by the energies of the vito-chemical life element. Though the plant embraces electro-magnetic energies, they are not the energies which produce the phenomena of vegetable life. The same is true of animal life. The animal is, primarily, governed by the spiritual life element. While the organs of digestion and the bone, blood and tissue of the physical body manifest the energies of electro-magnetism and vito-chemical life, these energies, however, do not give rise to nor control animal life. The animal brain and the individual animal will and desire are expressions of a higher life element. In the same way, the kingdom of man is, primarily, controlled by the energies and potencies of the soul element. It is true, that human life embraces all of the elements, energies and affinities of the lower life elements. The involuntary operations of electro-magnetism and vito-chemical life serve the physical body. The spiritual life element furnishes the physical appetites and passions.

Man, however, is something vastly more than all of the elements, energies and affinities which go to make up mineral, vegetable and animal life. He is something more than an organism which manifests the activities of the electro-magnetic and the vito-chemical life elements. He is also something more than an animal intelligence, consciously seeking to gratify physical appetites and passions. He is more than mere bone, blood and tissue. He is more than appetites and passions.

Man is a Living Soul, with powers, capacities and necessities beyond all of the life which lies below him. He is the rational entity who inhabits and operates a physical body. He is not the maker, but he is the governor of that body. He has the necessary force to increase or decrease its natural demands. He has the power to indulge or deny its appetites and passions. He has the power to injure or improve his physical body. He has the power to leave it at any moment he desires.

These are the facts which show that the higher, individual, intelligent activities of human life are primarily, governed by the soul, the operating entity.

When the intelligent Ego operates upon the still higher plane plane through the mediumship of the physical body, the physical brain, and the physical sensory organs, those resultant activities are classified as "physical."

As, for example, all forms of voluntary physical exercise and muscular effort, including physical sport and physical labor.

When the intelligent Ego operates directly upon the spiritual plane through the mediumship of the spiritual organism, the spiritual brain, and the spiritual sensory organs, those resultant activities are classified as "spiritual." These are the activities in which a master engages when he voluntarily uses his spiritual sensory organs to study spiritual phenomena, or when he leaves the physical body to travel upon the spiritual plane.

When the intelligent Ego operates upon the still higher plane of pure intelligence the resultant phenomena can only be classified as "psychical." They are so classified in that they transcend those activities which are unmistakably physical and those which are unmistakably spiritual in their nature. There are certain operations of intelligence which appear to take no note of matter, either physical or spiritual; as for example, the exercise of pure reason and of abstract thought. Neither is there anything in the mere consciousness of loving, nor in the mere conception of moral principles which we can associate with, or mistake for, either physical or spiritual activities. The activities of love and the development of morality clearly transcend the limitations of physical and spiritual activities as classified by science.

Thus, it appears that Nature assigns to the soul, as well as to the body and spirit, a class of activities peculiar to its own realm.

While each of the three natures of man presents a class of activities peculiar to its own plane, it must be remembered that the operating Ego gives rise to, and governs each and all of the activities themselves. To the soul, therefore, and not to the body, we must look for the factors and causes of the higher evolution of man.

By a normal and steady development upon the physical plane the Ego strengthens and improves its physical instrument, refining it in particle and increasing its vibratory action. By inertia or inactivity the physical organism and the physical powers are left unimproved and undeveloped. In this case man remains as close to the animal plane as is possible for a human being. If the Ego perverts the physical laws of being he not only fails to improve, but he even falls below the brute physically and morally. If, on the other hand, the Ego indulges in abnormal physical development, he overshadows the powers of both spirit and soul, His spiritual intuitions are dull, his moral perceptions are blunted.

For illustration, the average prize fighter develops an abnormally strong physical body with correspondingly strong appetites and passions. This he does at the expense of both of his higher natures.

What is true of physical development is analogously true of spiritual development. A man may subject the physical nature to rigid asceticism. He may render his physical organism entirely passive. He may control the physical appetites and eliminate the physical passions. He may thus develop the spiritual faculties. He may revel in the sights, sounds and beauties of the spiritual plane. He may hold daily communication with "spirits." That very process, however, abnormal in its severity, may weaken the physical body and destroy the physical brain.

What is true of the body and spirit is analogously true of the soul itself. It is possible for the Ego, during earthly life, to concentrate its powers unduly upon the plane of pure intelligence. In this case, we find an individual who entirely neglects the physical activities of life for the pleasure he finds in the activities of pure abstraction and speculation. Such a man becomes the speculative philosopher, or the average metaphysician. He is an intelligence who expends his energies in dealing with mere ideas. To such a man the struggle for nutrition is of little moment. His physical appetites and passions have little influence over his life. Wealth and worldly position are disregarded. The practical duties of life are neglected. Even the æsthetic pleasures fail to touch him. Such a man is as narrow and unequal in his development as the prize fighter or the Yogi.

Still further, the Ego may concentrate all of its energies in any one of the several activities which represent the psychical plane. For it must be understood that the soul, like the body and spirit, has its many distinct powers which are to be used for different purposes.

When we speak of the powers of the body, we refer to the several members and organs, each of which performs a distinct duty and function in physical life. When we speak of the powers of the soul, however, we refer to those distinct activities of intelligence which lie above physical and spiritual manifestation. We refer to activities which are intangible to both the physical and spiritual senses. We refer to activities which are comprehensible by intelligence only.

The psychical powers are broadly divided into two classes, one of which we term the purely intellectual. The other we term the purely moral. The first class of activities has to do with the acquirement of knowledge. The other has to do with the exercise of love. The one class embraces the phenomena of reason, memory, comparison, calculation, abstraction and introspection. The other class embraces the phenomena of love, friendship and altruism.

An individual may cultivate or neglect the powers of the soul, just as he may cultivate or neglect the powers of the body or spirit. Everybody knows that a man may cultivate the muscles of one arm and neglect those of the other. He may train one hand to skillful labor, leaving the other unskilled. He may excel as a pedestrian and lack strength in both his arms. A man may cultivate one or more of the physical sensory organs, neglecting the others.

For example, a musician may develop a fine sense of sound and yet be unable to distinguish the primary colors. He may cultivate the sense of taste and neglect that of hearing. The musician, the painter, the epicure, are examples of cultivation in the several senses of hearing, sight and taste.

In the same way the intelligent Ego may develop or neglect its higher powers. A man may give himself to abstraction, meditation and introspection. At the same time he may wholly neglect the activities of love, friendship and altruism. He may acquire knowledge and ignore both morality and justice. On the one hand, he may expend the soul's energies in love, friendship and altruism, wholly neglecting the exercise of reason and the acquirement of knowledge. On the other hand, the soul may pervert its powers into evil ambition, greed, jealousy, envy and hate. In this case we find that character of degeneracy which is far more deplorable than a diseased physical body or a degenerate physical brain. Here is an example of self-invited moral degeneracy, a sin which Nature pardons but slowly. From this it must be understood that an individual may be as unevenly developed, or as abnormally developed, or as perverted and degenerate in the psychical nature, as he may be in body or spirit.

These are the infinite inequalities of development in each nature of each individual, which give rise to what we know as Individuality and Personality. The individuality of a man is the manifestation of the Ego, the intelligent soul. The word stands for the impression which such an Ego or intelligence makes upon another Ego or intelligence. The personality of a man is the material manifestation of the Ego or soul, as we view it through its physical or spiritual instrument. The word stands for that impression which the external appearance, habits and manners of one individual make upon another individual.

Thus, individuality has to do with the psychical nature only, while personality has to do with the material manifestation of the individual.

True development embraces an equal and steady improvement in all of the members and organs of the physical body, in all of the members and organs of the spiritual body, and in all of the powers and capacities of the soul.

Out of such development, only, does the equilateral triangle rise.

Only by a steady and equal activity in all directions, upon each plane of being does man properly develop his triune nature. He thus attains to that admirable condition of self-poise, physically, spiritually and psychically, which is so readily recognized, but so little understood. Such a man gives us the example of one who is strong, brave, sympathetic, wise, just and merciful.

With this explanation, the reader may be able to measure the task of Nature which aims to so develop every human being.

With this understanding of the relation and offices of the trinity, it must be admitted by men of all religions, philosophies and sciences, that the average earthly man has not attained to such development. The equilateral triangle is, therefore, the ideal and symbol of what we may become, rather than of what we really are.

Man, as we usually find him, represents every conceivable stage of development except the perfect. He represents every kind of triangle except the equilateral.

Figure 3 may be said to represent the primitive man, scarcely risen from the purely animal conditions. It represents a man whose life is lived almost entirely upon the physical plane, or in the enjoyment of sensations received through the physical organs of sense. Both the spiritual nature and the soul fall far short of their proper development. There is, however, a certain degree of symmetry in the figure, in that the two higher natures are equally deficient. Such a man possesses neither keen spiritual intuitions nor good intelligence. He has gross ambitions, appetites and passoins, which he lives to gratify. Such a figure represents man in the lower grades of civilization.

Figure 4 is but a slight variation upon figure 3. The base line is the same, the strongest of the three. Here, however, the psychical nature, though low in development, is yet stronger than the spiritual. This indicates a certain exercise of the intelligence. Such a man is equally gross with the other, with no finer intuitions but better reasoning powers. He is a man of equally strong appetites and passions, but has a better knowledge of the things of this life. This figure might well represent the average Indian chief who rules in council.

Figure 5 is another slight variation upon figure 3. The difference here, however, consists in the better development upon the spiritual plane. Such an individual will have keener spiritual intuitions. He or she will be slightly less gross in physical habits, if not in moral perceptions. This person will incline to a love of beauty and adornment, but will exhibit very little intelligence in such tastes and adornments. This figure might well represent the Indian squaw who, though as coarse as her chief, and even less intelligent, yet exhibits a certain degree of æstheticism in her decorative work with skins, beads, feathers, shells and grasses.

Figure 6 represents a man whose life is lived largely upon the physical plane, with strong appetites and passions. This long base line and long psychical line, together with the short spiritual line, indicate a strong intelligence operating upon the physical plane. Such a man about equally divides his life between purely physical and purely intellectual enjoyments and occupations. He prides himself upon his rationality. His ambitions and aspirations are mainly intellectual, in spite of his strong physical proclivities. In this man the spiritual organism is overshadowed by the physical, and he receives few intuitions of the spiritual world. He is, therefore, more naturally a physical materialist, or an agnostic. Robert G. Ingersoll fairly represents this type.

Figure 7 is the direct opposite of the preceding figure as to the lines representing the two higher natures. It has, however, the

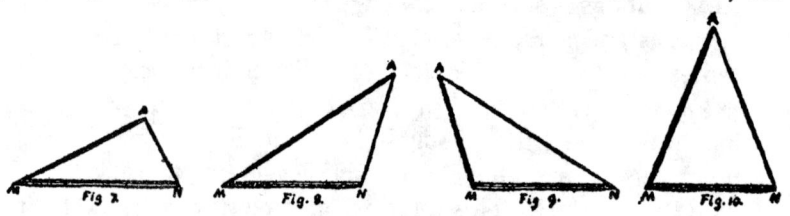

same heavy base line. This figure represents a person with a spiritual organism unusually sensitive. The shortness of the psychical line and the direction of the spiritual, indicate a rather low grade of intelligence and of moral perception. The tendencies of such a person are mainly physical. While he has luxurious tastes and desires, his appetites and passions largely govern him. His intuitions being strong, he is continually admonished to higher things. The intelligence and the moral forces being weak, he is unable to rise above his physical desires. Such a man we know as well meaning, but weak, sensual and foolish. Such men do not acquire enough fame, except by accident, to be cited as examples. George IV, of England, is the best illustration that occurs to the writer.

It has been said that "Hell is paved with good intentions." If this be true, it is apparent that this type of man has contributed a large share of that paving.

Here is a figure (Fig. 8) which represents an individual with only a medium physical organism, an average psychical development, but strong and sensitive spiritual organism. This represents a person who is guided by the impulses and emotions, rather than by the rational judgments. He is æsthetic in his nature, a lover of the arts, though lacking the intellectual development necessary to become the artist. Such men as this often become religious evangelists, but never philosophers or skeptics. The great evangelists are generally fitting representatives of this type of man.

We might say in passing, that this figure fairly represents the large majority of women.

Figure 9 represents one whose psychical activities predominate over both the physical and spiritual organisms. The lines here indicate what we would term a purely intellectual development. Such an intelligence would more naturally incline to the professions of science, law or philosophy. This represents a mind too strong for its environment. This is the type of man who often dies from the results of over mental work.

Figure 10 represents a rare, yet abnormal, type of development. Such an individual has a very frail physical body and weak physical powers. He is, however, highly developed in both the psychical and spiritual nature. Such a man has strong spiritual intuitions, checked and governed by a fine intelligence. His impulses never control him. He invariably consults reason. Such development indicates superior moral perceptions and a keen sense of justice. There is, however, little physical strength, consequently the physical appetites and passions are below normal. Such men are naturally students, philosophers and religious teachers. Emerson and Whittier fittingly represent this type.

The foregoing ten figures represent but ten distinct types. Ranging between these, however, are countless variations from the normal or perfectly balanced type. These countless variations might be represented by an equal number of triangles, varying only in the degree of unequal development.

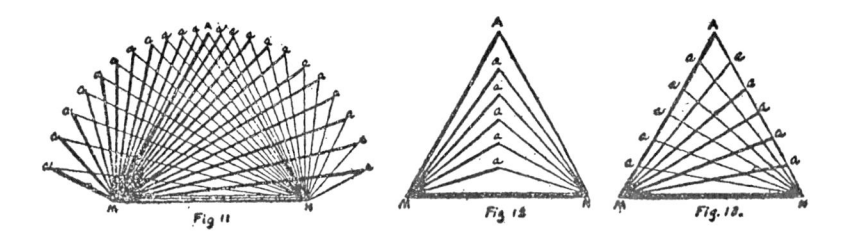

For illustration, figure 11 shows that with the same base line, there may be almost countless variations in triangles whose upper angles, a, a, a, etc., fall outside of the equilateral triangle, MAN.

Figure 12 shows another series of variations with the same

base line. This suggests the countless triangles which may be built on the same base line, each triangle having two equal sides and two equal angles, and the apex of each triangle falling within the equilateral triangle.

Figure 13 shows still another series of variations with the same base line.

From the few illustrations given here, it must be seen that the variations in the triangle are practically infinite in number. Nor can the student fail to see how truly they represent man as we know him.

The figures from 14 to 16 are triangles showing the variations which may occur where the line MA is common to all.

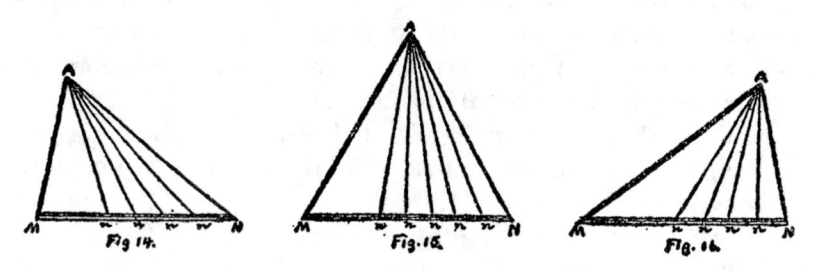

Fig 14. Fig. 15. Fig. 16.

The figures from 17 to 19 represent triangles which have the common line, NA.

These illustrations must suggest even to the most thoughtless:

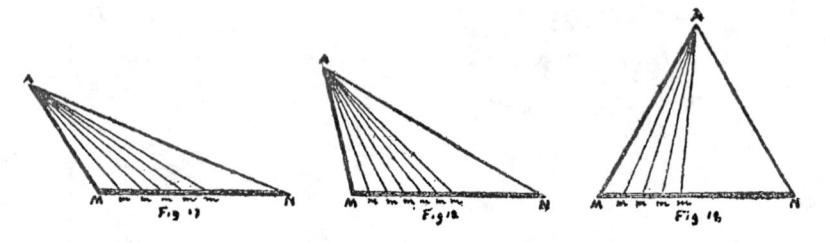

Fig 17. Fig 18. Fig 19.

(1) The infinite variations of individual character.
(2) The difficulties of reaching a perfect balance between the triune natures of the individual.

Science, it must be understood, does not designate a perfectly balanced man, as a perfect man. It does not even define him as a Completed Individual. It must be remembered that the normal balance of the three natures may occur in an individual of very low general development. That is to say, a man may represent any sized equilateral triangle.

For illustration, see the figures 21 to 25 inclusive. Each figure of this series is an equilateral triangle with lines similarly shaded. Each represents a man normally developed. Each, however, represents a different degree of development, or what we might define as a different degree of maturity. Each differs from the other in stature and power of body, spirit and soul. The first figure of the series indicates what we naturally designate as an under-sized individual, but one who is worthy as far as he goes. Such an individual is weak in body, in spirit and in soul. The last figure of the series, on the contrary, represents a man whose physical, spiritual and psychical powers are unusually strong, fine and brilliant. This type of man becomes the hero among the common people.

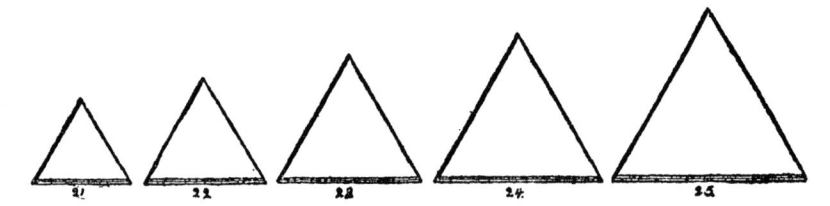

Even among so-called great men, such types are rare. Conspicuous, however, are the few whom their fellow men accept as standards of manly development. Among these few we may name our own greatest of men, Abraham Lincoln, and England's "Grand old Man," Mr. Glastone.

Such as these meet the requirements of physical proportion and strength, of spiritual intuition, of rational judgment and of moral sensibility.

The fact that there are any such as these proves the possibilities in human nature. It suggests also the mighty undertaking of Nature which aims at even a higher development of these harmoniously balanced individuals.

The task of Nature in bringing the individual into a unity of the trinity, is a gigantic one. The far greater task, however, is to establish a perfect vibratory relation between the triune natures of two individuals of opposite polarity.

This is an achievement toward which the Great Intelligence and the individual intelligence move in supplementary lines.

When such relation is established the fundamental principle in Nature is satisfied, viz., that principle which impels every entity to seek vibratory correspondence in another like entity of opposite polarity. When such a relation is established the fundamental principle of human life is satisfied, viz., that principle which impels one intelligent soul to seek happiness in another like soul of opposite polarity.

In this achievement lies the Completion of the Individual.

Thus, the struggle for completion, even to the end, is conducted by universal intelligence through mathematical necessities, and by the individual intelligence because of ethical necessities.

It has already been shown that the human family, in its individual inequalities and abnormalities, represents an infinity of diversity. It must now be recalled that this infinity of diversity is sharply divided into masculine and feminine. It must be remembered that Nature is forever seeking to establish vibratory correspondence or equilibrium of forces between these masculine and feminine halves of humanity. By keeping Nature's purpose (which is equilibrium of forces), and the purpose of the individual (which is happiness), clearly in mind, it may be better conceived why evolution is a process requiring unmeasured time. It will be more easily understood why the individual so often fails to achieve his purposes. It will be better understood why true marriage is so rarely accomplished during this earthly life.

For the purpose of illustrating the mathematics of marriage, we must revert to the symbol previously used to represent the triune nature of man, viz., the equilateral triangle.

All that has been illustrated or said concerning the individual triangle has been with reference to the subject in hand. The present purpose is more especially a consideration of the relation of two triangles to each other. No such explanation were possible expect the student has a clear conception of the triune nature of the individual who is seeking a perfect relation with another individual having a triune nature.

To avoid confusion, however, the masculine and feminine symbols must differ in outline. For this reason the shaded triangle will be employed to represent man, while the dotted triangle will be used to represent woman.

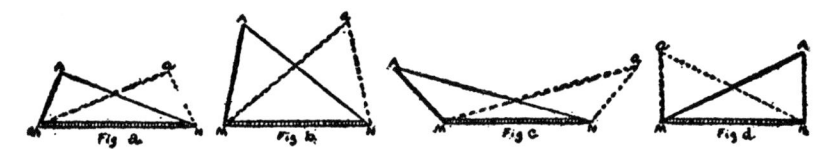

Figure (a) of the present series represents two unequal triangles having in common only the base line. As already explained, the triangle, MAN, represents man, while the triangle MaN represents woman. By referring back to figures 3 and 4 there will be no difficulty in reading the marriage relationship illustrated in figure (a). Here are represented two people of equally low general development. Their one line of correspondence or sympathy, is the physical. The man has a somewhat better psychical development, while the woman is the more sensitive upon the spiritual side. This means that while these two are equally gross as to the physical life, the man has the better rational intelligence, while the woman has the keener spiritual intuitions. Except in physical life, however, they have no sympathies. Such persons have little aspiration above the physical. Such a marriage appears to be purely animal. It is indeed, mainly a marriage of physical passions, appetites and instincts. This, perhaps, represents the commonest type of mismating among people of low development and strong physical natures. Such persons continually mistake these physical correspondences and sympathies for love. They are misled into unions which scarcely suggest the true relation of marriage.

In figure (b) is represented another very common type of mismating. The only difference between this and figure (a) is the superior development which obtains in the two higher natures of each. While both are pushing the lines of their lives upward, they are not the same lines. The individual relationship is not improved. In fact, it indicates greater possibilities for discord and unhappiness. Both of these people have aspirations far above the physical. They have not, however, the same aspirations. The height and inclination of the masculine line, NA, indicate a man well developed in rational lines, while the height and inclination of the feminine line, Ma, indicate a woman of very sensitive spiritual organization.

Both of these individuals have strong physical natures with correspondingly strong physical appetites and passions. The physical, then, is the one line of coincidence. This physical correspondence, however, yields nothing but temporary satisfaction to people thus developed in the higher natures. Each has risen to higher demands and higher necessities than the physical. Unfortunately they are not the same demands and necessities.

The order of masculine intelligence here represented would incline the man to the pursuit of practical knowledge, such as the study of law, medicine and the sciences. The order of feminine intelligence here represented would incline the woman to that which is artistic and æsthetic. She would occupy her intelligence, mainly, with matters of luxury and comfort, and with ideas of beauty in dress, decoration, etc.

The man would pride himself on his "rationalism" and good sense. He would have few spiritual intuitions himself, nor patience with those who had. The woman, on the contrary, would live in her impulses and emotions and trust to her "impressions" for guidance.

While these people are well mated physically, they are far less companionable than a man and woman of lower development who look entirely to physical gratifications for their pleasures. The man would crave companionship in his intellectual life and aspirations. He would need a sympathetic co-worker in the acquirement and use of practical knowledge. The woman, on the other hand, would crave companionship in her higher æsthetic life. She would need a sympathetic soul to share her emotions, impulses and impressions. She requires appreciation for her endeavors to beautify the home, to adorn herself, and to make a figure in society.

For such a pair there is nothing in marriage but perpetual misunderstanding and irritation and disappointment. To such a man this woman would appear as irrational, sentimental and shallow. To such a woman this man would appear as cold, unsympathetic and narrow.

Both may be honest, but neither can be happy in a relation which furnishes but one line of coincidence.

Figure (c) represents a curiously assorted couple, harmonious however, only on the physical plane. The man here represented shows an unusual psychical development—in the wrong direction. Here is a good intelligence prostituted to base uses. It runs parallel with the earth, rather than in an ascending line. This indicates a mind occupied with selfish ambition and with sordid gain. This figure would well represent the money-getter, the great financier who makes his millions in total disregard of honesty, of justice and of common humanity. Such a man may have little education. He is without natural refinement. He is, nevertheless, hard-headed, practical, capable, and a power in the business world. This is the type of man who boasts of being "self-made." This is, indeed, a fact which he mistakenly fancies is to his credit.

The woman here represented shows an unusually strong spiritual development—in the wrong direction. The lines indicate an extremely sensitive spiritual organism, but a very low grade of intelligence. Such a woman, if wholly uneducated, would be superstitious as well as cunning. She would be given to a consideration of signs, omens, presentiments, etc. Such a woman, if educated, is naturally religious, but her religion is as narrow as the figure which represents her. Such religion would mean faith without reason, and piety without principle. Such a person would make a good nun or a good church woman without being able to give one reason for her vocation. Such a woman is impressional, sensitive and unreasonable. Her "religion" is merely an expression of keen spiritual intuitions. It is not a rule of life, based upon rational conceptions of life.

Such a man and such a woman are, in fact perverted from the normal balance. Their development, as indicated by the lines, is in the wrong direction. While such a woman would easily become the religious bigot, the man would just as naturally be a materialist of the most radical type. In matters pertaining to physical life and to material gain, they would find common ground. When it came, however, to matters of "opinion" and to matters of "faith" there would be perpetual and irreconcilable differences.

Who of us, if we jog our memories, but can recall some couple who fairly approach this type? It is unfortunately, a very common type of matrimonial failure.

Figure (d) represents still another type of mismating. In this case it is the man who possesses the more refined spiritual organism. It is the woman who has the higher and better grade of intelligence. Here, as in the other illustrations, the physical side of the union is harmonious, but with those already given as types from which to study relations, the reader will have no difficulty in interpreting this and other double triangles, without suggestion from the writer.

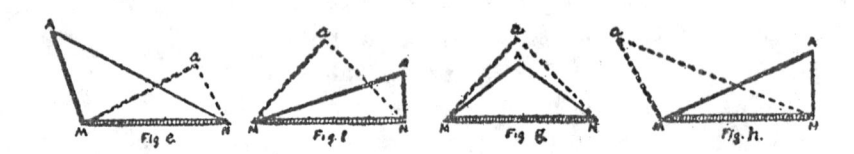

Figures (e), (f), (g) and (h) represent other marriages, or rather partnerships between men and women, where the correspondence is on the physical plane alone.

Figures (i) and (j) are drawn to illustrate the countless variations which may occur with any given form of triangle. Figure (i) is a masculine triangle, suggesting his possibilities in the

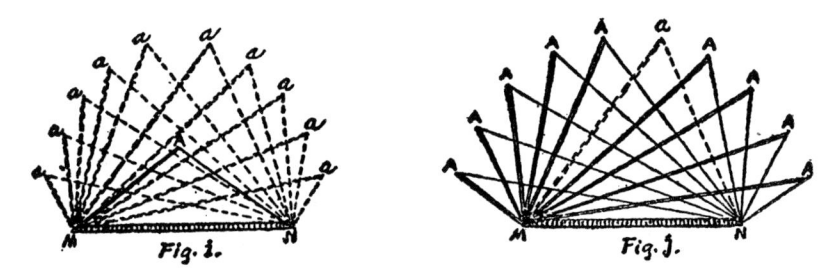

Fig. i.   *Fig. j.*

selecton of a mate. Figure (j) is a feminine triangle, suggesting the possible mistakes in the same selection.

Figures (k), (l) and (m) illustrate mismatings between people

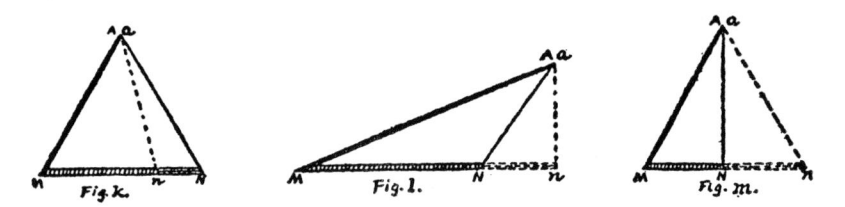

Fig. k.   *Fig. l.*   *Fig. m.*

who are harmonious upon the spiritual plane only. Neither the physical nor the psychical lines coincide. As a result, the bond of union here is in the realm of spiritual intuition, of artistic powers, and æsthetic tastes. Such a pair must look for their happiness through mutual sympathies upon the spiritual plane.

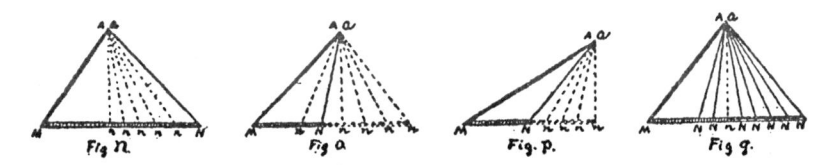

Fig. n.   *Fig. o.*   *Fig. p.*   *Fig. q.*

Figures (n), (o), (p) and (q) suggest the variations which might occur in a mating where the spiritual natures alone coincide.

Figures (r), (s) and (t) illustrate the most unfortunate, and at the same time, a very large class of marriages. Here are represented the unions of men and women who have absolutely nothing in common. In neither of these conjoined triangles are there

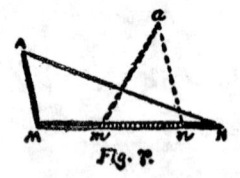

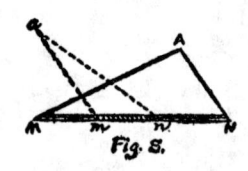

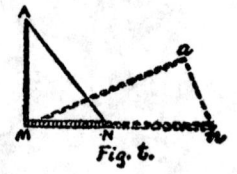

Fig. r.   *Fig. s.*   *Fig. t.*

two lines which coincide. This means that here are represented legal partnerships having none of the elements of natural union. It means relationships without correspondence or sympathy in any one department of being, physical, spiritual or psychical.

In such marriage there is no common ground. There is the bond of neither physical passion, of spiritual intuition, of intellectual pursuits, nor of moral principles.

These figures fairly represent the conventional, mercenary, political and diplomatic marriage. They stand for every character of motive except those of passion, sympathy or love. These are invariably unhappy marriages, without hope of adjustment. In such marriage universal intelligence teaches individual intelligence the error of his way through the discords and disappointments evoked by his own acts.

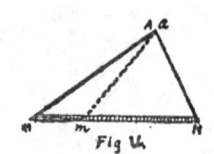

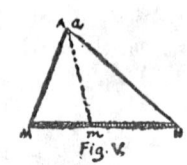

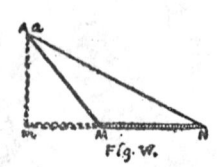

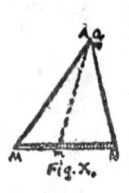

Fig. u.   *Fig. v.*   *Fig. w.*   *Fig. x.*

Figures (u), (v), (w) and (x) represent marriages which are perfect upon the psychical plane alone. This, however, is a relation in which there is hope of final adjustment.

Given, a man and a woman in whose highest natures the perfect harmonic obtains, and you have a relation in which there is a basis of development. Such a pair are naturally and indissolubly bound. There may be differences in the degrees of refinement and vibratory action of both their bodies and spirits. They may be unequal in both physical and spiritual development. In this case the tendency is to bring those physical and spiritual conditions into a harmonious adjustment. When the inharmony of two individuals lies only in the differences of the souls' instruments, those differences will disappear. The soul which has the power to improve and refine either body, and to increase its own activities in this or that direction, can effect any change desired. When once the soul has recognized its perfect mate, it has then the inspiration and therefore the power to work through all conditions which constitute a barrier.

This law of Nature is refered to by Bulwer in his masterpiece, "Zanoni."

Zanoni, the Master, discovers his own in a simple, child-like woman, a gifted artist, but a creature of impulse and intuition, not one of learning and judgment. While she adores the man, Zanoni, her superstitious mind is awed and terrified by the wisdom of a Master. She, the bigoted child of the church, saw only sorcery in his great psychic powers, while he was exerting those same powers to save her from disease and from death. When at last her childish fears have brought him to the guillotine, Zanoni vaguely refers to this difference in their conditions. He assures her, however, that she will one day know him as he is, but not, he says, "until the laws of our being become the same."

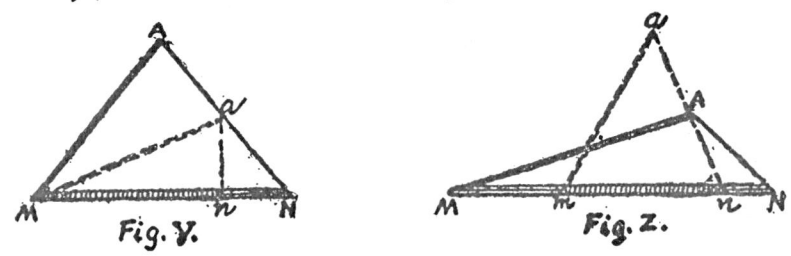

Fig. Y.   Fig. Z.

Figures (y) and (z) represent yet another type of discordant marriage. Thus far the figures have represented individuals of an average general development. This is the rule the world over, that people of the same race, nation and social class, intermarry.

392 HARMONICS OF EVOLUTION

It will be observed that all the previous figures represent mismatings of people who have attained to the same general level in life. Other and more radical violations of the law may occur. A still more deplorable misalliance is possible.

In figures (y) and (z) is indicated, not merely inharmony upon each plane of life, but a general inharmony in point of average development.

For example, figure (y) might represent the marriage of a highly developed Caucasian to a middle class Chinese woman; while figure (z) might represent the marriage of a delicate, educated, cultured, white woman to an Indian of the plains. Such a relation as this means degeneracy to the more highly developed of the pair. It is an unnatural relation for the less developed. It is a misery to the one and stupid dissatisfaction to the other. This is a crime against Nature which Nature but slowly condones.

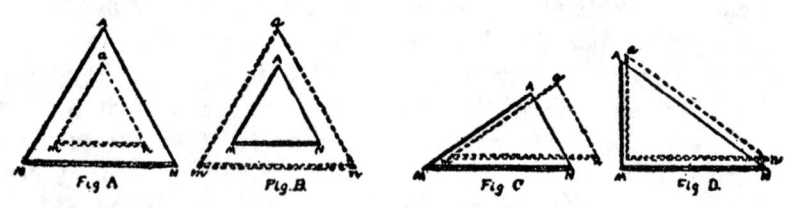

Fig A   Fig B   Fig C   Fig D

Still another form of mismating is revealed in figures (A) and (B). Here is simply disproportion as between two normally developed individuals. Such a marriage is a failure, but more especially to the individual of the larger stature. In this union is disappointment, rather than open conflict; for two well balanced people, however great the difference in their general development, will be considerate and amenable to reason. In every such case the lesser of the two is the happier. As far as the lines of life coincide the lesser finds response. The larger nature, however, lives alone. For him or for her there is no companionship. The lesser never has enclosed and never can enclose the greater, whether the proposition involves two geometric triangles or two human beings.

True marriage may occur between two individuals of any degree or character of development. This fact is illustrated in figures (C) and (D). Here we find true marriage, that is, union which fulfills the natural law of marriage. These figures, however, do not represent the highest possibilities in marriage. They do not represent, either in form or stature or effects or influence, what marriage may mean to the individual and to the world. Marriage, as here represented, is, nevertheless, a true relation, physically, spiritually and psychically. It is a true relation in that the individual natures of the one find perfect response in the triune nature of the other. Such a pair will have realized their ideal. No man's ideals can transcend the best and the highest intuitions of his own soul. Such a pair will find in each other a perfect companionship. The demands of each may be limited. They will, however, be the same demands. Such a pair will seek no further for satisfaction, for sympathy or for companionship in any department of life.

For the purposes of illustration, it is necessary to let the individuals in these pairs of related triangles, stand alone. To join them, as in the other illustrations, would make but one figure. As all lines coincide, the feminine dotted triangle would disappear in the straight lines of the masculine symbol. This, by the way, is another suggestion as to what occurs in true marriage. We would then have but a single figure which is the proper symbol of true marriage.

There now remains but one general type of marriage for illustration. This is the rarest in human society. This is the ideal toward which all marriage systems are naturally, though slowly, tending.

Figures (E), (F), (G) and (H) represent the perfect unions of two individuals. Each figure represents the union of two individuals who are harmoniously balanced in every nature, physical, spiritual and psychical. While all of these figures represent perfect union they represent unions of different value. Here is not merely the union of two harmonious beings, but several harmonious unions of different grades. While each pair represents a perfect individual relation and an individual completion, the several pairs do not by any means represent the same quality of companionship. They do not represent the same physical, spiritual or psychical strength and stature. They do not represent the same degree of energy, nor of power, nor of influence. They do not represent the same measure of learning, nor the same capacity for loving. They do not represent the same degree of happiness to the individual soul, nor the same degree of benefit to the world at large.

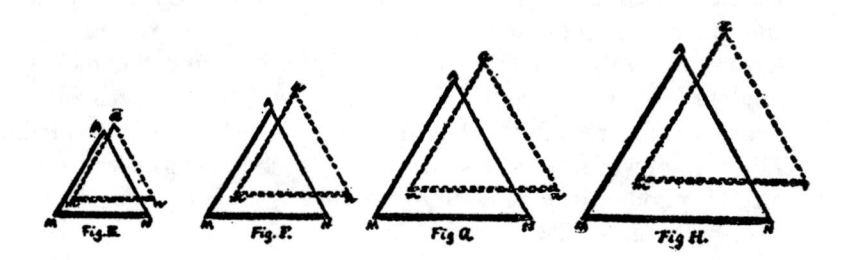

This chapter is not properly closed without one other suggestion. That suggestion must be very brief, however, for it touches a very large question outside of the limits of this work.

The reader may ask, How is society to be benefited by this individually perfect marriage, and by the happiness of one man and one woman?

For answer, he is referred once more to Nature, which accomplishes the evolution of man only through individuals.

The law of heredity is as inexorable as the law of motion and number. Figures (I) and (J) are given as mere suggestions, as to the effects of true and false marriage, upon the child, the family and society.

Figure (I) represents the perfect marriage relation and its mathematical results.

Figure (J) represents the unnatural and inharmonious marriage relation and its mathematical results.

It will be observed that all of the triangles which radiate from the perfect equilateral triangle, as a center, are also equilateral triangles, and that their combination presents a perfect composite figure.

In figure (J) the radiating triangles, with their unequal sides and unequal angles, also illustrate the laws of heredity. They suggest the inharmony of family and society, which necessarily

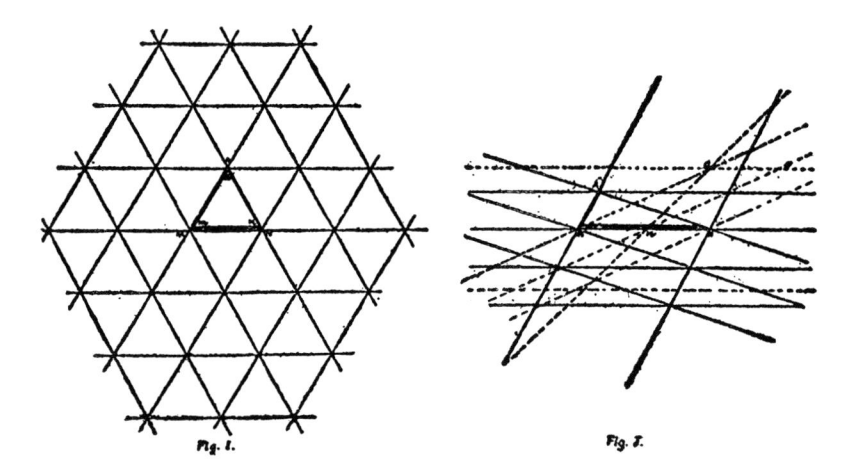

Fig. I.   Fig. J.

result from an imperfect center. This figure, however, is a very fair illustration of our present stage of marital, family and social development.

Thus it is that every man and every woman becomes a factor in the world for social order or for social chaos, as he or she fulfills or fails to fulfill the law of True Marriage.

The Harmonics.

Universal intelligence governs the mathematics of marriage.

Individual intelligence governs the Harmonics of Marriage, and enjoys the music, or suffers from the discords which marriage itself produces.

It has already been shown how universal intelligence seeks to complete the individual upon the purely mathematical principle of vibratory correspondences. It is now to be shown how the individual intelligence completes himself through purely harmonic relations with another individual.

From all that has been said of the fundamental principle of

affinity, it must be understood that when this principle operates through material substance, it means merely vibration of matter. When it operates through individualized intelligences, however, it constitutes a harmonious activity which we call love. It seems unnecessary to repeat that the same principle which produces vibratory correspondences and equilibrium in material substances, induces harmonic or love relations in human life.

Such, however, is the law.

The mathematical principle in marriage has been but briefly touched. This is a subject of interest, mainly to science. We now approach the Harmonics of Marriage, which are of interest, mainly to the individual and society.

To fashion a well balanced man, physically, spiritually and psychically, does not exhaust Nature's resources. Neither does it fulfill the purposes of Nature and the individual. The completion of the individual is not yet accomplished. By a unity of the trinity are established only what science defines as "primary" harmonics, viz., an interior harmony of body, spirit and soul. This is the state or condition of individual poise and power. It is not necessarily a state of completion or happiness. This, however, is the state of being which enables the individual to more easily arrive at the next higher range of harmonics. That next higher range is defined as "secondary" harmonics. It refers to harmonics set up by exterior conditions, viz., by and through perfect vibratory correspondences with another individual of opposite polarity.

Natural Science is, therefore, right when it declares that love is the highest activity of the soul, and that rational happiness is its highest attainment. If the reader doubt this, let him review history. Let him study the activities of men and the love relations of life. Let him consult his own soul, analyzing his own highest aspirations and ideals. Let him discover the main-spring of his own life. If it be not love and loving, he must confess that he falls below his own ideal.

This present work is no more than an attempt to suggest the love principle in Nature. To actually show the power of this principle in human development, would be an endless task. The simple record of human development, without the aid of either science or philosophy, is enough to convince the intelligence that love is the most potent energy in human affairs.

The scientific skeptic is right when he points to the havoc created by "that physical passion miscalled love." He must admit, however, if fair and impartial, that the heroism and faith and endurance of love are the forces which yield to neither time, distance, nor circumstance. If honest with himself, he will confess that love for some one other than himself is the inspiration of his own life and efforts.

The writer is aware that this position contravenes physical science, as well as the theories of certain learned pessimists and degenerates. Physical science, engrossed with the physical vestiges of descent, overlooks the equally patent facts of intellectual and moral ascent. A science, however, which fixes its limitations at the physical, must be prepared to see intelligence turning elsewhere for explanation of that which is not physical. Philosophy which reads man through deformity, disease and degeneracy, must be prepared to find man seeking elsewhere for explanation of that universal impetus toward health, intelligence, morality and love, which we name "evolution."

We are told by men who study the human affections and human intelligence through the digestive organs and through disease and abnormalities, that sex is a physical device for reproduction, that love is essentially lust, and that happiness is a delusion. To such as these we are indebted for statements to the effect that the love relations of men and women are responsible for all of the strife, jealousy and unrest of society.

The eternal principle of harmonics refutes such deductions.

Nature, history and experience show, instead, that all of the discord, jealousy and crime associated with sex, are due, not to love, but to the absence of love.

What these scholars really mean is that the struggle for happiness among the undeveloped and ignorant is a process involving unmeasured strife and discord. They mean, in reality, that the effort to achieve the soul's happiness through the physical passions, has been a prolific source of folly, sorrow and crime.

To debased passion which is preëminently selfish, and not to love which is absolutely unselfish, must be referred all of the strife and discord which mark the evolution of sex. Misplaced affection may leave a temporary sorrow in the soul. Unwisdom in love relations may entail embarrassment. Separation from the beloved may mean loneliness; but never yet, in the affairs of men, did love evoke unhappiness or commit a crime. Ungoverned passion and blind jealousies have made havoc in human affairs. Love, however, from the beginning, has never been anything but love; and "love suffereth long and is kind."

The attempt to define love, except as the highest activity of the soul, would amount to an absurdity. We cannot go behind the word itself. There are no synonyms for the word love. There are no other words which would convey any clearer understanding as to the activity itself. Each individual must measure the meaning and value of "love" by his own observations, his own experiences, or his own ideals.

There is, however, one correction or explanation in this connection, that seems necessary. True philosophy clearly distinguishes between "love" and "altruism." These words must always be used with a distinction as to meaning and value. The interchangeable use of these words is unscientific, and therefore erroneous and misleading. Love is not altruism, nor is altruism love. Love and altruism may or may not exist in the same individual. Though representing the same principle they are not the same thing. The difference between them is the distinction between an activity that is individual and personal, in its object and effects, and one that is impersonal in both object and effects.

Love demands a particular individual to love. Altruism considers the rights and welfare of all individuals. Love represents an attraction, an emotion, and a desire which can obtain only between individual intelligences. Love necessitates a definite object upon which to expend its energies. Altruism represents an intelligent perception of the rights of others and an honest desire to discharge one's own obligations to humanity. Altruism may well be defined as a static condition of love. It represents a love nature highly developed, but it does not represent loving itself. Altruism is simply the impulse and capacity for loving without particular objects to love. It indicates, however, the nature which is especially equipped for loving individuals.

In the "Ascent of man," Mr. Drummond continually uses the words altruism and love interchangeably. This inaccuracy confuses his entire work. He also defines "Love or Altruism" as essentially self-sacrifice, pity and compassion. Several years ago the same author published a smaller book, but a greater one. In this earlier treatment of "The Greatest Thing In the World," he unconsciously gave the world a true definition of love.

He asks this question: "Why do you want to live to-morrow?" He answers it by saying: "It is because there is some one who loves you, and whom you want to see to-morrow and be with, and to love back. There is no other reason why we should live than that we love and are beloved."

Here is a definition of love and the true philosophy of life, condensed to a paragraph. When Mr. Drummond wrote this, altruism was not in his mind. Neither was he considering the claims of reproduction. On the contrary, he was thinking of something very definite and very dear to the individual soul. It is safe to say that no intelligent man or woman will take issue with this reading of love or of life's values.

When Mr. Drummond wrote this he did not have in mind either self-sacrifice or pity or compassion. Instead, he was thinking of something in which he and every other intelligent being delights. He was thinking of something that means inspiration and exhilaration to the individual soul. He was thinking of that in Nature which brings to each of us the sense of fulfillment, and not a sense of sacrifice.

Self-sacrifice, pity and compassion are expressions of the love nature. They are essential elements of altruism. These, however, do not constitute either love or loving. These are not the elements which make up the perfect love relations of life. It does not matter whether those relations are between man and woman. between parents and children, or between friends. Self-sacrifice, pity and compassion are not the things we crave for ourselves. They are not the offerings which we ourselves bring to love. True, love is surrender, but it is not sacrifice. It is tenderness, but it is not pity. It is unselfishness, but it is not compassion.

Love is always ready for sacrifice. It has always a reserve of pity and compassion; love, however, which makes this life worth living, is none of these, nor all of these combined.

The single purpose of this work is to explain the love principle in Nature. It is not to discuss altruism. It is, therefore, limited to a consideration of that character of phenomena which science classifies as love. That is to say, an activity which is essentially individual and personal in its nature. The individual love relations of life are many. They are limited in number and value by nothing except the opportunities and capacities for loving.

The natural order of creation determines the order of the evolution of love. The natural order begins with the affinities of male and female, and of man and woman. After this comes the affinity of mother and child. Then comes the bond of father and child, and of brother and sister. Beyond these are the gradually evolving affinities of family, which we call the blood relationships. This family bond is very largely a physical bond. After the family, come the affinities of friendship, representing spiritual instead of physical attractions. Beyond friendship is that impersonal consideration for all men, which we define as altruism. This is the bond of our common humanity.

Thus, it appears that Nature assigns man and woman to the leading rôle in the evolution of love. Into their hands is committed the task of discovering the love principle, of exemplifying it, and teaching it to the world.

Science points to this relation as the key to the higher evolution of man. Here also it discovers the source of all social discord as well as of all social harmonics. Here is the basis of human degeneracy as well as of human development.

Is it possible for science to explain this activity of love so that it shall be fixed in the mind as a principle in Nature rather than a poet's dream? Is it possible for philosophy to so clearly indicate the natural pathway of love that young men and young women will be induced to seek their happiness in accordance with natural law?

It is believed that the activity of love may be fairly explained by analogy. It is only necessary to find some familiar phenomenon which expresses the same general principle in Nature. That perfect analogy is found in music. This is the one familiar, tangible, earthly activity which constitutes the perfect analogy to the activity of love.

Music, like love, has its purely mathematical side, a mere matter of vibration. It has also, like love, a harmonic side, which covers the relation of vibrations to each other. Finally, as with love, it has the side of effects, viz., the value of those vibratory harmonics when they are reported to the intelligent Ego. These effects constitute musical sounds, or music.

Both activities, therefore, of music and love, rest upon the same general principle of vibration. Both are made up of harmonic ratios in vibratory action. Both produce effects upon human intelligence which are pleasing and desirable.

The difference between musical harmonics and the harmonics of love is as great as the difference between atoms and men. The difference in the value and effects of music and in the value and effects of love, can only be conceived by comparing the activities of atmospheric waves, with the activities of intelligent souls. The one activity represents only the vibrations of the atmosphere falling upon the human ear. The other represents impulses and emotions of a self-conscious soul responding to the impulses and emotions of another self-conscious soul.

Music represents harmonic relations between waves of unconscious physical atmosphere. Love represents harmonic relations between two intelligent souls. When musical harmonics fall upon the ear the Ego experiences a pleasurable sensation. When the harmonics of love exist between two souls each enjoys an emotion which is translated as happiness.

If knowledge of the science of music is deemed an accomplishment, how much greater an achievement is knowledge of the science of love. If a man must know the vibratory theory of music before he can become a composer, how much more necessary that he should know at least the rudiments of the science of love before he defines himself as a lover.

One may love music without knowing either the theory or the practice. So one may desire love without knowing either the principle or the practices.

On the other hand, one may know the science of music without being a musician. That is, one may master the vibratory theory of sound waves, may know the value of every note, and at the same time not be able to run the scale. So the student of human nature may master the scientific theory of love, may fully comprehend the principle of those higher harmonics, without himself ever having experienced the joy of loving.

So also, a man may be a natural musician, born with a "musical ear." Such an individual without a scintilla of the technical knowledge of music, may sing or play fairly well. Nobody, however, will insist that a technical knowledge of music would interfere with, or detract from the natural gift. In the same way, a man or woman may be a natural lover, born with a love nature highly developed. To such as these loving is a natural and necessary condition.

Who that knows human nature, who that has seen the effects of wasted and misplaced love, but will admit that a rational knowledge of the scientific principle of love would be a benefit to mankind? Loving "by ear," or loving in ignorance of all the natural laws of love, is so common a thing as to become a by-word and a reproach. "She loved not wisely but too well," is the epitaph of how many reputations?

A loving nature coupled with an undeveloped intelligence is the natural victim of selfishness and lust. The very fact that woman is fundamentally conditioned to the stronger love nature and the stronger desire for love, makes it doubly her duty to study the laws involved. "Knowledge is power," and the woman who is able to discriminate between friendship, passion and love, either in herself or in another, has the power to avoid all unhappy complications in the love relations. The woman who knows scientifically and rationally what love is and what it is not, need never become the victim of her own desires for love.

To the man or woman who knows the theory of music the science of love will be clearly intelligible. In the same way the analogies between the effects of music and the effects of love will be more readily comprehended by the lover. Only the unhappy soul who knows neither the theory of music nor the practices of love will seriously dispute the findings of the higher science upon this question.

Science has determined that sound involves the following processes, viz.:

(1) By concussion or agitation, atmospheric waves or vibrations are set in motion.

(2) These atmospheric waves or vibrations, traveling through space, strike upon the tympanic membrane of the ear, and are thence conveyed to the internal organs of the sense of hearing.

(3) The impression thus made upon the organs of the internal ear is transmitted to the brain by way of the auditory nerve, and is there recognized by the intelligence as sound.

The intelligence of man recognizes in Nature two distinct general classes or kinds of sounds, viz.:

(1) Musical sounds.

(2) Sounds which are not musical.

A musical sound involves three distinct properties, viz.:

(1) *Pitch,* or that property which distinguishes a musical tone as high or low.

(2) *Volume,* or that property which distinguishes a musical tone as loud or soft.

(3) *Quality,* or that property which distinguishes a musical tone as harsh or mellow, pleasing or offensive to the sense of hearing.

Science has proven:

(1) That the pitch of a musical sound is governed by the number of vibrations per second.

(2) That the volume is governed by the distance covered by the oscillations of the vibrating body or substance.

(3) That the quality of a musical sound is governed by the shape of the vibrations or sound waves.

For illustration: Open the lid of a piano and you will observe that its strings vary in both length and size, beginning with the longest and largest string which is about six feet long and almost a quarter of an inch in diameter, and ending with the shortest and smallest which is less than one foot long and no larger than an ordinary thread.

Now strike the end key on the left as you face the instrument and it will set the largest and longest string in motion. The sound thus produced is the lowest tone of the piano. Now strike the end key on the right and it will set the shortest and smallest string in motion. The sound produced is the highest tone of the piano.

If you could only count them, you would find that the longest and largest string oscillates or vibrates about fifty times per second, while the shortest and smallest vibrates about six thousand times per second. Thus, the slower the vibrations the lower the pitch, and vice versa.

Again, strike the end key to the left very hard, and watch the longest string closely. You will observe that it vibrates at first back and forth over a considerable distance, and that the tone is very loud. The longer you watch it, however, the shorter is the distance covered by the vibrations and the softer the tone becomes, until the string ceases to vibrate and the soul dies out entirely. This proves that the volume of a musical sound is governed by the distance covered by the oscillations or vibrations of the vibrating body or substance.

If you could only see the waves of sound you would observe that the sounding board of the piano forms them into shapes which correspond with its own particular form or shape. It is difference of form in the sound waves that produces the different qualities of tone which we so easily discover in different instruments or voices.

Let us turn now for a moment to the analogies of human life.

As pitch in music is governed by the number of vibrations per second, so the true pitch of a human life is governed by the rapidity of all of its activities. The man whose physical organism is coarse, whose spiritual sensibilities are blunted, and whose moral nature is low and degraded, represents the lowest pitch of human life. He, however, holds the keys of his own development in his own hands. He may raise or lower the tone of his life in exact proportion as he refines his body, cultivates his spiritual sensibilities, and elevates his moral perceptions.

Again, as the volume of a musical sound is governed by the distance covered by the oscillations of the vibrating body, so the volume of a human life is determined by the distance over which its influence may be felt. The amount of energy, physical, spiritual and psychical, thrown into a human life, also determines the distance to which its influence may be felt and its volume extended.

And finally, as the quality of a musical sound is governed by the shape or form of the vibrations or sound waves, so the quality of a human life is determined by the form or manner in which its activities are exerted. Human character is the common expression for quality in human life. Here, again, the key is in the hands of each individual. He may make his life harsh or sweet toward his fellow man, pleasing or offensive, as he chooses to develop his character.

In all these properties of human life, man is both instrument and performer at the same time. The music of his life is what he makes it. It is the result of his own intelligence. If he would attain to the sweetest and the loftiest harmonies of life he must make of himself a perfect instrument and he alone must have absolute command of the keyboard of his own existence.

For the purpose of obtaining additional facts of Nature, however, let us turn to the piano again. Begin with middle C. Strike the key and hold it with the finger. While the C string is vibrating strike successively the notes running up the scale and carefully observe the effect each one has upon the vibrating note C. It will be observed that the first note, D, produces a most unpleasant effect. It is thoroughly discordant when sounded with C. The next note, E, produces a very pleasing effect which seems to be a musical harmony. The next note, F, is not so pleasant. The next note, G, is particularly pleasing in its effect. It is in very close harmony. The next note, A, produces a strange effect. It is not so pleasant in its relation to C. It has a somber or saddening effect. The next note, B, produces the most unpleasant effect of all. It is painfully discordant.

Now strike the next note, which is C, an octave above the first note. Here, again, a most interesting result follows. Its vibrations blend so perfectly with those of the lower C, that the most acute ear can scarcely distinguish the fact that more than one string is vibrating. The effect is one of unison rather than of harmony. It is the same throughout the entire key-board of the piano. If all the seven C's of the instrument are set vibrating at the same time the effect is still one of seeming unison.

Mathematically, there is an explanation for this, as follows:

Science has discovered the fact that the numbers representing the vibrations of octaves sustain to each other the ratio of 1 to 2.

For illustration, 17 vibrations per second will produce the lowest tone, C, which the human ear can distinguish. 17x2, or 34 vibrations per second, will produce the next C an octave above. 34x2, or 68 vibrations per second, will produce the next C an octave still higher. 68x2, or 136 vibrations per second, will produce the next octave above, and so on.

Thus, 17x2x2x2x2x2x2x2x2x2x2x2 equal 34,816, which is the number of vibrations per second necessary to produce the highest tone C, which the human ear can distinguish.

Thus, it appears that the human ear can distinguish as musical tones, a range of only about eleven octaves, or 78 notes of the regular ascending scale. Just why the number of vibrations of any given tone multiplied by 2 will produce an octave above, is not so easily explained. It is simply a fact in Nature and as such must be recognized.

Just why the first and third, or the first and fifth of the scale struck together will produce a pleasing harmony, while the first and second or the first and seventh struck together will produce the most painful discord, are facts not easy to explain. They are facts of Nature, however, and as such must be recognized. The **pleasing effect of the first and third, and of the first and fifth, is** doubtless due to the vibratory ratios fixed by Nature between those particular notes of the scale. A different ratio between the first and second and the first and seventh is doubtless responsible for the unpleasant effect they produce upon the sensitive ear of the musician.

Human life illustrates this same law of relationship, this same law of sympathy. Wherever we find a man and a woman whose lives seem to be in perfect unison there is represented the perfect marriage relation. They stand as distinctive among the human race. They are thus distinguished because Nature has so provided that the perfect unison of all the elements in man and woman produces effects not common in society. Such a harmonic unison and such a response between the body, spirit and soul of a man and a woman produce musical effects which cannot be concealed. Mutual love which creates the most intense and exquisite music of life, results in a condition of happiness to which the great world is a stranger. Such a pair are the wonder and admiration and envy of the less fortunate. Absolute love and perfect happiness are so rare in the experience and observation of mankind that such a relation appears to be a gift of the gods. In truth, however, it is simply the fulfillment of the same law that impels one C string to respond to the other.

Take another pair (and of this class there are many in married life), whose relation is pleasant though not perfect. Here we have a relation analogous to the first and third and the first and fifth of the scale of music. There is a certain harmony though not a unison. This illustrates the relation of friendship. This is not love. The ethical state which such a pair experiences may be likened to the state which pessimism defines as happiness, viz., "the absence of pain." There are others, and a very large majority of married pairs, whose relations are most painful and irreconcilable discords. This unhappy state also depends upon an unfortunate ratio of relationships. These discords correspond to the vibratory relation of the first and second or the first and seventh of the musical scale.

It may be of interest and benefit to experiment a little further. We will now take two pianos tuned to the same pitch, placed on opposite sides of the same room. Sit down at one of the instruments and place your foot upon the loud pedal. Ask some friend to go to the other piano and strike middle C. Hold your ear close down over the key-board and the instant C is struck upon the other piano you will hear the same string of the piano before you respond with a clear and distinct tone. Now ask your friend to stike A, and immediately you will hear the A string of your own piano respond. Why is this? Because the vibratory ratings of the same strings are the same.

It is a principle in Nature that wherever two different objects have the same vibratory rate, if one is set in vibratory motion the other will respond to it. This is not only true of musical strings, but it is equally true of everything else in Nature. This is due to the harmonic relations throughout all Nature.

Given two musical strings keyed to the same pitch located near to each other, and it is impossible to set one in motion without causing the other to immediately respond. Why? Because such is the law of sympathy between them. The analogy is equally true and far more beautiful in the law of human life. It is the key to the perfect marriage relation. A man and a woman whose lives, physically, spiritually and psychically, are in perfect accord, can no more resist this law of sympathy than two strings keyed to the same pitch under the conditions above suggested.

But while you are at the piano try another experiment. Place your foot on the loud pedal and strike middle C again very softly. While it is vibrating have your friend strike the first D above middle C on the other piano very hard. You will find that your C string will stop vibrating and you will cease to hear it. On the other hand, you will hear your D string set up a strong vibration in response to the D struck by your friend upon the other piano.

You are now prepared to ask why your C string ceased to vibrate so quickly. It is because in all cases of discordant notes, the ratio of vibration between them is such that they neutralize each other. In other words, the waves cross one another in such manner as to destroy their force. The result is paralysis of the weaker tone.

The reader will have no difficulty in finding the unhappy parallel in human life. The marriage relation of all times has furnished the painful analogy. A man and a woman whose lives sustain to each other a vibratory ratio analogous to the first and second or the first and seventh of the musical scale, can produce nothing but discord. Every activity of the one is neutralized by that of the other. Life is only a terrible discord resulting in total paralysis of all that tends toward love or happiness. The overwhelming necessity for calm study and a rational understanding of the true marriage relation is best indicated by the records of the divorce courts and the number of unhappy and mismated men and women everywhere.

In view of all that has been said concerning the three-fold or triune nature of man and the frequent use of the triangle as an illustration, it must not be inferred that man, in his harmonic nature, is only an instrument of three strings. On the contrary, each of his three natures, physical, spiritual and psychical, represents a distinct and separate key-board, covering all the range of tones and harmonics possible upon its particular plane.

For illustration, the physical nature of man which constitutes one side of the triangle, is an instrument of itself, upon which may be played all the tones and harmonics possible to physical nature. Every physical sensation, impulse, desire, emotion or passion is a different string upon the harp of physical nature.

Thus, the physical side of man's nature alone represents all the possibilities of the fullest and most complete orchestration. The same is true of his spiritual nature, the difference being that it lies upon a higher and finer plane of life, covering a different range of harmonics. In the normally developed man, however, the orchestration of both his physical and spiritual natures is upon the basis of a perfect harmonic relation between them. The same is true of his psychical nature. While it touches the heights of harmonic possibilities, it is in perfect accord with the other two orchestras of his being.

Imagine the harmonic possibilities of three such orchestras

combined in a single instrument with a performer who could properly operate them all.

Thus, man, even when we consider him alone, represents infinite harmonic possibilities.

Let the human mind run on, however, until it is weary with contemplation of the infinite sweetness, grandeur and power of harmonics inexpressible, and it has only touched upon the border-land of those realities which flow from a perfect union of man and woman in the highest conception of true marriage.

Here, every string of human life in its three-fold orchestration, finds its responsive harmonic in another life.

We must now consider another phase of musical sound which involves a principle of great importance to science, as well as to the true philosophy of individual life based upon science.

It has been shown that musical sounds are the result of vibratory activity. Harmony in music is based upon the relations between musical sounds. Harmony, therefore, must also be traced back to the same vibratory activity which produces musical sound. Harmony, however, produces a sensation through the ear of the listener which is very agreeable and pleasurable. The pleasure we receive from musical harmonics must, therefore, depend upon the same vibratory activity which produces the musical sounds.

When viewed in their scientific aspect, therefore, musical sounds, musical harmonics, and all the pleasures and joys of the soul's response to music, are but Nature's expression of an intense vibratory activity.

Whoever has experienced the deep delight of listening to the symphonic harmonies of a grand orchestra under the direction of a master of music, will better understand and appreciate the principle it is here designed to make clear.

Picture the scene as it appears to the eye from the auditorium of a large theater. The orchestra is upon the stage. Count its members. There are one hundred different performers. They are playing upon as many different instruments. Each man's mind and energies are bent upon performing his particular part as it is written upon the score in front of him. He does not know what his neighbor is doing except as he catches the sound of his instrument. Watch him closely. You will see that he is all alive with an intense activity. He is working as if his life were at stake. Now look at his neighbor and you will see the same evidences of intense individual activity. Now take in at a glance the entire orchestra as it is in the midst of a most difficult presto, crescendo movement, approaching a final and brilliant climax. Every member, from the first violinist down to the drummer, is exercising all of his energies to properly execute his particular score. Even the impresario, the master of music, is beating the air with his baton as if he were endeavoring to annihilate a band of invisible demons.

The picture is one of the most intense energy and activity. It is one of individual activity. It is one of general activity. It is one of combined activity. But what are the results? Musical harmonies and pleasurable sensations.

Thus, it is observed that musical harmony is not a static condition. It is the result of the most intense activity.

So it is in the love relations of man and woman. Love is not a static condition. It is the highest activity of the soul. It sets in motion every instrument and every member of the triple orchestra of man's three-fold nature. It finds its response in the harmonic activities of the triple orchestra of the triune nature of woman. What are the results? The harmonics of life, the happiness of perfect love.

Thus, science observes, investigates and demonstrates the harmonics of marriage as it does the harmonics of music. The man of average intelligence is able to comprehend that music represents the principle of vibratory correspondence in operation. It requires, however, a higher order of intelligence to comprehend, much less to demonstrate, that love is the same principle in operation.

Such, however, is the law.

With this understanding of the principle and processes involved in the harmonics of love, it may sound paradoxical to say that love is a state of rest.

Literally, the activities of love are the farthest removed of all activities from a state of rest or inertia. In a purely ethical sense, however, the harmonics of that activity induce a condition of mind which is termed by the poets as rest. The word "rest," in this connection, therefore, really stands for that perfect equilibrium and harmony of activities which obtain in a reciprocal love relation. It represents that condition where all activities of all the elements in a man and woman constitute a perfect harmonic.

Thus, the "rest" which love confers upon the soul represents, in reality, a state of the most intense activity. It is, in fact, a tremendous increase of energy and accomplishment in every department of being. That increase of activity is, however, so free from friction and so reinforced by a mutual response as to produce upon the intelligent soul only the sense of relaxation, freedom and rest.

That "divine unrest," of which the poets sing, is simply the absence of the true harmonic relations in life. The charm of fine music is the sense of relaxation and rest which follows upon its perfect harmonies. Music is the refuge of tired souls. It rests body, spirit and soul from the friction of daily living.

In countless lives music is the only substitute for love. How many lonely men and women are there who, deprived of love, seek natural and needed rest in the harmonies of sound? Indeed, music is the natural consolation of lonely, loveless lives.

Music and love correspond in their general effects. Both musical harmonies and the harmonies of mutual love produce upon intelligence, under various conditions, a sense of exhilaration, of recreation, of relaxation and rest. But these are only the general effects. Both music and love have infinite moods and variations, with infinite shadings in effects. Those moods and variations, however, must represent the natural law of harmony. The standards of music and the standards of love are as fixed and immutable as the law of motion and number which governs both activities.

Chinese music is but another expression of the perversions in Chinese character. It is not music. It is discord. It is simply

noise which contravenes every principle of harmony. To instruct that nation in natural harmony of sound alone would tend to establish therein a normal marriage system. Neither a man nor a nation that delights in discord can appreciate the natural harmonies of love.

Neither music nor love is the result of arbitrary arrangement. They are not creatures of man-made customs and laws. Neither are they illusions nor habits of thought. On the contrary, they are definite activities governed by universal principles. They are verities in exactly the same sense that electro-magnetism and the vito-chemical life element are verities.

Thus, it is seen that the harmonics of marriage depend upon conformity to the eternal principle of affinity. Such marriage, therefore, simply represents that state or condition wherein all of the physical, spiritual and psychical activities of two human beings constitute a perfect harmonic as to pitch, volume and quality.

THE ETHICS

The general intelligence of Nature seeks equalization of forces through vibratory correspondences and completion of individuals.

The intelligent soul seeks its own equalization or happiness through harmonic relations with other individuals.

Thus, the law of motion and number has a general, material, mathematical design; while the intelligent soul has a particular, individual, ethical motive and purpose in view.

The ethical purposes of an intelligent soul are far removed from a mere matter of vibratory action.

When a great orchestra enters upon the rendition of a symphonic poem it has an object in view which is distinctly apart from agitating the atmosphere. Nobody would say that these musicians had come for the purpose of setting up vibrations of the atmosphere. We must acknowledge, however, that this is exactly what they do, and all that they do, in a purely physical sense. Nobody would say that the intent of that audience centered in the law of vibration. We must admit, however, that this is the literal fact.

What, then, occasions this meeting between orchestra and audience? It is clearly not a question of interest in vibratory law. It is not to learn how music is produced. It has, in fact, nothing to do with the science of music. The motive of a great concert is very far removed from the mere matter of vibrations of the atmosphere, or from the relation of sound waves to each other.

That which inspires the musician and attracts the audience is simply and solely a matter of effects.

The object which the musician has always in mind is the effect he is producing. His object is to set up sound waves of such pitch, volume and quality, so related to each other as to delight his audience. The audience, on the other hand, is sitting in a perfectly receptive mood, with the single motive and intent of being delighted.

In its final analysis, then, the meaning and value of sound waves are the sensations and impressions which they produce upon listening intelligence.

To expect that the lover should keep this vibratory principle continually in mind, would be like insisting that an epicure should consider the processes of alimentation whenever he sat at a dainty feast. It will be admitted, however, that a scientific knowledge of the processes of digestion would tend to prevent gluttony, as well as enable the gastronome to cater to a refined taste without disquieting consequences.

The lover seeks his beloved with the same intent that moves the audience to seek the musician. What the lover desires, and all that he desires, are the ethical effects of love; those exhilarating and inspiring sensations, impressions and emotions which he is to share with another. He cares nothing, generally knows nothing of the vibratory principle which governs the impulses, passions and affections which he experiences. He has no knowledge of the vibratory conditions which loving induces. He has no remote idea that by the very act of loving he changes the vibratory conditions of both his physical and spiritual organisms. The lover, first and last, is seeking his own Happiness, which is bound up in the Happiness of one other human being.

It is not until a philosopher becomes the lover that the lover concerns himself with the science of love.

So the average man everywhere is individually concerned with the effects of love instead of its mathematical processes. The object of intelligence is universally its own ethical content. Intelligence investigates the mathematics only as a final means to its desired ethical ends.

Where, and under what conditions, or in what relations, shall I as an individual, find content, peace and Happiness?

War, art, science, law, literature, religion, philanthropy and "society" represent activities in which individual intelligence is attempting the solution of this question. The important consideration in this instance, is whether the lover, after all, is not the wisest of all experimenters.

This philosophy of individual life accepts the perfect marriage relation as the necessary condition of individual Happiness, an essential experience in the development of individual character, and in the normal progress of the soul. This position, it will be observed, maintains happiness as the normal state or condition of the soul, and unhappiness as a departure from that state.

If the normality of Happiness be doubted let the reader study mankind and himself.

What principle in Nature is it that impels an individual to conceal his unhappiness from the world as if it were a deformity or a disease?

Unhappiness is as clearly an abnormal condition as are physical deformity and mental aberration. The commonest facts of daily life prove that there is a normal ideal of the soul as well as of the body, that there is a normal standard of psychical intelligence, harmony and Happiness, just as there is one of physical strength, courage, health and beauty.

The failure to attain this ethical ideal is a deeper humiliation to pride than failure to reach the physical standard. If this were not the case of Nature how shall we explain the fact that the **unhappy, like the deformed, diseased and criminal, shrink from critical observation and employ every art and artifice to conceal their misfortune?**

Without conscious reasoning the soul perceives its natural right to happiness. This knowledge comes first by intuition and next by independent reason. To be physically perfect, mentally strong and ethically happy is the normal ideal towards which humanity moves. There is another common but significant fact which bears out this deduction as to the normality of Happiness and its dependence upon the love relations. If we measure the value of a thing by the sorrow its loss occasions then love is surely the "greatest thing in the world." There is no loss that compares to the loss of the beloved one. There is no unhappiness like that of disappointed love. There is no form of poverty that an individual so skillfully conceals or so reluctantly confesses as the poverty of a loveless life. A man will admit financial straits or physical disease, he will confess his ignorance and thwarted ambitions. He will not, however, if he have natural pride, confess that he has lost the object of his love. Under such affliction he shrinks from pity as he does from scorn. He has but one desire, to hide his poverty, and to deceive his friends as to the state of his soul.

To hide this misfortune and poverty in the higher nature, men and women daily and hourly live lives of deception. Who can estimate the number of unhappy marriages deliberately entered upon by men and women for the sole purpose of concealing previous disappointments.

The mere sight of a disappointed and of a successful lover is a lesson in the law. The one arouses almost the same sense of pity and commiseration that we feel in the presence of physical deformity and disease. The other instantly gains our sympathy. We pass him smiling involuntarily, reflecting back his joyousness. If the disappointed lover but drop his mask for a moment he excites either ridicule or pity. He is a social failure, a depressing object to his friends and a burden to himself.

But all the world loves a lover. It loves him for his radiance. He represents to the soul, ethically speaking, what physical perfection and beauty do to the eye. The world loves a lover because, for the moment he is the visible, living ideal of every other soul. He is our own desire tangibly realized in the flesh. Even the skeptic and cynic, decrying love as lust and happiness as delusion, envy that ecstasy which lifts the lover above the plodding men of earth.

Even the onlooker realizes that the lover lives in a world of his own. The unloved and unloving are always keenly conscious of the vast gulf which rolls between their own estate and his.

If the mere sight of the lover so clearly suggests his separateness from other men, what is that experience to the lover himself? He who has realized the transforming power of a perfect love already occupies a new heaven and a new earth. To him all things have been changed in the twinkling of an eye. Even the face of inanimate nature appears changed and glorified to his sense of vision. The difficulties and perplexities of his life have vanished. What was discouragement is now hope. What was in doubt is now cleared up. Tasks have become opportunities, and failure is a word he appears to have forgotten.

Every faculty of body, spirit and soul responds to the new conditions of harmony. Physical weariness, spiritual inertia, and mental indolence vanish, to be replaced by a new and bewildering strength, buoyancy and activity. To himself, if not to his friends, he suddenly appears to possess wealth and power and knowledge. For the first time in his life he knows the true meanings of the words, life, liberty, wealth and Happiness. For the first time he knows himself to be at one with all Nature. He wonders that he never before realized the loveliness of this world. Unsuspected beauties flash upon him everywhere. He feels himself at peace with all mankind. He discovers new virtues in his friends and acquaintances. He suddenly realizes the joy of mere existence. Nor is this in any sense imagination. It means merely that the lover becomes a perfect note in the higher harmonics of Nature. He has risen to conscious sympathy with the highest laws of being. He is become a seer and interpreter of truth. He boldly proclaims that God is Love and Love is God.

All things, humanly speaking, are possible to the lover. The courage, the endurance, the patience and the suffering of faithful love are reread and retold from generation to generation. The bare facts of history irrespective of science constitute unanswerable proof that man and woman alone have wrought out the true Love Story of the World. These are the unanswerable proof that love is born of the soul and not of the body.

Nothing in the great drama of human life so quickly and so deeply rouses the soul to sympathy as the sight of the mutual, loyal love of a man and a woman. It does not matter whether that romance is enacted in the lowest or in the highest social life.

Who that loves happily or has witnessed the perfect relation, can question or deny that transformation in his own life or the transfiguration he has perceived in others truly mated? Only the man and woman who love really live. Only such as these are exercising the highest faculties of the soul. Only such as these experience that rare exhilaration of body, spirit and soul which constitutes the highest earthly Happiness. Only these have found the key to the higher life. Only these have proved that life is worth the living.

If this harmonic or Happiness principle, and this ideal of perfect love were not so firmly rooted in the soul, what disastrous results had followed these modern, "rational" doctrines of physical materialism. Even religion, teaching faith, hope, patience and compassion, fails as a philosophy of Happiness here and now. Physical materialism is pessimism. The doctrine of degeneration is the doctrine of despair, while Orthodoxy stands for resignation rather than inspiration to the individual.

Happiness is the normal destiny of the soul. It is, therefore, neither the outcome of physical passions, which are blind, impatient and fitful, nor a state of negation. It is not a delusion of the mind.

On the contrary, it is as much a verity as matter. It has for its base the same principle that gives warmth, color, life and music to this physical world. It is the normal state of the intelligence, just as health is the normal state of the body. That subtle but distinct exhilaration which distinguishes the happy individual is based upon an actual condition of the soul, just as physical intoxication is based upon an actual condition of the physical nervous system.

"Beaming," "radiant," "illuminated," "transfigured" are the words familiarly used to describe this psychical phenomenon as it manifests itself in the physical countenance of man. These words describe conditions which are as much facts of psychical nature as the words "height," "weight," "strength" and "beauty" describe conditions of physical nature.

This exhilaration called Happiness is just as truly an expression of natural law as molecular action, or growth, or life, or love itself. Indeed, it is an expression of the same law—in a higher realm. Happiness is therefore, just as properly the subject of scientific investigation as the law of gravitation or of heat or of light. The Struggle for Happiness is just as truly a human activity as the struggle for nutrition and the struggle for reproduction.

What then, according to science, is the final and vital issue in the struggle for individual Happiness?

This question has been generally answered in those chapters dealing with the natural law of selection, the struggle for completion, the struggle for Happiness, and the mathematics and harmonics of marriage. It may, however, be more specifically answered from still another point of view.

Science, familiar with the intellectual struggle and the ethical ideals of man in two worlds, finds that the struggle for completion and Happiness appears, in the last analysis, as the struggle for intelligent companionship.

Actually proving this as a matter of natural law has consumed ages on the part of both Nature and science. It has cost untold energy and effort and suffering to the individual who finally must admit that it is the intelligent soul and not the body of man which is forever seeking satisfaction. On the part of science it has consumed ages of study and experiment. It has required all means and methods known to intelligence. It has required the deepest thought of the wisest scholars to finally determine that the highest relation and the highest Happiness are essentially the intellectual response of soul to soul. All this has been necessary to final acceptance of this perfect individual ethical relation as a prerequisite in the full development of individual character and as the necessary gateway to still higher achievement.

In a dim way every living soul realizes his need of companionship for his intelligence. This, however, is a need which very slowly defines itself as such to the intelligence itself. In a dim way every unmated soul feels its incompleteness and its isolation. The necessity for overcoming or satisfying this vague but persistent need gives rise to that restless and often irrational condition which we define as "society."

In the lowest reaches of "society" the individual seeks to satisfy his intelligence almost entirely through physical appetites and relations. Even here, however, he must have only intellectual equals, if he hopes for pleasure. Even here his demand is unconsciously for an intelligent companion. In a higher stage of physical refinement where the spiritual nature has asserted itself, the effort is for æsthetic sympathies as well as physical affinities. Here also, the real effort is to find another intelligence which enjoys the beauties and harmonies of Nature from the same point of intellectual development as well as physical refinement.

It is only among the highest types that the struggle for intelligent companionship becomes a self-conscious and rational effort on the part of the individual. As a result, companionship means vastly more and *is* more to men and women of high degree than it is to those of lower range. The demand is now, self-consciously, the demand for response and fellowship in the rational activities and the ethical ideals of one's own soul.

In one respect this psychical demand conforms to all lower demands, that is, it represents, primarily, the operation of the law of polarity. This statement may, on the first thought, provoke criticism. It is, nevertheless, natural law, provable and proved under every test. This psychical demand, scientifically stated, means the demand of one intelligence for another of such pitch, volume and quality as will constitute a perfect harmonic. This perfect harmonic, as already elucidated, necessitates the complementary energies of a positive and those of a receptive intelligence.

This, then, a companion for intelligence, is what every individual is, in reality, seeking in his struggle for completion. This, then, is the guaranty of individual Happiness. Nothing other than this satisfies the soul. Nothing lower completes it. Nothing different confers permanent and rational Happiness.

This is the fiat of the Great Intelligence which guides Nature in all departments.

Universal history and experience corroborate these deductions as to the harmonics of individual life. The individual who knows history or society or himself, knows that the value and beauty and charm of life are made or marred by the individual relations which he deliberately assumes, or which are thrust upon him by stronger will or circumstance. So long as the closest relations of life are discordant, the soul knows neither rest nor peace nor Happiness. This is the fact, no matter what may be the individual possessions or powers or knowledge or honors.

Who that is wise seeks his Happiness except through the love relations?

It is not, however, until human intelligence exhausts all other means provided by Nature, that it finally comes to this conclusion. After the soul has long and vainly soughts its ethical needs through physical passions and æsthetic pleasures, it comes to realize that the joys of existence lie far outside of the domain of the purely material.

In the lowest rounds of life the struggle for Happiness appears, indeed, as the unhappiest of struggles. Brutal brawling and sports and debaucheries absorb all of life's energies which are not required in the struggle for nutrition. This degraded phase of revelry and sport merges into the "good time" so persistently sought by honest ignorance. In this grade of life innocent folly and trivial sports replace the brutalities and immoralities of vicious ignorance. Later on we have "society" more refined, more cultured, but scarcely wiser than the common people. Here is the struggle for Happiness with infinite waste of energy, vitality and intelligence. In the name of "pleasure" life is given up to mere recreation, entertainment and amusement. "Society" represents the struggle for Happiness through more refined physical and spiritual conditions. It represents epicureanism and æstheticism.

There is, however, a world of still higher ideals where the intelligence in terms of its essential nature seeks its own content. This is the realm of purely intellectual and moral activity; the world that embraces science, art, philosophy and philanthropy. This is the world whose inhabitants seek Happiness in knowledge, labor, thought, and in service to their fellow men. This includes the religious life in which the soul seeks surcease of sorrow. Here are the dreamers and ecstatics who, wrapt in visions of a future Happiness, take little thought for achieving it in this present life.

It is only after long trial that the soul discovers that its ultimate serenity is not secured through the senses, nor by intellectual occupations, nor in altruistic labors. It comes to acknowledge that the individual Happiness of a rational being rests, primarily, upon its individual relations with other rational beings like itself. Not until all other expedients have been tried and abandoned does man confess that the life of the soul is a life of love. Not until then does he realize that the charm of life lies solely and only in his individual relation to some one other human being.

This is the marvelous truth which has been in process of proving ever since man set out upon the path of self-completion and Happiness.

Loneliness is the commonest and the heaviest cross that is borne in this earthly life. A sense of isolation as to one's inner life and motives and aspirations is an almost universal impression. The conviction that one is misunderstood is an everyday experience. To feel one's self without sympathy, even in one's own family, is so common as to excite neither wonder nor protest.

One man says that he is lonely in "spirit," another that he is alone in his "heart life," still another that he is alone in the "life of the soul." No matter what terms one employs, they all mean the same thing. Each means that the essential intelligence, the "I," is alone in its rational and ethical existence.

To be misunderstood and alone is accepted by the average individual as the state of Nature.

This sense of loneliness increases as the Ego ascends the scale of development. There comes a time in the life of the soul when rational intelligence assumes control, when it discards the fleeting satisfactions of the senses, as the elements of Happiness. There comes a time when æsthetic pleasures, intellectual occupations and even altruistic labors are no longer mistaken for the primary sources of Happiness. This is the period when the soul realizes its essential need for an individual companionship in its highest activities, when intellectual and ethical fellowship become the first necessities of existence. To such an individual physical association, without this higher response, is loathsome; and mere æsthetic sympathies yield but a momentary pleasure. Neither the treasures nor triumphs of this world confer permanent joy upon such a being.

"It is not good for man to be alone."

The ancient poet gave voice to Nature. His pronunciamento has been established throughout the ages as the voice of God himself. This has become an article of faith confessed by the whole human family.

The inspirations which underlie the most enduring works of man had their source in this cry of the intelligent soul for true companionship. The most sublime in human accomplishment stands for this hope and expectation of the soul. Art and literature are but echoes of this universal refrain. The loftiest in poetry, the sweetest in music, the loveliest in color and form, are contributions to the true Love Story of the World. They are the shadows of other men's desires. They are confessions of our own. This hunger of the intelligence runs in minor key through everything that is worthy in human achievement, and the loftier the ideal of the artist the loftier his production.

This inspiration, so clear and distinct to genius, is also the unconscious motive of the plodding lives below the level of genius. And if we but analyze the individual need that inspires the enthusiasm of the religious devotee we shall find that his hope and expectation do not differ from the hope and expectation of all other men. The search for the Personal God, is unconsciously the search of the soul for its mate.

This solitary life of the soul is proof against every distraction or occupation that intelligence can devise. It yields to neither the honors nor the pleasures of the world. A man may rule a kingdom and be absolutely alone. He may have indulged himself in every phase of revelry and entertainment, and in every right of power, remaining, however, isolated in his real life. A woman may live in the close relation of wife or mother, she may pose as a social queen, yet she may never have known the first joys of real companionship.

For this immeasurable loneliness of intelligence there is an alleviation, if not cure, which lies outside of the individual relation. This is honest toil, whether it be with hands or brain.

The sadness of this isolation, on the other hand, may be immeasurably increased by idleness, wrong association and reckless misalliance.

This craving for an intelligent response to one's own intelligence, and rebellion against this inner solitude of the soul, constitute that subtle factor which leads men and women to violation of the social conventions. The necessities of the intelligence, however, lead only to infringement of custom, whereas the passions of the body induce violation of legal restrictions.

Marital infidelity, the world over, is more of soul than of body.

One eminent divine of New York estimates that there are two hundred and fifty thousand married persons in that city who are legally unfaithful. If this be true, of the physical relation, who will undertake to estimate the number of those who are estranged in the higher realm of the intelligent soul life? How light would be the task of numbering the few pairs who are loyal in body, in spirit and in soul.

If this inference appears extravagant, let the reader stop and count the number of ideal marriages which have come under his own personal observation. Let him count the number of happy married men or women whom he actually knows as such in his own little world.

A school of philosophy which declares that there is a perfect marriage relation attainable under mathematical laws of Nature, runs the risk of popular ridicule, if not of legal suppression. It is, nevertheless, true that such a relation is demonstrable under the law of vibration, and that the struggle for individual Happiness ends in that accomplishment. In this relation and in this alone, that "divine unrest" of which the poets sing, is forever stilled.

The higher science, studying this psychical struggle for Happiness in two worlds, declares that by far the greater number of individuals are seeking that end along other lines than the true principle of harmonics. Comparing the motives, ideals and lives with the measure of happiness that men secure through their so-called successes, it is seen that the ideal of Happiness too often conflicts with the universal Happiness principle. Taking into account this principle of harmonics in Nature, this modern philosophy of the ancient school declares that this long-sought Happiness of the individual rests, primarily, upon the love principle and upon his ability to find completion under that principle through permanent union with his perfect affinity.

This reading of the law furnishes the basis of propositions new in philosophy. These are propositions which must change the life of any man or woman who adopts them as the working formula of his or her daily life. The philosophy of individual life, built upon a principle of harmonics instead of a principle of competition, declares as fundamental doctrines:

(1) Rational Happiness is the moral destiny of the soul.

(2) Such Happiness rests, primarily, upon the individual relation of man and woman.

(3) The attainment of such Happiness is the first duty, as well as the highest privilege, of rational beings.

Man and woman must work out this marital problem along lines conforming to the elements they represent. There have been and will be sins of omission and commission on both sides. This, indeed, is the only path possible for the evolution of a rational and moral relation between these complementary intelligences.

The past half century marks an epoch in the struggle for completion, especially on the feminine side. This unprecedented incursion of woman into higher and hitherto forbidden fields of educational and commercial life has its evolutionary meaning. It marks that strengthening of the feminine nature which guarantees higher rational and moral achievements, increased activity, added usefulness, and larger Happiness. Without losing the essential grace of womanliness, she is gaining in will force, and in reasoning powers. She is gaining in breadth of intelligence and in direction of purpose.

The best womanhood of to-day has earned that purely intellectual equality with man she has so long desired. She has fitted herself to be the companion of man's intelligence, as well as his wife, or his mistress, or even the mother of his children.

The popular science of the twentieth century will not study woman as simply a "female" or a "mother." It will study her as a *woman*. The best manhood of the new age will accept this woman at her own valuation. It will accept her as explained and illustrated by herself.

The best types of both sexes to-day enjoy an equality and fellowship in the higher lines which have no parallel in history. This new relation represents the struggle for completion at one of its critical periods. Not only a new century, but a new balance between man and woman has been struck.

Though this new era marks a closer harmony, it does not alter the relative natures or positions of these two powers. Man, if he does not degenerate, must continue the superior will force and master of this material world. Woman, if she does not deteriorate, must remain as the spirit of peace, the guardian of the love relation, and the inspiration of the ethical life of the world. The long and wearisome quest of human intelligence for its own completion has an ending in the course of Nature, just as surely as the efforts of physical nature had an ending in the perfect physical organism. Whenever man and woman meet upon those terms of equality and reciprocity which Nature intends, they will have experienced marriage which is not made by courts of law. They will know a fellowship of which "society" is barren.

All of life's energies are then exactly balanced and fully employed. All activities are harmonious activities. To every demand of every element is response and reinforcement. Thought answers thought. Principle strengthens principle. Will and desire are one. Ideals are verities. The soul has entered into rest.

Man has been well named "The Hunter for Truth," since we are indebted to his aggressive intelligence for our accumulation of rational knowledge and scientific fact. Woman, on the other hand, the conserver of the spiritual relation and the developer of ethics, may well be named "The Searcher for Love." For whatever her weakness of body or will, whatever her sins of omission or commission, the world must admit that it looks mainly to woman for the preservation and improvement of the love relations, for the status of home and the harmony and purity of society.

In the final summing up of life's purposes it will be found that masculine and feminine intelligence have served as equal factors in the higher evolution of man.

In this last analysis it will be found that the Hunter for Truth has discovered the rational road to Happiness, while the Searcher for Love has guarded the relation in which Happiness is found. The Hunter for Truth and the Searcher for Love finally meet in the perfect relation. Truth and Love shall be made one. The law will be fulfilled and Happiness established in wisdom.

* * * * * *

Thus, in the light of a higher science, men and women exist for living and learning and loving, and not merely (as contributions to species) for feeding and breeding and battle. They exist for individual action, improvement and accomplishment, and not merely to reproduce. They exist for a self-conscious completion in each other and for individual knowledge, power and Happiness here and hereafter.

Physical materialism, interpreted by certain learned rationalists and distinguished degenerates, is promulgating theories of sex, and of love, and marriage which menace both general progress and individual Happiness. The trend of physical science is clearly to establish men and women as the mere agents of nutrition and reproduction. The effort is to establish marriage in the interests of progeny, rather than for the well-being and Happiness of the individual.

Such doctrine is a contravention of Nature. The struggle for self-completion does not and can not (primarily) take progeny into account. The present necessities of the individual must take precedence of benefits to possible offspring. Neither passion nor sympathy nor love, ever has been nor will it ever be evoked in the interests of the unborn. Marriage has but one proper, natural impulse, inspiration and motive, viz., the well-being and Happiness of the individuals concerned.

Neither science, law nor religion can abrogate the primal and natural laws of intelligent life. Neither the passions of the body nor the responses of the soul can be governed through appeals in behalf of progeny. Whenever and wherever marriage can be raised to Nature's standard of harmonics, the interests of reproduction are guarded. When individual choice is based upon an individual fitness, the child is the direct beneficiary. When marriage violates the principle of harmonics, the consequences of that violation are visited upon children "to the third and fourth generation."

That recent popular admonition, "for the sake of the unborn," sounds well, reads well. It is safe to say, however, that not one man or one woman in ten thousand actually considers the unborn when his or her own affections or ambitions or interests are engaged. On the contrary, one familiar with the temper of this age, knows that there is a growing distaste for reproduction. The larger number of men and women marrying to-day are hoping to conserve their own individual ambitions and individual Happiness by escaping the obligations and cares of parentage. Nobody doubts that a large proportion deliberately plan to rear no children.

The individual man and woman are so centered upon their own Happiness, so moved by their own impulses, needs and desires, that the rights of the child are ignored until it arrives. Then love, the sublime love of maternity and paternity, invests the helpless offspring with its rights and privileges and—when too late—mourns for the defects and diseases to which it has been condemned.

If posterity depended upon the actual desire for reproduction, over-population would never be a menace. It is safe to say that no woman would voluntarily take upon herself the pains, penalties and sacrifices of maternity for the fourth and fifth, much less the tenth and twelfth time.

Reproduction must be accepted for just what it is—a physical function and a moral duty which conserve the interests of the race, rather than the individual interests. It must be treated as an office of vast importance to the race, but incidental to the life of the individual, and subordinate, in his eyes, to his own completion.

Giving to men and women new and higher ideals of marriage is the one any only method of securing the best interests of reproduction. When the will and desire of man and woman are directed toward higher marriage, reproduction is already safeguarded and the interests of the race are secured. The destiny of man and woman is no more fulfilled in offspring than the destiny of the child is fulfilled in the act of its birth. The lives of parents are usually sealed books to their children; and how soon the child learns to live his own life, to follow his own ideals and work out his own purposes.

"I find my Happiness in my children" is the familiar expression, which literally means "I have not found it in marriage." In such cases the word "consolation" should be used instead of the word "Happiness."

Happiness is a matter of intelligent companionship, of sympathy and confidential co-operation, with another individual of our own plane of development. Children stimulate the love nature and develop altruism. They teach patience and sacrifice. They waken the sense of personal responsibility. They are objects of anxiety, too often of unrest and sorrow. They are objects of love, of pleasure, possibly of pride, but they are not the fulfillment of destiny, nor the primary source of individual Happiness.

It is not meant that perfect marriage shall defeat reproduction. It means instead reproduction guarded and ennobled by the proper relation of parents. It means children as a contribution to the love life of man and woman, rather than a burden. The child of such union is the inheritor of prenatal harmony. He is the harmonious expression of will and desire when perfectly united. In such a relation unwelcome children would not be born, nor more of them than could be properly maintained. In the perfect marriage relation the child is the privilege of love and not the penalty of lust.

With such recognition of the individual rights of man, woman and child, science would no longer promulgate a theory of sex selection based upon physical fitness and the interests of reproduction, nor "a practically free selection with children reared by the state."

Looking to the perfect marriage relation and individual Happiness as expressions of natural law, we find that our poets and singers have been the realists. We perceive that the skeptics are the intellectual visionaries. We know that the pessimists are the moral degenerates.

Looking to the whole of Nature, rather than a part, it appears that love between man and woman is neither lust nor disease, and that Happiness is not a delusion nor a dream. On the contrary, skepticism is abnormal and pessimism is a perversion of intelligence.

To such as seek rational evidences of the law, to such as have the Intelligence, Courage and Perseverance to conform to that law, Nature says: "Affinity is the law of laws. Harmony is a mathematical possibility. Peace is a natural principle. Hope is the healthy attitude of the soul. Faith has a foundation in cold fact. Living is a science. Love is the fulfilling of the law. Happiness is the fundamental reality of existence."

CHAPTER XXI.

INDIVIDUAL SOLUTION.

If perfect marriage is a scientific possibility, how shall I proceed to discover the individual to whom I am naturally allied?

What of those who have married unhappily without previous knowledge of the law?

What of those who discover the true mate after mismating has occurred?

If perfect marriage is Nature's intent, and the essential condition of individual earthly happiness, what of marriage in general? Must men and women cease from marrying until they discover the perfect mate?

These questions would naturally occur to one who has followed this primary work of the philosophy of individual life.

A general answer to the first question is to say that the ability to marry in accordance with the true law of marriage depends, primarily, upon a rational comprehension and scientific knowledge of those principles and elements which go to make up the perfect relation. This first question is undoubtedly the most important one which any individual could ask. Its importance lies in the fact that it implies an individual solution of the chief problem in the philosophy of life. It suggests an individual application of the principles involved in the true relationships of life.

It must, therefore, be clear that the first prerequisite is a definite knowledge of the elements which go to make up the true relation.

If the reader has carefully followed the work to this point, he will have in mind the fact that each individual is a triune being, that the elements of this triune being are physical, spiritual and psychical, and that the complete relation involves correspondence in all three of these elements.

Affinity, or correspondence, may obtain between two individuals, in any one or two, or in all of the three natures composing the trinity. The perfect union, however, demands response in each nature, and those several correspondences have different values in their relation to individual happiness; that is to say, the comparative values of the harmonics of marriage are as follows:

(1) Psychical.
(2) Spiritual.
(3) Physical.

This, it will be observed, reverses the usual order of the trinity, as it has been discussed in this work. This means merely, that the most important harmonic in the individual relation (looking to Happiness) is the psychical. Next in value is the spiritual, and least in importance is the physical. The indices of the true relation are:

(1) Psychical harmony, perfect response and agreement in the rational intelligence and in ethical principle, two individuals who think from the same point of view and act by the same moral standards.

(2) Spiritual sympathy, an equal culture and refinement, æsthetic tastes in common, similar appreciation of the beauties of Nature and of art, similar standards as to the pleasures and conduct of society.

(3) Physical affinity, a mutual desire for the physical presence, a mutual pleasure in the physical touch and the physical caress.

The ability of an individual to find his own depends upon his ability to determine when all of these conditions have been fulfilled. It depends upon his ability to discriminate between physical passions, æsthetic sympathies, and a rational and ethical agreement. The investigator must bear in mind that true natural selection involves the intelligent processes of both intuition and reason. It necessitates both an involuntary or impulsive selection and a voluntary and rational choice on the part of the individual. That is to say, it embraces the involuntary sanctions of spiritual intuition and the voluntary mandates of reason.

Intuition indicates where correspondences exist. Reason, however, must determine whether those impulses or passions or affections are complete, embracing each element, or only existing upon one plane or another of the trinity.

When it is said that true marriage must have the sanction of both intuition and reason, it is meant that the act must be a matter of impulse, feeling, love, and also a matter of reason, judgment and reflection. It must be remembered that Nature supplies man with his impulses and his intuitions. Man, however, must supply his own reasoning. Intuition and impulse indicate the operation of the law of harmonics. It requires, however, an individual and definite knowledge of harmonics to determine the nature and the value of our intuitions and impulses. Man intuitionally seeks harmony for his triune nature. Science alone enables him to determine whether the impulses and attractions he feels are evidences of completion or partial completion.

Thus, the individual solution in reality depends upon individual development. The nearer the individual approaches the equilateral triangle the greater are his chances for intelligently securing his own happiness. The gross and the stupid must depend mainly upon their physical impulses for guidance. The physically refined and spiritually sensitive have keener spiritual intuitions upon which they may depend. It is not, however, until the individual intelligence arrives at an independent and rational conception of the law that he can hope to form a rationally happy alliance.

It is admitted that happy marriage may occur in total ignorance of the scientific principles involved. Such cases, however, are rare; nor can such happiness be called rational, since it cannot account for itself. The conditions which equip an individual to intelligently search for his own require, first, physical refinement, next, spiritual refinement, and finally, a high degree of rational and moral development. Indeed, it is only through such general development that human intelligence may even conceive the true ideal of this relation.

The rule and guide in the marriage problem are to seek the individual who responds, first, upon the higher plane of intelligence, next, in spiritual sympathies, and finally, in physical passions and proclivities.

The man and woman who do not represent such union may legally marry, but they are never mated. Only the man and woman who fellowship and co-operate in the higher, rational and ethical pursuits, hold the key to true marriage and to rational happiness.

"As a man thinks so is he." This is a concise statement of the law.

When two rational beings think alike they are alike. They are already indissolubly bound. To think alike is to live, aspire, feel and act upon the same general principles. This is fellowship which guarantees permanent understanding, establishes confidence, fixes faith, and banishes solitude. This is the union which wipes out the numeral "1" and absorbs the personal pronoun "I." This is the relation in which man and woman may aspire and work and accomplish. This is the alliance which means health, progress and happiness.

This is the only marriage which has the sanction of the Great Intelligence.

The second and third questions have been covered in those chapters on legal marriage and divorce. This philosophy recognizes the necessity for law and legal restraints in the marital relation. It therefore has no remedy for error, except the proper fulfillment of self-assumed obligations. The purpose of philosophy is rather to prevent error and its inevitable penalties. Both Nature and law provide the opportunity for natural selection. If that selection proves imperfect, legal and moral obligations still remain. When that mismating results in a family the question ceases to be an egoistic one. It is then purely altruistic, and must be dealt with in view of the rights of that family, and not upon the ground of individual happiness. This applies also wherever the unhappily married discover the true relationship.

Such a pair can no more check that natural response than one C string can check its response to another C string. This, however, does not necessarily mean either legal violations or the repudiation of moral responsibilities.

In such cases where men and woman have misused the great opportunity of youthful freedom, they must pay the penalty of that act according to reason and conscience. They must not solve this problem through impulse and feeling. According to the particular circumstance, this may or may not mean a lifetime of duty to principle. If so, those separated lovers should perform their several duties ungrudgingly and cheerfully.

To one who knows the philosophy of right action and right thought, this seeming misfortune does not mean despair. To one who knows the fact of life after death and the freedom of love in that higher life, that knowledge furnishes a perpetual hope. It furnishes the inspiration of a right life under every present hard condition. That which is our own will come to us in time, and however severe the penalties of our own errors, the very fact of that great discovery and the certainty of future companionship should sustain any loving soul through whatever duties it is called upon to fulfill.

To answer the fourth and last question, it need only be said that this work is intended as the presentation of a rational philosophy; a rational treatment of evolutionary processes. It looks merely to educational methods and gradual improvement. It contemplates neither arbitrary interference with the natural law of selection, nor with the established order of things. This philosophy is essentially evolutionary and not revolutionary.

Marriage is the vital process in the struggle for completion. It is the necessary school of experiment in which individual intelligence acquires rational knowledge of the laws involved. Despite its shortcomings, our present marriage code represents the best efforts of the general intelligence to rationally guard the natural law of selection. When jurists see the necessity of better guarding the act of marriage itself, one of the purposes of this work will have been accomplished. The legal code must look more closely to the questions of age, heredity and health, to length of acquaintance and financial responsibility.

The acceptance of this philosophy of the true relationship, therefore, means neither abrogation of natural law nor disregard of legal statutes. It would, however, mean a higher ideal of the marriage relation. It would elevate the practices of married life. It would quicken the sense of moral responsibility to children and society. It would lessen the tide of premature, hasty and irrational marriages. It would lessen the number of conventional and mercenary alliances. It would diminish the number of deformed and diseased children. It would insure an infinitely greater proportion of harmony in the individual marriage relation. It would mean fewer divorces, happier men and women, and a more stable government.

How shall the attainment of this true relation be accelerated?

The answer to this question rests, first, with the individual, next, with society.

It is the province of science to disclose the facts of Nature. It is the province of philosophy to illustrate principles. It is the privilege of the individual and society to accept or reject those facts as the rule and guide to action.

It remains with the individual to reject, or accept and exemplify the law. It remains with educators to condemn or to instruct the young in accordance with this philosophy.

Science and philosophy discharge their duty when they present the facts, elucidate the principles, and furnish the key to personal demonstration.

The True Altruist.

The natural occupation of the completed individiual is Altruism.

The end of the struggle for self-completion is the true beginning of the intelligent struggle for others.

The attainment of rational happiness is the end of selfishness. **The satisfaction of egoism is the opportunity of Altruism. The fulfillment of personal desire is the dawning of impersonal sympathy.**

True Altruism is that state or condition of the soul in which all of its energies and activities are centered upon the needs and requirements of our common humanity. It is that stage of development where the well-being and the advancement of others become the normal occupation of intelligence. It is that stage of experience where pleasure, recreation and entertainment are found in labor for others. It is that point of individual life beyond which our happiness consists in the transmission of the truths we have learned and the benefits we have enjoyed.

True Altruism is that state of being in which the intelligent soul increases its happiness through what it may bestow rather than through what it may gain. It is that state in which will and desire are concentrated upon giving instead of acquiring.

To the completed individual Altruism is a joy and a privilege. It is neither a duty nor a sacrifice.

Altruism thus interpreted, is the opposite pole of egoism. It means the substitution of "you" and "yours" for "I," "me" and "mine."

"I," "me" and "mine" are the dominant notes in human society. From the cradle to the grave we are victims of the personal pronoun. We live in it ourselves and our neighbors thrust it upon us. Absorbed in this personal pronoun and in the great personal problem, men, women and children, the wide world over, live, think and labor for "I," "me" and "mine." So self-centered are the most of us that these personal pronouns of our neighbors are but half heard and immediately forgotten.

So exacting is Nature, so intense the struggle for individual completion, that only in completion can we turn from "I" to "you," or merge the interests of "mine" into "thine."

This intense egoism is an unconscious and innocent selfishness. Absorbed in this vital struggle for happiness, nobody realizes his own egoism. To accuse such an individual of selfishness, would be to surprise and wound him, and in a sense to misjudge him. The man who gets "outside of himself" is the unusual man, but he is as welcome as sunlight anywhere and everywhere. The artist who can sink himself in his art is ungrudgingly praised by the world. Whoever loses himself in a common cause, or in his love of humanity, is the man we would canonize.

The incomplete individual, man or woman, is not prepared for true Altruism.

This does not mean that Altruism is not practiced in the world. It does not mean that generous impulses, noble sacrifices and splendid giving are lacking. It does not mean that the egoistic toilers of the earth are without sympathy, charity and generosity.

The philanthropies of times past and of the present forbid such suggestion.

The world is full of kindness, pessimism to the contrary. Generous impulses abound. Charity is everywhere. The average man experiences altruistic impulses in many forms. He is moved by distress. He deplores misery, crime and poverty. He has compassion upon even the unworthy. He is impelled to relieve conditions which induce unpleasant emotions in his own breast. His Altruism comes in the nature of a duty or a sacrifice or a contribution to his own ethical content.

This does not mean that Altruism has become his actual occupation, nor that his benefactions spring from an unmixed motive —the simple joy of giving.

When it is said that the incomplete individual is not prepared for Altruism no more is meant than to say—no man or woman is prepared to give all of his or her energies to the world so long as part of those energies must be consumed in the egoistic struggle for completion. It mearely means that no one is prepared to practically live and teach the philosophy of love and of intelligent happiness except he or she is grounded in the love principle, except he or she is individually and rationally happy.

Every man and every woman is a factor for health or disease, for harmony or discord, for happiness or unhappiness. Every human being radiates his own conditions, physical, spiritual and psychical. The radius of personal influence is only limited by personal power and by the counter influences of other individuals.

Every rational being is responsible for the character of influences which he exerts upon his fellow man. A vicious man will arouse the vicious instincts of other men. An immoral man lowers the moral tone of his associates. A quarrelsome individual breeds discord. One melancholy member clouds an entire family circle. One fretful, peevish soul irritates every other soul in its neighborhood. An exhibition of selfishness provokes other people to withhold their generosities. Selfish unhappiness is as contagious as smallpox. One selfishly unhappy individual is a source of positive evil and moral degeneration, as far as the poison of his personal influence extends.

The state of true Altruism is as definite a state as that of selfishness or viciousness or immorality. It is also farther reaching in its influence and richer in effects that any other known state of being.

To arrive at the state of true Altruism requires that the soul should have individually demonstrated the principle of harmonics and personally attained to a rational happiness.

The individual who is rationally happy has reached the end of personal desire. He is neither oppressed nor distracted by individual wants. He has ceased to make demands upon Nature or society for his individual happiness. In this natural cessation of personal demands egoism dies its natural death. It has not been strangled by austerities nor crushed out by religious superstitions nor covered up by social conventions. Happiness is simply unselfishness in its literal sense. It is spontaneous Altruism.

The completed individual, the perfectly happy man, has no other choice of occupation than a work for humanity. It must be remembered that the state of completion is the state of the most intense psychical activity. Intelligence must have occupation. The happy man, as well as the unhappy, must find employment for his energies and capacities. The divine law of labor encloses the completed individual and urges him to other achievements. Such a man or woman is alive with splendid enthusiasm; and now, seeing life beyond the narrow limit of self, maps out accomplishments undreamed of by those in the midst of the egoistic struggle for happiness. To such as these, action is necessity, while inertia and idleness are as impossible as egoism and selfishness.

To such as these Altruism is a practical occupation, as well as recreation and pleasure. To such an one, humanity is his family, the world is his field and to do good is his religion. This is an Altruism which makes universal brotherhood a splendid possibility, and an eternal hell a hideous impossibility.

What other occupation is open to an individually completed life, than the splendid task of imparting its gains to others? What other motive can inspire the really happy man except the desire to make other people happy? It is a law of Nature that impels us to bring others to our own condition of mind. The universal principle of harmonics impels every individual to seek to bring his neighbor to his own intellectual and ethical state of being. The man and woman, mutually attuned to the higher harmonics of love, and released from egoistic considerations, are thenceforth impelled by every law of psychical development to impart their own condition to their fellow man.

Being happy themselves, they are irresistibly moved to minister to the happiness of the world. They are irresistibly inspired to pass on to others that knowledge and those principles which shall change discord into harmony, fear into hope, loneliness into companionship, and sorrow into joy.

Ask any rational man what he would do if he were absolutely happy. He will invariably reply, "I would make other people happy." Ask the rationally happy man in what he finds his greatest pleasure, and he will tell you, "In doing what good I can in the world and helping other people to be happy."

Just here is the radical point of departure between the ancient and this modern interpretation as to the place, value and purpose of the individual in Nature. An individually happy earthly life was apparently the last and least consideration in the ancient philosophies. Indeed, most of the teaching, and much of the practices of its devotees, are calculated to inspire the mind with the insignificance of the individual and the presumption of his desire for individual happiness. The natural and happy love life is made to appear as something quite foreign to, if not actually opposed to the "higher life of the soul." The desire for, and the determination to seek such a relation, are made to appear rather as the temptation of the "lower nature" than a legitimate part of so-called "Spiritual Illumination."

Impressed by such doctrine, the would-be "mystic" comes to consider marriage and individual love with its duties and joys, as a mere phase, a mere passing experience of the soul which is seeking to "lose itself in the Universal." This attempt to reach the higher altitudes of altruism by ignoring the natural necessities of the individual, results in a doctrine of "Impersonality" that is contrary to Nature, and deadening to the individual faculties. This is a doctrine which rests securely upon such terms as "selflessness," and whose aim is defined as "oneness" with a universal, uncomprehended, and incomprehensible Ultimate.

This attempt of the modern "occultist" to govern his life by ancient interpretations of man and his destiny, results in curious encounters between ancient mysticism and modern common sense.

The clear-headed western skeptic may be forgiven his occasional criticism of what he designates as "Oriental Fads." Such a man may well question the wisdom of an American woman who, born and reared in this western atmosphere of religious, social and legal equality for women, abjures Christianity, joins an Indian sect, takes vows of celibacy and poverty, and dons a yellow robe, that she may be better able to practice the golden rule.

Such a critic, if a thinker, would know that celibacy is a contravention of Nature. If he were a scientist he would know that the celibacy of the highly developed defeats racial improvement. If he were an economist, he would know that "vows of poverty" mean the shifting of personal responsibility for maintenance upon an already overtaxed public. If he were a practical statistician, he would calculate the physical, industrial, and moral results, if nine out of every ten citizens of this great republic vowed themselves to celibacy and poverty, leaving the other one-tenth to replenish the earth and furnish subsistence for the whole.

In these particular interpretations of individual rights and responsibilities this philosophy contravenes ancient dogmas and practices. It unequivocally affirms that true altruism rests upon neither celibacy nor poverty. It declares, on the contrary, that he who seeks the approval of Nature's God and the blessings of his fellow man, defeats that end by seclusion, celibacy and poverty.

A philosophy which hopes to satisfy modern progressive intelligence must banish the unnatural and therefore unholy martyrdom of priest, monk and nun. It must replace selfish seclusion with an active usefulness in the world. It must replace the degenerative effects of austere celibacy with the ennobling impulses of a natural love life. It must substitute the self-respect of individual effort for the self-degradation inseparable from "vows of poverty."

In brief, it will banish an altruism wrung from somber doctrines, from cheated hearts, and povety-stricken lives. It will set forth, instead, that splendid altruism which overflows in the soul who has self-sought and self-earned an individual independence and an individually happy love life.

Happiness is the most irresistible teacher of goodness on earth or in heaven. The example of one rationally happy being is farther reaching and more enduring than volumes of precept. Such a man or woman is the most potent factor for healthful development to be found on earth.

As a man thinks, so is he, and so does he labor. The soul who has climbed to the summit, who has reached the goal, who has attained his heart's desire, is the one and only mortal properly equipped to teach the gospel of happiness to an ignorant and sorrowing world. He is the only individual rightly conditioned to furnish both example and precept.

Thus it is that only such as have reached the stage of an individual completion, are prepared to give themselves wholly and joyously to the task of teaching mankind. The man and woman who have together proved the law of love, and whose individual happiness is completed in each other, have no other will nor desire than Altruism.

Such a pair might well say, "Our mission is to teach the science of love as a law of fulfillment here and hereafter. Our pleasure is to present a philosophy of life, which, if accepted, shall hasten the fulfilling of the law in other lives. Our highest purpose and happiest occupation are to serve you and yours in this present earthly life, and also in the life to come."

To Know, to Dare, and to Do.

This philosophy of individual life is, therefore, something more than a compilation of scientific fact or an array of intellectual opinion, to be lightly scanned and soon forgotten. On the contrary, it is an array of fact and a declaration of principle which call for immediate investigation and immediate action.

This is a philosophy of action, as well as of introspection. It means the doing of that which is practical, as well as a contemplation of that which is ethical. It calls for the exercise of reason and the practice of principles, as well as the indulgence of the emotions and development of the æsthetic tastes. It looks to knowledge and goodness, as well as to culture and refinement. It means more than thinking or speculating or believing. It is living and learning and doing. It is a *life,* not a creed.

This philosophy is essentially a philosophy of this life, rather than of a life to come. It fixes upon a noble earthly life as the gateway to the splendors of a higher life. It includes a practical effort to refine the physical body and to control abnormal physical appetites and passions. It means cultivation of the spiritual faculties, opening to life a perception of the order and harmony and beauty of the spiritual side of Nature. It means psychical development, and equal and steady exercise of the intelligent soul in the acquirement of knowledge and the practices of love.

As previously explained, the philosophy here presented represents the modern masters of the law. It therefore stands for a later and larger interpretation of Nature, of life, of love and of duty.

This is not a philosophy of negation, of self-suppression, self-sacrifice, nor resignation. On the contrary, it is distinctly a philosophy of affirmation, self-development, self-importance and self-satisfaction. It is the philosophy of fulfillment and not of resignation.

This interpretation of man, of his relation to Nature, and to his fellow man, leaves no basis for stoicism. It removes all ground for extreme austerities and self-chastisement. Indeed, it teaches none of those sad and benumbing doctrines of the ancient East which belittle the importance of physical and material development, which renounce the sweetness of individual love, and point to an absorption in the universal intelligence as the final "Place of Peace."

The philosophy of this later time is, indeed, the dawning of a new day in the intellectual and ethical life of the world.

Later authorities wisely hold that stoicism and austerities, if ever necessary, belong to an age long past. Stoicism, indeed, is Will without Desire. It is courage without hope. It is learning without wisdom, sacrifice without purpose, effort without gain,—it is man without woman, and life without love.

Physical austerities are less important than the exercise of psychical powers. The refinement of the physical body is better achieved through the right activities of the soul than through processes of digestion. Rigid austerities have no place in the life whose rule and guide are temperance and self-control.

This philosophy accepts this earthly life, and this physical body with all of its functions, as a necessary, important and legitimate part of the destiny of the soul. It, therefore, enjoins an earthly life well sustained and well rounded in all of its activities and relations. The individual is not admonished to "lose himself in the Universal," but rather to find himself in a particular world of actualities. Earthly life is not presented here as an illusion of the senses, but as a very real and tangible opportunity for the intelligent soul.

Instead of stoicism it teaches courage. It inculcates unselfishness rather than sacrifice. It commands temperance and not asceticism. It enjoins patience instead of resignation; for true living is a state of progress and fulfillment, irrespective of the external conditions of this earthly life.

If this initial volume of the law seems to ignore ultimates, it is because those individual relations and activities this side of ultimates have been too much neglected. Where it fails to dwell upon a remote "Nirvana" it enlarges upon the immediate present.

If it does not portray the final "Place of Peace," it does point to a present home of happiness. Earth life is a privilege and not a penalty. The purpose of earth life is not to find a heaven, but to make one. Religion is not a matter of duty to God, but of duty of man to himself and to his fellow man.

Whoever has the Intelligence to *know,* the Courage to *dare,* and the Perseverance to *do,* may understand, accept and prove this philosophy. To such an one the yoke is easy and the burden is light.

Such is the true philosophy of life, the philosophy of present action and of future hope. Nature thus interpreted and life thus lived, illuminate the word "evolution" with a new meaning and a new purpose.

Such a philosophy restores Nature to its true position as the Father-Mother of life, intelligence and love. It advances individual life to new and splendid possibilities. It looks to individual happiness as to the normal destiny of the soul. It clears the mists from physical science, and unravels the confusion of a moral philosophy based upon physical materialism. It extends the limitations of Nature and of recognized science from the physical to the spiritual planes of life and activity. It gives to science a motive for its knowledge and to religion a reason for its faith.

It extends the sphere of man from a world of physical functions and physical activities, to a world of spiritual functions and psychical activities. It extends individual destiny beyond a present contribution to species, or a future contribution to the soil of Mother Earth. It lays the terrors of loneliness and death. It banishes the shadow of annihilation. It opens to the soul unmeasured possibilities. It guarantees an individual completion through individual love, and a permanent happiness here and hereafter.

It makes of each individual man and woman a natural heir to all the beneficences of Nature and of Nature's God.

END OF VOLUME I.